# METHODS IN MOLECULAR BIOLOGY



For further volumes:
http://www.springer.com/series/7651

For over 35 years, biological scientists have come to rely on the research protocols and methodologies in the critically acclaimed *Methods in Molecular Biology* series. The series was the first to introduce the step-by-step protocols approach that has become the standard in all biomedical protocol publishing. Each protocol is provided in readily-reproducible step-by-step fashion, opening with an introductory overview, a list of the materials and reagents needed to complete the experiment, and followed by a detailed procedure that is supported with a helpful notes section offering tips and tricks of the trade as well as troubleshooting advice. These hallmark features were introduced by series editor Dr. John Walker and constitute the key ingredient in each and every volume of the *Methods in Molecular Biology* series. Tested and trusted, comprehensive and reliable, all protocols from the series are indexed in PubMed.

# KRAS

## Methods and Protocols

Edited by

**Andrew G. Stephen and Dominic Esposito**

*NCI RAS Initiative, Cancer Research Technology Program, Frederick National Laboratory for Cancer Research, Frederick, MD, USA*

Humana Press

*Editors*
Andrew G. Stephen
NCI RAS Initiative, Cancer Research Technology Program, Frederick National Laboratory for Cancer Research
Frederick, MD, USA

Dominic Esposito
NCI RAS Initiative, Cancer Research Technology Program, Frederick National Laboratory for Cancer Research
Frederick, MD, USA

ISSN 1064-3745 ISSN 1940-6029 (electronic)
Methods in Molecular Biology
ISBN 978-1-0716-3821-7 ISBN 978-1-0716-3822-4 (eBook)
https://doi.org/10.1007/978-1-0716-3822-4



Cover Caption: The cover illustration shows the covalent attachment of sotorasib (cyan) to cysteine 12 (yellow) and binding in the switch 2 (teal) pocket of KRAS G12C. Switch 1 is depicted in mauve, and GDP is represented in green. The PBD structure 6OIM was used to generate this representation. This illustration was prepared by Joe Meyer, Scientific Publications, Graphics and Media, Frederick National Laboratory for Cancer Research.

This Humana imprint is published by the registered company Springer Science+Business Media, LLC, part of Springer Nature.
The registered company address is: 1 New York Plaza, New York, NY 10004, U.S.A.

Paper in this product is recyclable.

# Preface

It is now over 50 years since the transforming potential of RAS genes was first discovered. During that time, RAS genes have been established as one of the most frequently mutated family of oncogenes in human cancer. Described as the "beating heart" of signal transduction, RAS proteins function as molecular switches which relay extracellular signals to intracellular signaling molecules that are responsible for cellular growth and proliferation. Oncogenic mutations in RAS proteins cause the molecular switch to be locked in the on-state, contributing to the uncontrolled proliferation of cancerous cells. Of the three RAS isoforms, KRAS is by far the most frequently mutated, with over 130,000 patients diagnosed every year with a cancer driven by a KRAS mutation. Over the last several years, renewed focus on the structural, biochemical, and signaling properties of KRAS has resulted in the approval in 2021 of sotorasib to treat G12C-mutated non-small cell lung cancer (NSCLC). While it is a remarkable achievement that these patients now have viable therapeutic options, NSCLC patients with KRAS-G12C mutations represent only 16% of all patients with KRAS mutations. Therefore, more work is needed to develop new drugs to target other KRAS mutant alleles. The objective of this book is to provide RAS researchers with a handbook containing the relevant methods that will enable them to lead the charge to identify new drug candidates for KRAS-driven cancers. The methods in this book cover a broad spectrum of protocols, such as those for high yield expression of isotopically labeled KRAS for NMR studies to approaches that quantify engagement of novel molecules that bind KRAS in live cells. The book highlights methodologies that focus on protein production structure, and characterization; targeting KRAS; biochemical assays; cell-based assays; and cell models. In these chapters, experts from the KRAS community have provided descriptions of the methods with step-by-step instructions that will enable researchers to perform these protocols. We are grateful to all the authors for their time and expertise in contributing to this book. Our hope is that these methods will be instrumental in the further development of clinically approved KRAS therapeutics.

*Frederick, MD, USA*

*Andrew G. Stephen*
*Dominic Esposito*

# Contents

# Contributors

J. Aaron Crapster • *Vibliome Therapeutics LLC, Bozeman, MT, USA*
Maria Abreu-Blanco • *NCI RAS Initiative, Cancer Research Technology Program, Frederick National Laboratory for Cancer Research, Frederick, MD, USA*
Patrick Alexander • *NCI RAS Initiative, Cancer Research Technology Program, Frederick National Laboratory for Cancer Research, Frederick, MD, USA*
Makoto Araki • *Department of Biochemistry, Meiji Pharmaceutical University, Tokyo, Japan*
Trent E. Balius • *NCI RAS Initiative, Cancer Research Technology Program, Frederick National Laboratory for Cancer Research, Frederick, MD, USA*
Lauren Beaumont • *NCI RAS Initiative, Cancer Research Technology Program, Frederick National Laboratory for Cancer Research, Frederick, MD, USA*
Pedro Andrade Bonilla • *NCI RAS Initiative, Cancer Research Technology Program, Frederick National Laboratory for Cancer Research, Frederick, MD, USA*
Konstantin Budagyan • *Drexel University College of Medicine, Philadelphia, PA, USA; Fox Chase Cancer Center, Philadelphia, PA, USA*
William Burgan • *NCI RAS Initiative, Cancer Research Technology Program, Frederick National Laboratory for Cancer Research, Frederick, MD, USA*
Alexa C. Cannon • *Drexel University College of Medicine, Philadelphia, PA, USA; Fox Chase Cancer Center, Philadelphia, PA, USA*
Mayukh Chakrabarti • *NCI RAS Initiative, Cancer Research Technology Program, Frederick National Laboratory for Cancer Research, Frederick, MD, USA*
Albert H. Chan • *NCI RAS Initiative, Cancer Research Technology Program, Frederick National Laboratory for Cancer Research, Frederick, MD, USA*
Jonathan Chernoff • *Fox Chase Cancer Center, Philadelphia, PA, USA*
John Columbus • *NCI RAS Initiative, Cancer Research Technology Program, Frederick National Laboratory for Cancer Research, Frederick, MD, USA*
Gabriel Cornilescu • *NCI RAS Initiative, Cancer Research Technology Program, Frederick National Laboratory for Cancer Research, Frederick, MD, USA*
Robert A. D'Ippolito • *NCI RAS Initiative, Cancer Research Technology Program, Frederick National Laboratory for Cancer Research, Frederick, MD, USA*
Caroline J. DeHart • *NCI RAS Initiative, Cancer Research Technology Program, Frederick National Laboratory for Cancer Research, Frederick, MD, USA*
John-Paul Denson • *NCI RAS Initiative, Cancer Research Technology Program, Frederick National Laboratory for Cancer Research, Frederick, MD, USA*
Marcin Dyba • *NCI RAS Initiative, Cancer Research Technology Program, Frederick National Laboratory for Cancer Research, Frederick, MD, USA*
Dominic Esposito • *NCI RAS Initiative, Cancer Research Technology Program, Frederick National Laboratory for Cancer Research, Frederick, MD, USA*
Scott Eury • *NCI RAS Initiative, Cancer Research Technology Program, Frederick National Laboratory for Cancer Research, Frederick, MD, USA*
Nicole Fer • *NCI RAS Initiative, Cancer Research Technology Program, Frederick National Laboratory for Cancer Research, Frederick, MD, USA*

WILLIAM GILLETTE • *NCI RAS Initiative, Cancer Research Technology Program, Frederick National Laboratory for Cancer Research, Frederick, MD, USA*
KENJI KONTANI • *Department of Biochemistry, Meiji Pharmaceutical University, Tokyo, Japan*
KARI KOPRA • *Department of Chemistry, University of Turku, Turku, Finland*
ARTHUR LAGANOWSKY • *Department of Chemistry, Texas A&M University, College Station, TX, USA*
ERIK K. LARSEN • *NCI RAS Initiative, Cancer Research Technology Program, Frederick National Laboratory for Cancer Research, Frederick, MD, USA*
ROGER MA • *NCI RAS Initiative, Cancer Research Technology Program, Frederick National Laboratory for Cancer Research, Frederick, MD, USA*
ANNA E. MACIAG • *NCI RAS Initiative, Cancer Research Technology Program, Frederick National Laboratory for Cancer Research, Frederick, MD, USA*
FRANK MCCORMICK • *NCI RAS Initiative, Cancer Research Technology Program, Frederick National Laboratory for Cancer Research, Frederick, MD, USA*
JENNIFER MEHALKO • *NCI RAS Initiative, Cancer Research Technology Program, Frederick National Laboratory for Cancer Research, Frederick, MD, USA*
ANI MICHAUD • *Promega Corporation, Madison, WI, USA*
RAKESH SATHISH NAIR • *Department of Cell and Molecular Pharmacology & Experimental Therapeutics, Medical University of South Carolina, Charleston, SC, USA; Hollings Cancer Center, Medical University of South Carolina, Charleston, SC, USA; Ralph H. Johnson VA Medical Center, Charleston, SC, USA*
ABIGAIL NEISH • *NCI RAS Initiative, Cancer Research Technology Program, Frederick National Laboratory for Cancer Research, Frederick, MD, USA*
NORITAKA NISHIDA • *Graduate School of Pharmaceutical Sciences, Chiba University, Chiba, Japan*
DWIGHT V. NISSLEY • *NCI RAS Initiative, Cancer Research Technology Program, Frederick National Laboratory for Cancer Research, Frederick, MD, USA*
JOHN P. O'BRYAN • *Department of Cell and Molecular Pharmacology & Experimental Therapeutics, Medical University of South Carolina, Charleston, SC, USA; Hollings Cancer Center, Medical University of South Carolina, Charleston, SC, USA; Ralph H. Johnson VA Medical Center, Charleston, SC, USA*
KATIE POWELL • *NCI RAS Initiative, Cancer Research Technology Program, Frederick National Laboratory for Cancer Research, Frederick, MD, USA*
DANA RABARA • *NCI RAS Initiative, Cancer Research Technology Program, Frederick National Laboratory for Cancer Research, Frederick, MD, USA*
MEGAN RIGBY • *NCI RAS Initiative, Cancer Research Technology Program, Frederick National Laboratory for Cancer Research, Frederick, MD, USA*
MATTHEW B. ROBERS • *Promega Corporation, Madison, WI, USA*
GRACE M. SCHEIDEMANTLE • *NCI RAS Initiative, Cancer Research Technology Program, Frederick National Laboratory for Cancer Research, Frederick, MD, USA*
ELENA SCOTT • *Department of Chemistry, Texas A&M University, College Station, TX, USA*
ICHIO SHIMADA • *RIKEN Center for Biosystems Dynamics Research, Yokohama, Kanagawa, Japan; Graduate School of Integrated Sciences for Life, Hiroshima University, Higashi-Hiroshima, Japan*
REBIKA SHRESTHA • *NCI RAS Initiative, Cancer Research Technology Program, Frederick National Laboratory for Cancer Research, Frederick, MD, USA*

DHIRENDRA K. SIMANSHU • *NCI RAS Initiative, Cancer Research Technology Program, Frederick National Laboratory for Cancer Research, Frederick, MD, USA*
BRIAN P. SMITH • *NCI RAS Initiative, Cancer Research Technology Program, Frederick National Laboratory for Cancer Research, Frederick, MD, USA*
MATTHEW J. SMITH • *Institute for Research in Immunology and Cancer, Université de Montréal, Montreal, QC, Canada; Department of Pathology and Cell Biology, Faculty of Medicine, Université de Montréal, Montreal, QC, Canada*
ANDREW G. STEPHEN • *NCI RAS Initiative, Cancer Research Technology Program, Frederick National Laboratory for Cancer Research, Frederick, MD, USA*
EDWARD C. STITES • *Department of Laboratory Medicine and Yale Cancer Center, Yale School of Medicine, New Haven, CT, USA*
REGINA STRAKHOVA • *Institute for Research in Immunology and Cancer, Université de Montréal, Montreal, QC, Canada*
Y. STANLEY TAN • *NCI RAS Initiative, Cancer Research Technology Program, Frederick National Laboratory for Cancer Research, Frederick, MD, USA*
TROY TAYLOR • *NCI RAS Initiative, Cancer Research Technology Program, Frederick National Laboratory for Cancer Research, Frederick, MD, USA*
THOMAS TURBYVILLE • *NCI RAS Initiative, Cancer Research Technology Program, Frederick National Laboratory for Cancer Research, Frederick, MD, USA*
JAMES D. VASTA • *Promega Corporation, Madison, WI, USA*
VANESSA WALL • *NCI RAS Initiative, Cancer Research Technology Program, Frederick National Laboratory for Cancer Research, Frederick, MD, USA*
TIMOTHY WAYBRIGHT • *NCI RAS Initiative, Cancer Research Technology Program, Frederick National Laboratory for Cancer Research, Frederick, MD, USA*
MICHAEL WHABY • *Department of Cell and Molecular Pharmacology & Experimental Therapeutics, Medical University of South Carolina, Charleston, SC, USA; Hollings Cancer Center, Medical University of South Carolina, Charleston, SC, USA*
BINGFANG XU • *NCI RAS Initiative, Cancer Research Technology Program, Frederick National Laboratory for Cancer Research, Frederick, MD, USA*
MING YI • *NCI RAS Initiative, Cancer Research Technology Program, Frederick National Laboratory for Cancer Research, Frederick, MD, USA*
SANGHO YUN • *Department of Chemistry, Texas A&M University, College Station, TX, USA*
QINGCI ZHAO • *Graduate School of Pharmaceutical Sciences, Chiba University, Chiba, Japan*

# Chapter 1

# Progress in Targeting KRAS Directly


**Dwight V. Nissley, Andrew G. Stephen, Ming Yi, and Frank McCormick**



## Abstract

RAS research has entered the world of translational and clinical science. Progress has been based on our appreciation of the role of RAS mutations in different types of cancer and the effects of these mutations on the biochemical, structural, and biophysical properties of the RAS proteins themselves, particularly KRAS, on which most attention has been focused. This knowledge base, while still growing, has enabled creative chemical approaches to targeting KRAS directly. Our understanding of RAS signaling pathways in normal and cancer cells plays an important role for developing RAS inhibitors but also continues to reveal new approaches to targeting RAS through disruption of signaling complexes and downstream pathways.



**Key words** RAS, GTP hydrolysis, Oncogenic alleles, Mutation frequency in cancer, Targeting RAS


## 1 Introduction

Over the past decade, the field of RAS biology and the optimism for therapeutic intervention against RAS-driven cancers have undergone seismic shifts driven in part by the application of emerging technologies and fresh approaches to targeting this major oncogenic driver. A basic understanding of the role of RAS in MAP kinase signaling was elucidated decades ago with the discovery of mutants that impair GTP-GDP cycling and the RAS regulatory proteins (GAPs and GEFs) that modulate this binary switch. The elegance and simplicity of a small protein that fluctuates from active to inactive via hydrolysis of a common nucleotide belies the myriad of regulatory mechanisms, and layers of feedback and crosstalk complexity, which influence cell growth through RAS-gated signaling cascades.

## 2 RAS Mutations Are Present in Many Cancers

Oncogenic mutations in *RAS* genes cause impairment of GTP-GDP cycling, resulting in increased signaling. Mutations impact this cycle through multiple mechanisms including impaired


Andrew G. Stephen and Dominic Esposito (eds.), *KRAS: Methods and Protocols*, Methods in Molecular Biology, vol. 2797, https://doi.org/10.1007/978-1-0716-3822-4_1,

intrinsic hydrolysis, abrogation of the GTPase-stimulating activity of GAPs, or increased nucleotide exchange which favors rebinding of GTP. The phenotype of mutations is proportional to the amount of time that RAS exists in the GTP-bound (on) state where the orientation of switch I (residues 30–38) and switch II (residues 59–76) permits the binding of effectors such as the RAF kinases and PI3Kα.

It is widely appreciated that RAS mutations are present in many cancers and that there is a correlation between tissue of origin and which RAS isoform is mutated. For example, pancreatic cancer is primarily associated with KRAS mutations, while melanoma is associated with mutant NRAS. The frequency of KRAS mutation in the pancreas, bowel, and lung has been widely cited [1, 2]. Figure 1 compiles information from multiple databases for cancers where RAS mutations are present at a frequency of 15% or higher.

**RAS Mutations Frequency in Human Cancers**

| | Biliary Tract | Bowel | Lung | Myeloid | Pancreas | Skin | Testis | Uterus | Vulva/ Vagina | Ovary/ Fallopian Tube | Cervix |
|---|---|---|---|---|---|---|---|---|---|---|---|
| KRAS | 20 | 49.7 | 20.1 | 5.5 | 78.7 | 1.3 | 14.4 | 16.7 | 8.6 | 13.9 | 13.8 |
| NRAS | 1.9 | 1.6 | 0.6 | 11.1 | 0.4 | 14.2 | 4.4 | 2 | 2.7 | 1 | 0.9 |
| HRAS | 1.4 | 0.2 | 0.2 | 4.8 | 0.1 | 3.4 | 0.7 | 0.7 | 5.1 | 0.1 | 2.4 |
| All RAS | 23.3 | 51.5 | 20.9 | 21.4 | 79.3 | 18.9 | 19.4 | 19.3 | 16.4 | 15 | 17.1 |

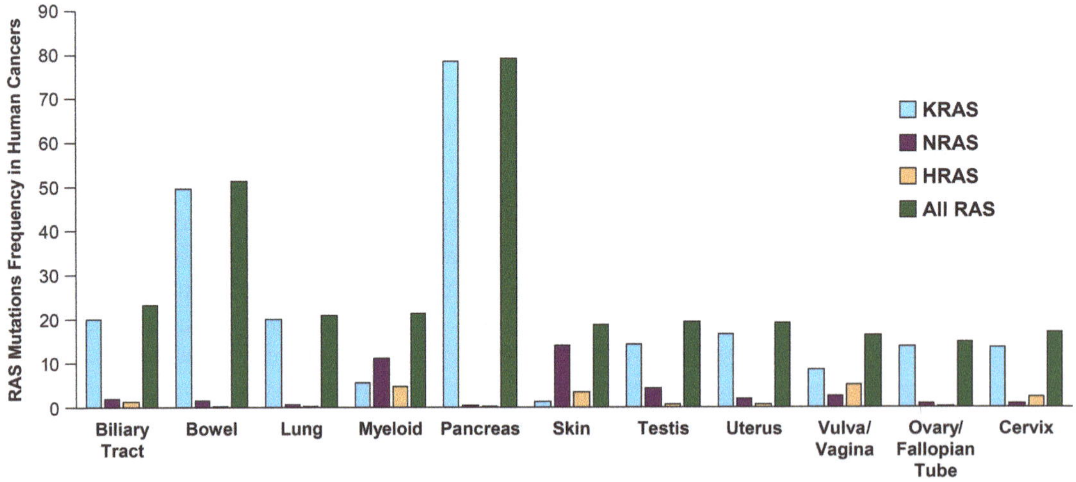


**Fig. 1** Distribution of RAS isoform (KRAS, NRAS, and HRAS) mutations across tumor types and frequency (%) of RAS mutations by isoform in each tumor type. RAS mutation frequency data were compiled from databases and public resource, including AACR Genie (Release 6.1-public), COSMIC (v90), and cBioPortal, TCGA Research Network: https:/www.cancer.gov/tcga and NCI's Genomic Data Commons (GDC). All human cancers that had a sample size greater than or equal to 300 and a total RAS mutation frequency greater than or equal to 15% are listed

With the exception of myeloid and skin cancers, where NRAS is the most frequent isoform, KRAS mutations predominate in other cancer types where RAS is a driver.

## 3 RAS Hydrolysis

RAS proteins cycle between an inactive GDP bound, "off-state" and an active GTP bound, "on-state" [3]. The cycling between these two states is essential for RAS to function as a molecular switch enabling the transmission of extracellular growth factor stimulation and the subsequent initiation of intracellular signal transduction. Rearrangement of the conserved switch regions in the on-state provides a binding site for effector proteins, such as RAF and PI3-kinase. When bound to GTP, two amino acids on KRAS, T35 and G60, form hydrogen bonds with the GTP γ-phosphate and hold switch I and switch II in an active conformation [4]. Hydrolysis of the γ-phosphate bond of the bound GTP results in the release of the switch regions and returns them to their flexible conformation when bound to GDP.

The proportion of RAS in the GTP-bound state is dependent on the hydrolysis rate and the GDP off-rate. The GDP off-rate is approximately 0.04 $\text{min}^{-1}$ ($t_{1/2} = 17$ min) (Waybright and Stephen, "unpublished data"), and the GTP hydrolysis rate is approximately 0.01 $\text{min}^{-1}$ ($t_{1/2} = 70$ min) (*see* Table 1). Certainly, the time scale of these transitions does not enable RAS to function as an efficient molecular switch and would leave it stuck in the off-state. However, the exchange of GDP for cellular GTP is catalyzed by guanine nucleotide exchange (GEF) proteins, such as SOS, which reduces the half-life of KRAS-GDP to just several seconds [5]. Insertion of an α-helix from SOS causes the displacement of switch I and distorts the conformation of switch II, which in combination opens the nucleotide binding pocket facilitating the dissociation of GDP [6]. Binding of cellular GTP, the concentration of which is approximately tenfold higher than GDP [7], causes the displacement of SOS and leaves RAS in the active state. Activated RAS is turned off by the stimulation of GTP hydrolysis through interaction with GTPase-activating proteins (GAP), such as NF1 or RASA1. Binding of GAP to KRAS-GTP stabilizes Q61 and therefore the nucleophilic water. In addition, placement of the arginine finger, the catalytically active amino acid from GAP, adjacent to GTP neutralizes the charge on the β-phosphate of GTP. NF1 stimulates the GTP hydrolysis rate by $10^5$-fold [8, 9], thereby rapidly converting RAS to the off-state. The regulation of GEF and GAP activity and localization is tightly coupled to receptor-tyrosine kinase activation and feedback inhibition [10].

RAS mutations that disrupt the GDP off-rate and GTP hydrolysis rate will inactivate this switching mechanism and result in

**Table 1**
**Summary of intrinsic GTP hydrolysis rate for RAS oncogenic mutants**

| Source | GTPase rate ($min^{-1}$) | | | | | | | |
|---|---|---|---|---|---|---|---|---|
| | WT | G12C | G12D | G12V | G13D | Q61H | Q61L | Q61R |
| John et al. (1988) [19] | 1.02E−02 | | | 2.00E−03 | | | | |
| John et al. (1989) [20] | 2.80E−02 | | | 3.80E−03 | | 1.90E−03 | | |
| Krengel et al. (1990) [21] | 2.80E−02 | | | 2.00E−03 | | | 1.30E−03 | |
| Bollag and McCormick (1991) [22] | 3.92E−03 | | | | | 3.60E−04 | 5.60E−05 | |
| Bollag et al. (1996) [23] | 1.30E−02 | | 5.60E−03 | | | | | |
| Wey et al. (2013) [24] | 1.26E−02 | 1.08E−02 | 8.40E−03 | 3.00E−04 | 1.14E−02 | | | |
| Smith et al. (2013) [14] | 8.80E−03 | | | 8.90E−04 | 3.10E−03 | | 1.10E−04 | |
| Burd et al. (2014) [25] | 2.70E−02 | | 1.38E−02 | | | | | 2.00E−05 |
| Hunter et al. (2015) [26] | 4.08E−02 | 2.94E−02 | 1.14E−02 | 2.52E−03 | 5.76E−03 | | 4.80E−04 | |
| Moghadamchargari et al. (2019) [27] | 2.42E−03 | | | | 9.20E−04 | 2.50E−04 | | |
| Rabara and Stephen (2021) [unpublished] | 8.57E−03 | 2.38E−03 | 2.63E−03 | 8.02E−04 | 6.41E−03 | 1.07E−03 | 3.20E−04 | |

GTPase rates collected from published results. Rate values vary due to measurement of three RAS isoforms, at varying temperature, and with different GTPase assays: HRAS, 37 °C [19]; HRAS, 37 °C [20]; HRAS, 37 °C [21]; KRAS, 25 °C [22]; KRAS, 37 °C [23]; HRAS [24]; HRAS, 37 °C [14]; NRAS, 25 °C [25]; KRAS, 25 °C [26]; and KRAS, 25 °C [27], Rabara and Stephen, 2021, "unpublished" (KRAS, 37 °C)

accumulation of RAS in the on-state. Oncogenic KRAS mutations are frequently observed at positions G12 and G13 and to a lesser extent as Q61 and A146. These oncogenic mutations can be categorized into those which are GAP insensitive, show fast GDP dissociation or a combination of both. Mutation of G12 to any of amino acid, except proline, resulted in cell transformation [11]. This result is easily explained based on the analysis of the structure of the RAS:GAP transition state complex that indicates replacement of the glycine 12 to any other amino acid would prevent the correct orientation of the arginine finger from GAP making these mutants GAP insensitive [8]. The steric effect of a proline at position 12 may still allow placement of the arginine finger to enable GAP-catalyzed GTP hydrolysis [12]. Mutation of G13, commonly to aspartic acid, results in a reduction in GAP-mediated GTP hydrolysis [13] but more importantly increases the GDP off-rate [14], causing a significant increase in the KRAS-GTP levels. Finally, less frequently observed, A146 mutations have a dramatically enhanced GDP off-rate, although retain GAP sensitivity (Rabara and Stephen, "unpublished data" and [15]) which suggests that loss of GEF sensitivity results in an oncogenic phenotype.

As described earlier, the level of RAS-GTP is determined by the ratio of the GDP off-rate and GTP hydrolysis rate. When KRAS oncogenic mutants lose GAP sensitivity, their intrinsic GTP hydrolysis rate will become limiting, dysregulating the switching mechanism, and leaving RAS stuck in the GTP state. Over the last 35 years, there have been many publications that measure the intrinsic GTPase rate of common oncogenic mutants [16]. However, differences in the RAS isoform, temperature, experimental methods, and conditions make comparing rates difficult (summarized in Tables 1 and 2). The decrease in GTPase rates (relative to wild-type RAS) for common KRAS mutations is presented in Fig. 2. Q61H/L/R and G12V all have a decrease in GTPase rate of more than tenfold compared with wild type. Rather surprisingly, G12C/D and G13D have a modest loss in GTPase activity of only threefold or less compared to wild type. This enhanced cycling of G12C has proven significant for therapeutic purposes as the first clinically approved covalent drugs (Sotorasib and Adagrasib) only targeted the off-state. The more dramatic loss of GTPase activity for Q61 mutant would predict that these mutant alleles would be locked in their active state. However, when measured, the levels of RAS-GTP were all lower than 50% [17]. Indeed, a recent study that measured the GTPase and GDP off-rates WT KRAS, G12V, G13D, and Q61L found the GTPase rates were faster and the GDP off-rates were slower in cells than those measured with recombinant proteins [18]. These data suggest that intracellular factors may

**Table 2**
**Summary of the relative decrease in the intrinsic GTPase rate for RAS oncogenic mutants**

| | Fold decrease in GTPase rate relative to WT | | | | | | |
|---|---|---|---|---|---|---|---|
| **Source** | **G12C** | **G12D** | **G12V** | **G13D** | **Q61H** | **Q61L** | **Q61R** |
| John et al. (1988) [19] | | | 5.1 | | | | |
| John et al. (1989) [20] | | | 7.4 | | | | |
| Krengel et al. (1990) [21] | | | 14.0 | | | 21.5 | |
| Bollag and McCormick (1991) [22] | | | | | 10.9 | 70.0 | |
| Bollag et al. (1996) [23] | | 2.3 | | | | | |
| Wey et al. (2013) [24] | 1.2 | 1.5 | 42.0 | 1.1 | | | |
| Smith et al. (2013) [14] | | | 9.9 | 2.8 | | | |
| Burd et al. (2014) [25] | | 2.0 | | | | | 1350.0 |
| Hunter et al. (2015) [26] | 1.4 | 3.6 | 16.2 | 7.1 | | 85.0 | |
| Moghadamchargari et al. (2019) [27] | | | | 2.6 | 9.7 | | |
| Rabara and Stephen (2021) [unpublished] | 3.6 | 3.3 | 10.7 | 1.3 | 8.0 | 26.8 | |

The GTPase rates shown in Table 1 have been normalized by dividing the WT rate by the mutant rate. This allows a comparison of the mutational impact on GTPase activity for the oncogenic mutants that is independent of the RAS isoform, temperature, and measurement technique

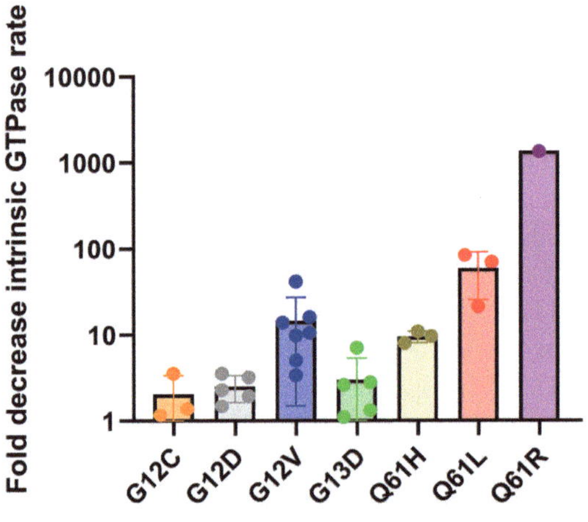


**Fig. 2** Relative decrease in the intrinsic GTPase rates for selected RAS mutants compared to wild type. (Source data listed in Table 2)

modulate these rates and impact the level of KRAS-GTP in cancer cells. Clearly further work is required to identify these factors and determine if these represent novel therapeutic opportunities for targeting mutant KRAS.

## 4 Approaches to Targeting RAS

The search to find compounds which bind directly to RAS proteins has been long and challenging. The well-documented lack of a deep-binding pocket presents a serious problem, but until recently, efforts to find compounds that bind directly to RAS, or to any protein, were limited by lack of a suitable assay to detect binding. The discovery of a fragment of RAF that binds to RAS in the GTP state presented an opportunity to screen for compounds that prevent RAF interaction by binding RAS [28]. However, the interface between RAS and the RBD contains no pockets to which a small molecule could bind.

Development of biophysical methods such as NMR and SPR enabled the detection of small molecule binding directly. SAR by NMR was used to identify fragments that bind to another difficult target, BCL-2. By stitching two low-affinity (131 μM, 60 μM) binders together, a compound was produced that binds at 0.32 μM, leading eventually to development of Venetoclax [29]. In 2012, a group at Genentech and another at Abbot labs published results of SAR by NMR analysis of KRAS [30, 31]. Both groups identified fragments that bind directly to KRAS, both at the same site. Unfortunately, this site is the SOS-binding pocket: compounds blocking SOS binding are likely to inhibit wild-type RAS signaling, while having little or no effect on oncogenic mutants. Furthermore, only one site was detected on the KRAS protein, precluding the approach of joining two independent fragments together to increase affinity. Nevertheless, these were significant steps toward drugs that bind RAS and inhibit its function.

The discovery of compounds that bind covalently to KRAS G12C, by Jonathan Ostrem and colleagues in Kevan Shokat's lab [32], was a landmark breakthrough that launched a stampede of G12C inhibitors into the clinic. Currently in November 2023, there are 11 G12C inhibitors, all of which bind at the "switch II" pocket in the GDP-bound form of KRAS G12C. Two of these, sotorasib from Amgen and adagrasib from Mirati, have already received FDA approval. This covalent approach solved two problems facing RAS inhibitors: covalent binding overcomes the problem of identifying high-affinity binders on a protein with very few pockets, and direct engagement of the activating cysteine at codon 12 confers unique specificity for the oncoprotein.

An early concern relating to this covalent approach was that compounds that bind weakly to the KRAS protein would need a relatively "hot" warhead to react with the G12C efficiently. A hot electrophile could have off-target toxicity. However, by good fortune, a lysine residue close to the reactive cysteine in the binding site activates the electrophile, so that a relatively weak warhead is sufficient [33]. The pocket to which the current generation of

G12C inhibitors bind is only accessible in the GDP state. At first, this seemed to be a potential liability, as the active form of KRAS is, of course, the GTP-bound state. Fortunately, KRAS G12C retains a relatively high intrinsic GTPase rate and therefore cycles into the GDP state sufficiently frequently to allow capture by the covalent compounds. Once bound, the KRAS G12C protein accumulates in the inactive state. However, in cells, the G12C inhibitors need to react before the KRAS protein switches back to the GTP-state, a process that is catalyzed by SOS. Therefore, upstream signals that increase SOS binding to KRAS G12C compete for drug engagement and limit efficacy.

Discovery of a compound that binds to KRAS G12D noncovalently with an affinity that is sub-nanomolar marks another important step forward [34]. This compound, MRTX1133, is potent against KRAS G12D cells with exquisite specificity for this allele. It is currently in Phase I clinical trials. In addition, drugs that target KRAS, but spare HRAS and NRAS, are being developed and present exciting opportunities as well as potential liabilities. One of these pan-KRAS drugs, developed by Boehringer Ingelheim, has been described recently and is expected to enter clinical trials in late 2023 [35]. The obvious liability of pan-KRAS drugs is the unknown consequences of inhibiting wild-type KRAS. Mouse models suggest this should not cause problems (discussed in [36]), but this concern can only be addressed in clinical trials.

The G-domains of HRAS, KRAS, and NRAS proteins are highly conserved, the first 80 amino acids being identical. However, residue 95 differs between isoforms and several KRAS G12C inhibitors engage this residue as part of the drug binding site. Early attempts to target KRAS specifically at the RAS Initiative at the Frederick National Laboratory focused on covalent attack on His-95. Compounds were identified that engaged this residue, but the chemical nature of electrophiles that react with histidine was not suitable for drug development. Nevertheless, these compounds, which were developed to target the GTP-form of KRAS, provided useful building blocks for future drug discovery efforts based on GTP-bound forms of KRAS.

Targeting the GTP-bound form of mutant RAS proteins has been more challenging than the GDP-bound forms but may provide significant advantages in the clinic. Resistance to G12C GDP inhibitors often involves increases in upstream signaling or increases in KRAS expression [37]. Upstream signaling generates active SOS, which, as discussed above, competes with GDP-state inhibitors for binding. If SOS wins this competition, the KRAS G12C protein cycles to the GTP form and drug cannot bind. Increased KRAS production can have two effects: newly synthesized KRAS is predominantly GTP-bound and therefore drug resistant. Also, a fraction of KRAS G12C is likely to escape from drug binding at some rate, and, obviously, the more KRAS protein is expressed, the more

KRAS G12C GTP is produced and able to maintain oncogenic signaling. In contrast, drugs that target the active GTP-bound state are expected to bind more efficiently in these circumstances and may therefore provide additional clinical benefit. Drugs developed by Revolution Medicines and the RAS Initiative with Bridge Bio will test this hypothesis in the clinic in the next few years.

While identification of compounds that bind directly to KRAS has been a major focus of drug discovery for the last 10 years, several other indirect approaches to target KRAS cancers are also being pursued on the assumption that multiple different forms of attack will be necessary to achieve significant clinical benefit and overcome resistance mechanisms. Inhibitors of enzyme downstream of KRAS, such as RAF, MEK, ERK, and PIK3CA, have not been effective to date, for various reasons [38]. Lack of therapeutic window is generally the major limitation, as these pathways are essential for normal cell signaling. At the NCI RAS Initiative, we have aimed at blocking the process by which KRAS activates its effectors rather than blocking the effectors themselves. These processes are different in KRAS cancers relative to normal cells, offering the possibility of a therapeutic window. For example, KRAS cancers depend on the activation of RAF1 exclusively, whereas normal cells can function normally without RAF1, by utilizing BRAF and ARAF [39]. RAF1 activation depends on SHOC2 in cancer cells expressing oncogenic KRAS, but not in cells expressing the wild-type KRAS protein [40]. KRAS depends on direct binding and activation of PIK3CA, whereas normal cells do not [41]. By analyzing structures of protein complexes, such as KRAS/RAF RBDCRD [42] and MRAS/SHOC2/PP1C [43], we have been able to identify potential new targets for therapeutic intervention.

## 5 Future Prospects

In the years ahead, research on RAS will focus on the effects of blocking KRAS in the clinic. Drugs targeting specific alleles of KRAS, different states of KRAS (ON vs OFF), pan-KRAS, and pan-RAS inhibitors are already in early stages of clinical trials and will be tested in protocols aimed at increasing potency, preventing or reducing resistance, and harnessing the immune system. Many more compounds targeting KRAS directly are sure to follow, as our understanding of KRAS structure and biochemistry continues to evolve. Analysis of data derived from clinical experience will likely inspire new approaches and combinations for future evaluation. In parallel, efforts to target KRAS effector pathways and other pathways that KRAS depends on will be tested in the clinic. This expectation assumes that substantial and durable clinical benefit from targeting KRAS will require multiple approaches, direct and indirect, as well as a better understanding of mechanisms of drug resistance and factors that affect clinical responses in individual patients.

Success in targeting KRAS cancers will likely inform new ways of targeting RAS family members, such as NRAS, and other GTPases whose roles in cancer and other diseases has been long appreciated, but not addressed therapeutically. Members of the Rac and Rho family of GTPases are obvious examples. Furthermore, advances in structural and computational biology will enable progress in understanding how complex signaling pathways function, particularly in local environments such as the plasma membrane, which has been particularly challenging in RAS biology.

The urgent clinical need to treat KRAS cancers has driven this ongoing revolution in RAS research, including the establishment of the NCI RAS Initiative to focus on this problem [44]. We still have a long way to go before we can declare "mission accomplished," but the path ahead and the steps that need to be taken are quite clear.

## Acknowledgments

This project was funded in part with federal funds from the National Cancer Institute, National Institutes of Health Contract 75N91019D00024. The content of this publication does not necessarily reflect the views or policies of the Department of Health and Human Services, and the mention of trade names, commercial products, or organizations does not imply endorsement by the US government.

### References

1. Prior IA, Hood FE, Hartley JL (2020) The frequency of Ras mutations in cancer. Cancer Res 80:2969–2974
2. Prior IA, Lewis PD, Mattos C (2012) A comprehensive survey of Ras mutations in cancer. Cancer Res 72:2457–2467
3. Gibbs JB, Sigal IS, Poe M, Scolnick EM (1984) Intrinsic GTPase activity distinguishes normal and oncogenic ras p21 molecules. Proc Natl Acad Sci U S A 81:5704–5708
4. Vetter IR, Wittinghofer A (2001) The guanine nucleotide-binding switch in three dimensions. Science 294:1299–1304
5. Huang WYC, Alvarez S, Kondo Y et al (2021) Relating cellular signaling timescales to single-molecule kinetics: a first-passage time analysis of Ras activation by SOS. Proc Natl Acad Sci U S A 118:e2103598118
6. Boriack-Sjodin PA, Margarit SM, Bar-Sagi D, Kuriyan J (1998) The structural basis of the activation of Ras by Sos. Nature 394:337–343
7. Traut TW (1994) Physiological concentrations of purines and pyrimidines. Mol Cell Biochem 140:1–22
8. Scheffzek K, Lautwein A, Kabsch W et al (1996) Crystal structure of the GTPase-activating domain of human p120GAP and implications for the interaction with Ras. Nature 384:591–596
9. Ahmadian MR, Hoffmann U, Goody RS, Wittinghofer A (1997) Individual rate constants for the interaction of Ras proteins with GTPase-activating proteins determined by fluorescence spectroscopy. Biochemistry 36: 4535–4541
10. Bandaru P, Kondo Y, Kuriyan J (2019) The interdependent activation of Son-of-Sevenless

and Ras. Cold Spring Harb Perspect Med 9: a031534

11. Seeburg PH, Colby WW, Capon DJ et al (1984) Biological properties of human c-Ha-ras1 genes mutated at codon 12. Nature 312:71–75
12. Der CJ, Finkel T, Cooper GM (1986) Biological and biochemical properties of human rasH genes mutated at codon 61. Cell 44:167–176
13. Rabara D, Tran TH, Dharmaiah S et al (2019) KRAS G13D sensitivity to neurofibromin-mediated GTP hydrolysis. Proc Natl Acad Sci U S A 116:22122–22131
14. Smith MJ, Neel BG, Ikura M (2013) NMR-based functional profiling of RASopathies and oncogenic RAS mutations. Proc Natl Acad Sci U S A 110:4574–4579
15. Poulin EJ, Bera AK, Lu J et al (2019) Tissue-specific oncogenic activity of KRAS(A146T). Cancer Discov 9:738–755
16. Burge RA, Hobbs GA (2022) Not all RAS mutations are equal: a detailed review of the functional diversity of RAS hot spot mutations. Adv Cancer Res 153:29–61
17. Donovan S, Shannon KM, Bollag G (2002) GTPase activating proteins: critical regulators of intracellular signaling. Biochim Biophys Acta 1602:23–45
18. Zhao Q, Fujimiya R, Kubo S et al (2020) Real-time in-cell NMR reveals the intracellular modulation of GTP-bound levels of RAS. Cell Rep 32:108074
19. John J, Frech M, Wittinghofer A (1988) Biochemical properties of Ha-ras encoded p21 mutants and mechanism of the autophosphorylation reaction. J Biol Chem 263:11792–11799
20. John J, Schlichting I, Schiltz E et al (1989) C-terminal truncation of p21H preserves crucial kinetic and structural properties. J Biol Chem 264:13086–13092
21. Krengel U, Schlichting I, Scherer A et al (1990) Three-dimensional structures of H-ras p21 mutants: molecular basis for their inability to function as signal switch molecules. Cell 62: 539–548
22. Bollag G, McCormick F (1991) Differential regulation of rasGAP and neurofibromatosis gene product activities. Nature 351:576–579
23. Bollag G, Adler F, elMasry N et al (1996) Biochemical characterization of a novel KRAS insertion mutation from a human leukemia. J Biol Chem 271:32491–32494
24. Wey M, Lee J, Jeong SS et al (2013) Kinetic mechanisms of mutation-dependent Harvey Ras activation and their relevance for the development of Costello syndrome. Biochemistry 52:8465–8479
25. Burd CE, Liu W, Huynh MV et al (2014) Mutation-specific RAS oncogenicity explains NRAS codon 61 selection in melanoma. Cancer Discov 4:1418–1429
26. Hunter JC, Manandhar A, Carrasco MA et al (2015) Biochemical and structural analysis of common cancer-associated KRAS mutations. Mol Cancer Res 13:1325–1335
27. Moghadamchargari Z, Huddleston J, Shirzadeh M et al (2019) Intrinsic GTPase activity of K-RAS monitored by native mass spectrometry. Biochemistry 58:3396–3405
28. Nassar N, Horn G, Herrmann C et al (1995) The 2.2 A crystal structure of the Ras-binding domain of the serine/threonine kinase c-Raf1 in complex with Rap1A and a GTP analogue. Nature 375:554–560
29. Shuker SB, Hajduk PJ, Meadows RP, Fesik SW (1996) Discovering high-affinity ligands for proteins: SAR by NMR. Science 274:1531–1534
30. Sun Q, Burke JP, Phan J et al (2012) Discovery of small molecules that bind to K-Ras and inhibit Sos-mediated activation. Angew Chem Int Ed Engl 51:6140–6143
31. Burns MC, Sun Q, Daniels RN et al (2014) Approach for targeting Ras with small molecules that activate SOS-mediated nucleotide exchange. Proc Natl Acad Sci U S A 111: 3401–3406
32. Ostrem JM, Peters U, Sos ML et al (2013) K-Ras(G12C) inhibitors allosterically control GTP affinity and effector interactions. Nature 503:548–551
33. Hansen R, Peters U, Babbar A et al (2018) The reactivity-driven biochemical mechanism of covalent KRAS(G12C) inhibitors. Nat Struct Mol Biol 25:454–462
34. Wang X, Allen S, Blake JF et al (2022) Identification of MRTX1133, a noncovalent, potent, and selective KRAS(G12D) inhibitor. J Med Chem 65:3123–3133
35. Kim D, Herdeis L, Rudolph D et al (2023) Pan-KRAS inhibitor disables oncogenic signalling and tumour growth. Nature 619:160–166
36. Simanshu DK, Nissley DV, McCormick F (2017) RAS proteins and their regulators in human disease. Cell 170:17–33
37. Dunnett-Kane V, Nicola P, Blackhall F, Lindsay C (2021) Mechanisms of resistance to KRAS (G12C) inhibitors. Cancers (Basel) 13:151
38. Moore AR, Rosenberg SC, McCormick F, Malek S (2020) RAS-targeted therapies: is the undruggable drugged? Nat Rev Drug Discov 19:533–552

39. Blasco RB, Francoz S, Santamaria D et al (2011) c-Raf, but not B-Raf, is essential for development of K-Ras oncogene-driven non-small cell lung carcinoma. Cancer Cell 19:652–663
40. Wang T, Yu H, Hughes NW et al (2017) Gene essentiality profiling reveals gene networks and synthetic lethal interactions with oncogenic Ras. Cell 168:890–903 e15
41. Gupta S, Ramjaun AR, Haiko P et al (2007) Binding of ras to phosphoinositide 3-kinase p110alpha is required for ras-driven tumorigenesis in mice. Cell 129:957–968
42. Tran TH, Chan AH, Young LC et al (2021) KRAS interaction with RAF1 RAS-binding domain and cysteine-rich domain provides insights into RAS-mediated RAF activation. Nat Commun 12:1176
43. Bonsor DA, Alexander P, Snead K et al (2022) Structure of the SHOC2-MRAS-PP1C complex provides insights into RAF activation and Noonan syndrome. Nat Struct Mol Biol 29: 966–977
44. Nissley DV, McCormick F (2022) RAS at 40: update from the RAS Initiative. Cancer Discov 12:895–898

# Chapter 2

# The Abundance of KRAS and RAS Gene Mutations in Cancer

**Edward C. Stites**

## Abstract

Mutant forms of the *RAS* genes *KRAS*, *NRAS*, and *HRAS* are important and common drivers of cancer. Recently, two independent teams that integrated cancer genomics with cancer epidemiology estimated that approximately 15–20% of all human cancers harbor a mutation in one of these three *RAS* genes. These groups also estimate *KRAS* mutations occur in 11–14% of all human cancers. Although these estimates are lower than many commonly encountered values, these estimates continue to rank *KRAS* and the ensemble of *RAS* oncogenes among the most common genetic drivers of cancer across all forms of malignancy.

**Key words** KRAS, NRAS, HRAS, Mutations, Epidemiology

## 1 Introduction

More than 40 years of research have investigated the role of the three *RAS* genes (*KRAS*, *NRAS*, and *HRAS*) in human cancer [1]. It has long been recognized that somatic *RAS* mutations are common in human cancer [2, 3]. The scientific literature frequently includes estimates for how common *RAS* mutations are found in human cancer, and the values presented vary widely. For example, one can find articles that cite the *RAS* mutation incidence in human cancer to be 33% [4, 5], 30% [6–13], 27% [14], 25% [15–17], 20% [18, 19], 19% [20, 21], and 15% [22–25]. Estimated values for *KRAS* mutations in human cancer also vary widely. For example, values of 30% [26], 28.7% [27], 20% [28], and 6.7% [29] can be found in the published scientific literature.

For a gene that has been so well studied, it might be surprising that the question of *RAS* mutation abundance in cancer has not been answered with more precision. With genomic approaches now implemented widely in cancer research and cancer medicine, the abundance of *RAS* gene mutations in the cancer patient population may seem like it would be a simple and straightforward question to answer.

Andrew G. Stephen and Dominic Esposito (eds.), *KRAS: Methods and Protocols*, Methods in Molecular Biology, vol. 2797,
https://doi.org/10.1007/978-1-0716-3822-4_2,

Recently, two different groups combined cancer epidemiology and cancer genomics to estimate the abundance of RAS mutations in human cancer. One group estimated that 15% of all human cancers have a *KRAS*, *NRAS*, or *HRAS* mutation and that the rates of each, respectively, is approximately 11%, 3%, and 1% [25]. Another group estimated that 19% of all human cancers have a *KRAS* (14%), *NRAS* (3%), or *HRAS* (1%) mutations by using a similar approach [20]. The discrepancy between 15% and 19% total *RAS* mutations (and 11% and 14% *KRAS* mutations) may be simply explained by the types of mutations included: the study estimating 15% excluded copy number variation, while copy number variation appears to have been included within the study that estimated 19%. Overall, the approaches and estimates appear to be in good high-level agreement, suggesting that 15–20% may be a more accurate range to use when estimating *RAS* mutation abundances.

However, there is quite a bit of nuance with respect to answering the question of how commonly *RAS* genes are mutated in human cancer. Further consideration of the nuance may be helpful for scientists and clinicians considering *RAS* (and other gene) mutations in cancer. This chapter will discuss estimates for *RAS* mutant abundances, approaches to estimate these values, and some of the challenges to obtaining accurate estimates.

### 1.1 Estimating RAS Mutation Abundances Before the Age of Cancer Genomics

Before cancer genetic and genomic databases became prevalent, review articles that aggregated data from a variety of studies involving tumor samples and/or cancer cell lines helped communicate that *RAS* gene mutations were common in a wide variety of cancers. For example, one highly cited review article [3] included a table titled "Incidence of ras mutations in human cancer" that summarized approximately 75 different measurements of the percentage of *RAS* gene mutations within specific tumor types, with 2230 total tumor samples summarized in the table. The table legend highlighted that some of the studies only considered *HRAS* codon 12 mutations, and some only considered *KRAS* codon 12 or 13 mutations. Even though no summary statistic was presented for the percentage of all of these samples that had a *RAS* mutation, this manuscript is commonly cited after many statements that describe the approximate percentage of human tumors that harbor a RAS mutation [6, 7, 9, 18]. Of note, these references estimated either 20% or 30%; within Table 1 of the cited review article [3], 525 of the 2230 samples, or ~24%, had a *RAS* mutation detected.

The Catalogue of Somatic Mutations in Cancer (COSMIC) is an important online database that aggregates information about cancer genetics at scale [30–32]. COSMIC began before cancer genomics became widespread. The first publication detailing COSMIC in 2004 discussed its original focus on four genes: *BRAF*,

*KRAS*, *NRAS*, and *HRAS* [30]. The data came from published articles that described the presence/absence of mutations in these genes. At the time of that first COSMIC publication, there were already 66,634 total samples [30]. Most of the tumor samples included in this database had information for only one of these four genes (51,988 of the 66,634 samples). The majority of the samples had some characterization of *KRAS* (35,716 of the 66,634 total samples). Thus, this resource was able to aggregate far more data than was found in traditional reviews and further highlighted the importance of mutations in these genes in a wide variety of cancers.

When using data from a review or a database, it is important to carefully consider the information that goes into the database. For example, studies that evaluate the presence/absence of *KRAS* mutations in cancer samples may only consider mutations that occur at the codon 12 and codon 13 hotspots in exon 2 and/or may limit analysis to a subset of the *KRAS* mutations [31, 33–35]. As the overwhelming majority of *KRAS* mutations occur at codon 12, this may be very well justified in the parent studies. However, when all of these studies are aggregated, the histograms that show the total number of mutations observed at each codon will naturally have disproportionately high peaks for codons that are both more common and more commonly assessed. There may also be reporting bias and bias for which studies are eventually accepted for publication [31]. Thus, a simple ratio between the number of mutations observed and the number of tumors analyzed may not actually be reflecting the proportion of the analyzed tumors that have a mutation in the considered gene, and the value may not be representative of what should more generally be expected for that type of tumor.

Another challenge is that there may be bias in which types of cancer are analyzed for *RAS* mutations. For example, *KRAS* mutations are known to be relatively common in pancreatic cancer, colorectal cancer, and lung adenocarcinoma and relatively rare in prostate and breast cancers. *KRAS* mutations additionally are relevant to treatment decisions in colorectal cancer, which may further motivate additional measurements of *KRAS* status in colorectal cancer and may lead to additional publications about *KRAS* mutation prevalence in colorectal cancer. A resource that aggregates data from a variety of sources can be useful for a variety of questions, and many reviews summarize this information to help convey the landscape of different cancers and which commonly involve *KRAS*, *NRAS*, or *HRAS* mutations [36, 37]. However, the ratio of samples included in COSMIC that includes a *KRAS* mutation may not accurately reflect the number of all cancer patients who have a *KRAS* mutation [31]; for example, it may yield a value of nearly 30%, whereas estimates that attempt to correct for biases in tumor-type sampling may find the prevalence of KRAS mutations in new diagnoses is more likely to be half to one third of that value [20, 25].

### 1.2 Cancer Genomics, Pan-Cancer Analyses, and Pan-Cancer Genomics Databases

One benefit of cancer genomics is that these studies should characterize and identify mutations in any exon of the gene. Although there can still be other sources of false positives and false negatives in genomic analysis, including the ratio of tumor to normal tissue in a sample, tumor heterogeneity, and sample quality [25], cancer genomics does seem to offer a more comprehensive characterization of the mutation status of the gene, as well as the assessment of many genes in parallel.

The age of cancer genomics includes many studies of individual forms of cancer. With the *RAS* genes long known to be common in a wide variety of cancers [2, 3], it was unsurprising for the *RAS* genes to commonly be found among the most mutated genes when individual forms of cancer were sequenced. For example, pancreatic adenocarcinoma sequencing found 93% had a *KRAS* mutation [38], lung adenocarcinoma sequencing found 33% had a *KRAS* mutation [39], melanoma sequencing found 28% had an *NRAS* mutation [40], and head and neck squamous cell carcinomas found 4% had an *HRAS* mutation [41]. Overall, cancer genomic information is quite useful for providing insight into the approximate level of *RAS* gene mutations in individual forms of cancer.

As the quantity of cancer genomics data increased for a variety of forms of cancer, "pan-cancer" genomic studies began to appear. Although the prefix pan- means "all" or "everything," pan-cancer studies do not truly include all forms of cancer. For example, an early pan-cancer genomics analysis only included 12 forms of cancer and did not include several common forms of cancer, such as prostate cancer, pancreatic adenocarcinoma, and melanoma [42].

Another important aspect to consider with a pan-cancer resource is that the relative numbers of tumor samples for the included cancer types may not be proportional to the relative burdens of those cancer types in the patient population. The number of samples with a mutation in a pan-cancer study will therefore be impacted by how the different included cancers are sampled. For example, an effort called the Pan-Cancer Atlas and that included approximately 10,000 tumor samples in its initial high-profile series of reports looked at 33 different forms of cancer [43]. Within the Pan-Cancer Atlas, the pan-cancer analysis by Bailey et al. [43] was an analysis of mutations in 9423 total samples, of which 311 (~3%) were glioblastoma and 510 (~5%) were lower-grade glioma. In the cancer patient population, these cancers combined comprise ~1% of all new cancer diagnoses [25]. Notably, *IDH1* mutations are important drivers of lower-grade gliomas (78% of the lower-grade gliomas analyzed in this study) and glioblastoma (5% of the glioblastomas analyzed in this study) but are less commonly encountered in most other cancers. Accordingly, *IDH1* mutations were found in ~6% of the samples from the Bailey et al.'s pan-cancer analysis, whereas they are estimated to be found in ~1% of all human tumors [25]. Thus, although the inclusion of a variety of cancer types within a pan-cancer analysis can offer many benefits for a wide variety of investigations, the proportion of samples with a mutation in such a collection of samples is not a good proxy for the proportion of all cancer patients who would also have that mutation. Nevertheless, it is fairly common to find estimates for mutations in the cancer patient population that refer to mutation proportions from pan-cancer cohorts [29].

The issue of proportionality is important to consider when one considers cancer genomics databases and resources that aggregate information from a variety of studies, including COSMIC [30–32], cBioPortal [44, 45], AACR Genie [46], and the TCGA [38–41, 43]. Although the data from these studies can be extremely useful for a variety of potential applications, it remains important to consider what the component data are, how they are desired to be used, and how well the question that would like to be answered will be addressed considering the nature of the data being considered.

### 1.3 Integration of Epidemiology and Cancer Genomics to Estimate RAS Mutation Abundances

Cancer epidemiology resources are available that catalogue the relative abundances of the different forms of cancer. For example, the NCI Surveillance, Epidemiology, and End Results (SEER) Program [47, 48] catalogues cases of cancer in the US patient population. The American Cancer Society (ACS) communicates an annual summary of cancer in the US population [49] that is based upon the NCI SEER data. Within the SEER data, cancer cases are categorized using both histology and anatomical location codes using terminology specified by the World Health Organization (WHO). Of note, there are more than 600 different malignant histology codes and more than 100 different anatomical location codes in the WHO International Classification of diseases for oncology, version 3 (ICD-O-3) [50]; each cancer is characterized by a pair of histology and anatomical location codes. These fine-grained subclassifications are commonly aggregated in analyses that use these data; for example, they are grouped by anatomical locations in the ACS report.

The lack of proportional representation between pan-cancer studies and pan-cancer resources can be identified by comparing the numbers of samples for each cancer type within a genetics or genomics data set with the number of cases of that cancer diagnosis in the cancer patient population within the epidemiological data [25]. One can envision a few approaches to develop a genomic resource where the number of samples for each form of cancer was proportional to the actual incidence of each form of human cancer. At one extreme, if every new cancer diagnoses had the corresponding tumor sequenced, the database would naturally have proportional representation. Such an effort does not currently appear practical for a variety of reasons, but an alternative would be assembling a cancer database where the number of cancer samples for each cancer type was proportional to the incidence of that cancer type. One issue with such an approach is that the large differences in incidence between the most common and least common cancers would result in a very small number of samples for rare cancers and a very large number of samples for common cancers. As the number of samples per tumor that accurately characterize the driver genes of a cancer type depends upon a variety of factors, such as background mutation rate [51], a proportionately representative database may introduce new problems. Alternative methods for estimating mutation abundances in the cancer patient population may therefore be useful.

One alternative approach to understanding the entire cancer patient population would involve weighting the information from the different cancer types in a manner proportional to the incidence of those cancers. Two different groups independently implemented this approach by utilizing epidemiological data to weight mutation frequencies as estimated by cancer genomics and cancer genetics [20, 25]. Here, epidemiology is used to characterize how the relative abundances of the different cancers and genomic and/or genetic data are used to characterize the relative abundances of the gene mutations in the different cancers. This approach is effectively finding the weighted average, where the epidemiological data provides the weights and the genomic data provides the values (percent of cases with a mutation) for which the weighted average is being taken [25, 52].

The mathematics of taking a weighted average is fairly straightforward. Almost every college student has performed a weighted average to calculate their grades in a class; in these scenarios, the weights for the different assignments, quizzes, and exams may be provided in a syllabus, and the grades are unique to each assignment, quiz, or exam. Although the math is straightforward, other factors complicate this approach for estimating population-level mutation frequencies. Inconsistent use of cancer nomenclature between genomic studies, as well as between cancer genomics and epidemiology, presents one significant challenge.

One approach to overcoming the cancer nomenclature involved reannotating cancer genomic samples and cancer epidemiology data to one common naming system. The naming system developed to enable integration of these data was chosen to be pathologically accurate, to correspond well to the types of samples that have been sequenced, and where the information provided with genomic studies enabled consistent subdivision across studies. This resulted in a system of cancer names of varying levels of granularity that spanned the cancer genomics data. After reannotation of the cancer genomic samples and the cancer epidemiology data to this naming system, cancer incidence proportions (from cancer epidemiology) and cancer mutation frequencies (from cancer genomics) could be combined to find the "weighted average" of *RAS* mutations in the cancer patient population.

An alternative, but similar, approach of combining epidemiology with cancer genetic data was included in an excellent review of *RAS* mutations in human cancer [20]. Within this article, estimates for *RAS* mutation rates came from combining epidemiology from the ACS Cancer Survey describing new cases per year in the USA [53] with *RAS* mutation rates from the COSMIC database (including both genomic and genetic studies) [31, 32]. This analysis utilized cancer nomenclature as used in the TCGA [43] and mapped them to the anatomical locations used in the ACS Cancer Survey [53]. For anatomical locations with more than one cancer type (i.e., the lung, with lung adenocarcinoma, lung squamous cell carcinoma, and lung small cell carcinoma), the proportion of each within each anatomical location was estimated and applied to the mutation rates for the individual cancer types.

Despite the differences between the approaches, both studies came to very similar estimates for the overall proportion of all cancers that harbor a *RAS* gene mutation [20, 25]. The Mendiratta et al. study estimated 14%, and the Prior et al. study estimated 19% of tumors harbor a mutation in *KRAS*, *NRAS*, and/or *HRAS*. With respect to the burden of *KRAS* mutations across all malignant tumors, the Mendiratta et al. study estimated 11% and the Prior et al. study estimated 14%. The two studies both estimate *NRAS* mutations are found in 3% of all malignant tumors and that *HRAS* mutations are found in 1% of all malignant tumors. One important difference that may explain the differences in magnitudes for *KRAS* is that the Mendiratta et al. study focused on mutations that result in a change to the amino acid sequence, whereas the Prior et al. study also included copy number variations. This may account for much of the difference in the estimates between the two complementary approaches.

### 1.4 Other Factors to Consider Regarding RAS Mutation Frequency Estimates

With the RAS community commonly stating the *RAS* mutations are found in one quarter to one third of all cancers, it may be seen as surprising that the integration of epidemiology with comprehensive genomic characterizations of cancer find *RAS* mutations in "only" ~15–20% of human cancers. There are several reasons that there may be a discrepancy. As the earlier portion of this chapter highlights, the field has been estimating *RAS* mutation abundances in cancer long before cancer genomics became so widespread. It is possible that estimates based on limited numbers of samples were misleading, and it is possible that values that were originally estimates and approximations began to be interpreted as accurate values due to the frequency with which they were encountered.

It is also possible that the estimates based on genomics are misleading. Consider that both the study of Mendiratta and Prior utilized cancer epidemiological data from the American Cancer Society that is based on cancer epidemiology in the United States. The relative incidence of cancers in the United States may not accurately represent the relative incidences of those same cancers in other countries [54]. If cancers with a high level of *RAS* mutations are more common in other parts of the world, the estimates based on the US cancer incidence levels would be low. Additionally, the rates of mutations in a gene may vary between countries [55–57]. Thus, if the genomic data comes from a different patient population than the population for which an estimate is desired, it may be another possible source of error. One can conceive of other possible sources of error or bias in the data as well [25]. Thus, although the epidemiologically and genomics-informed estimates appear to be reasonable and to have a strong logical and methodological foundation, it is likely that values will continue to be refined as data continue to increase and improve.

## 2 Conclusion

*RAS* mutations are common in cancer. Even though the new epidemio-genomic estimates of the abundance of *RAS* mutations across all cancers are less than historical, popular, estimates, an abundance of 15–19% of all cancers would rank the *RAS* oncogenes among the most common genetic drivers of cancer, and at 11–14% of all cancer diagnoses, *KRAS* would likely be the first or second most mutated proto-oncogene (with *PIK3CA* possibly/likely being slightly more commonly mutated across all cancers). With a critical role in cancer development and with the pharmaceutical targeting of RAS proteins becoming a clinical reality, research on *KRAS*, *NRAS*, and *HRAS* will continue to offer potential benefit to a large number of cancer patients.

### References

1. Chang EH, Gonda MA, Ellis RW et al (1982) Human genome contains four genes homologous to transforming genes of Harvey and Kirsten murine sarcoma viruses. Proc Natl Acad Sci U S A 79:4848–4852
2. Bos JL (1988) The ras gene family and human carcinogenesis. Mutat Res 195:255–271
3. Bos JL (1989) ras oncogenes in human cancer: a review. Cancer Res 49:4682–4689
4. Vigil D, Cherfils J, Rossman KL et al (2010) Ras superfamily GEFs and GAPs: validated and tractable targets for cancer therapy? Nat Rev Cancer 10:842–857
5. Li S, Balmain A, Counter CM (2018) A model for RAS mutation patterns in cancers: finding the sweet spot. Nat Rev Cancer 18:767–777
6. Khosravi-Far R, Der CJ (1994) The Ras signal transduction pathway. Cancer Metastasis Rev 13:67–89
7. Johnson L, Mercer K, Greenbaum D et al (2001) Somatic activation of the K-ras oncogene causes early onset lung cancer in mice. Nature 410:1111–1116
8. Kato-Stankiewicz J, Hakimi I, Zhi G et al (2002) Inhibitors of Ras/Raf-1 interaction identified by two-hybrid screening revert Ras-dependent transformation phenotypes in

human cancer cells. Proc Natl Acad Sci U S A 99:14398–14403

9. Shmulevitz M, Marcato P, Lee PW (2005) Unshackling the links between reovirus oncolysis, Ras signaling, translational control and cancer. Oncogene 24:7720–7728
10. Jin H, Wang X, Ying J et al (2007) Epigenetic silencing of a Ca(2+)-regulated Ras GTPase-activating protein RASAL defines a new mechanism of Ras activation in human cancers. Proc Natl Acad Sci U S A 104:12353–12358
11. Schubbert S, Shannon K, Bollag G (2007) Hyperactive Ras in developmental disorders and cancer. Nat Rev Cancer 7:295–308
12. Murugan AK, Grieco M, Tsuchida N (2019) RAS mutations in human cancers: roles in precision medicine. Semin Cancer Biol 59:23–35
13. Rukhlenko OS, Khorsand F, Krstic A et al (2018) Dissecting RAF inhibitor resistance by structure-based modeling reveals ways to overcome oncogenic RAS signaling. Cell Syst 7: 161–179 e14
14. Ryan MB, Corcoran RB (2018) Therapeutic strategies to target RAS-mutant cancers. Nat Rev Clin Oncol 15:709–720
15. Ward AF, Braun BS, Shannon KM (2012) Targeting oncogenic Ras signaling in hematologic malignancies. Blood 120:3397–3406
16. Saeed O, Lopez-Beltran A, Fisher KW et al (2019) RAS genes in colorectal carcinoma: pathogenesis, testing guidelines and treatment implications. J Clin Pathol 72:135–139
17. Hobbs GA, Der CJ, Rossman KL (2016) RAS isoforms and mutations in cancer at a glance. J Cell Sci 129:1287–1292
18. Downward J (2003) Targeting RAS signalling pathways in cancer therapy. Nat Rev Cancer 3: 11–22
19. Konstantinopoulos PA, Karamouzis MV, Papavassiliou AG (2007) Post-translational modifications and regulation of the RAS superfamily of GTPases as anticancer targets. Nat Rev Drug Discov 6:541–555
20. Prior IA, Hood FE, Hartley JL (2020) The frequency of Ras mutations in cancer. Cancer Res 80:2969–2974
21. Roskoski R Jr (2021) Blockade of mutant RAS oncogenic signaling with a special emphasis on KRAS. Pharmacol Res 172:105806
22. Davies H, Bignell GR, Cox C et al (2002) Mutations of the BRAF gene in human cancer. Nature 417:949–954
23. Kyriakis JM (2009) Thinking outside the box about Ras. J Biol Chem 284:10993–10994
24. Shimizu T, Tolcher AW, Papadopoulos KP et al (2012) The clinical effect of the dual-targeting strategy involving PI3K/AKT/mTOR and RAS/MEK/ERK pathways in patients with advanced cancer. Clin Cancer Res 18:2316–2325
25. Mendiratta G, Ke E, Aziz M et al (2021) Cancer gene mutation frequencies for the U.-S. population. Nat Commun 12:5961
26. Venkatanarayan A, Liang J, Yen I et al (2022) CRAF dimerization with ARAF regulates KRAS-driven tumor growth. Cell Rep 38: 110351
27. Saif MW, Shah M (2009) K-ras mutations in colorectal cancer: a practice changing discovery. Clin Adv Hematol Oncol 7(45–53):64
28. Foster SA, Malek S (2016) The RAS/MAPK axis gets stressed out. Mol Cell 64:854–855
29. Macheret M, Halazonetis TD (2015) DNA replication stress as a hallmark of cancer. Annu Rev Pathol 10:425–448
30. Bamford S, Dawson E, Forbes S et al (2004) The COSMIC (Catalogue of Somatic Mutations in Cancer) database and website. Br J Cancer 91:355–358
31. Forbes S, Clements J, Dawson E et al (2006) Cosmic 2005. Br J Cancer 94:318–322
32. Tate JG, Bamford S, Jubb HC et al (2019) COSMIC: the Catalogue Of Somatic Mutations In Cancer. Nucleic Acids Res 47:D941–D947
33. Bazan V, Migliavacca M, Zanna I et al (2002) Specific codon 13 K-ras mutations are predictive of clinical outcome in colorectal cancer patients, whereas codon 12 K-ras mutations are associated with mucinous histotype. Ann Oncol 13:1438–1446
34. Poehlmann A, Kuester D, Meyer F et al (2007) K-ras mutation detection in colorectal cancer using the Pyrosequencing technique. Pathol Res Pract 203:489–497
35. Neumann J, Zeindl-Eberhart E, Kirchner T et al (2009) Frequency and type of KRAS mutations in routine diagnostic analysis of metastatic colorectal cancer. Pathol Res Pract 205: 858–862
36. Karnoub AE, Weinberg RA (2008) Ras oncogenes: split personalities. Nat Rev Mol Cell Biol 9:517–531
37. Rodriguez-Viciana P, Tetsu O, Oda K et al (2005) Cancer targets in the Ras pathway. Cold Spring Harb Symp Quant Biol 70:461–467
38. Cancer Genome Atlas Research Network (2017) Integrated genomic characterization of pancreatic ductal adenocarcinoma. Cancer Cell 32:185–203 e13

39. Cancer Genome Atlas Research Network (2014) Comprehensive molecular profiling of lung adenocarcinoma. Nature 511:543–550
40. Cancer Genome Atlas Network (2015) Genomic classification of cutaneous melanoma. Cell 161:1681–1696
41. Cancer Genome Atlas Network (2015) Comprehensive genomic characterization of head and neck squamous cell carcinomas. Nature 517:576–582
42. Kandoth C, McLellan MD, Vandin F et al (2013) Mutational landscape and significance across 12 major cancer types. Nature 502:333–339
43. Bailey MH, Tokheim C, Porta-Pardo E et al (2018) Comprehensive characterization of cancer driver genes and mutations. Cell 173:371–385 e18
44. Gao J, Aksoy BA, Dogrusoz U et al (2013) Integrative analysis of complex cancer genomics and clinical profiles using the cBioPortal. Sci Signal 6:pl1
45. Cerami E, Gao J, Dogrusoz U et al (2012) The cBio cancer genomics portal: an open platform for exploring multidimensional cancer genomics data. Cancer Discov 2:401–404
46. AACR Project GENIE Consortium (2017) AACR Project GENIE: powering precision medicine through an International Consortium. Cancer Discov 7:818–831
47. Hayat MJ, Howlader N, Reichman ME et al (2007) Cancer statistics, trends, and multiple primary cancer analyses from the Surveillance, Epidemiology, and End Results (SEER) Program. Oncologist 12:20–37
48. Hankey BF, Ries LA, Edwards BK (1999) The surveillance, epidemiology, and end results program: a national resource. Cancer Epidemiol Biomark Prev 8:1117–1121
49. Siegel RL, Miller KD, Wagle NS et al (2023) Cancer statistics, 2023. CA Cancer J Clin 73:17–48
50. World Health Organization (2013) International classification of diseases for oncology (ICD-O). World Health Organization, Geneva.
51. Lawrence MS, Stojanov P, Mermel CH et al (2014) Discovery and saturation analysis of cancer genes across 21 tumour types. Nature 505:495–501
52. Mendiratta G, Jones MK, Stites EC (2022) How often is each gene mutated within the cancer patient population? Mol Cell Oncol 9:2065176
53. Siegel RL, Miller KD, Jemal A (2018) Cancer statistics, 2018. CA Cancer J Clin 68:7–30
54. Sung H, Ferlay J, Siegel RL et al (2021) Global cancer statistics 2020: GLOBOCAN estimates of incidence and mortality worldwide for 36 cancers in 185 countries. CA Cancer J Clin 71:209–249
55. Midha A, Dearden S, McCormack R (2015) EGFR mutation incidence in non-small-cell lung cancer of adenocarcinoma histology: a systematic review and global map by ethnicity (mutMapII). Am J Cancer Res 5:2892–2911
56. Gao B, Sun Y, Zhang J et al (2010) Spectrum of LKB1, EGFR, and KRAS mutations in Chinese lung adenocarcinomas. J Thorac Oncol 5:1130–1135
57. Novoplansky O, Jagadeeshan S, Regev O et al (2022) Worldwide prevalence and clinical characteristics of RAS mutations in head and neck cancer: a systematic review and meta-analysis. Front Oncol 12:838911

# Chapter 3

# Production of Isotopically Labeled KRAS4b

Troy Taylor and William Gillette

## Abstract

Isotopically labelled proteins are important reagents in structural biology as well as in targeted drug development. The field continues to advance with complex multi-isotope labeling. We have combined our experience in high-level soluble KRAS4b expression with protocols for isotope incorporation, to achieve reliable and efficient approaches for several labeling strategies. Typical experiments achieve nearly 100% $^{15}N$ incorporation, with yields in the range of 1.3–24.6 mg/L (median = 6.4 mg/L, $n$ = 53). Improvements in the growth parameters in the presence of deuterium reduce the standard time of fermentation from 5 days to 3 days by modifying the medium used during the weaning process. The methods described are compatible with multi-isotope labeling and site-specific labeling.

**Key words** RAS proteins, KRAS4b, Isotope, NMR, Deuteration, $^{15}N$

## 1 Introduction

As a core lab tasked with providing a variety of recombinant protein reagents across several diverse projects, our laboratory has spent considerable time exploring and optimizing approaches to produce recombinant proteins [1, 2]. A major component of that work is producing proteins with incorporated heavy isotopes (primarily performed using *E. coli*) for use in a wide variety of NMR-based structural studies [3]. A bottleneck in this process is the requirement for defined media to allow for incorporation of the desired isotopes, as this limits production levels. Also, *E. coli* requires adaptation to grow in the presence of deuterium, and this typically results in a long process with considerable hands-on labor. Despite these caveats, we were able to increase yields while also reducing both the labor involved and the time required to complete $^{15}N$ labelling. Additionally, by altering key aspects of the fermentation, we reduced the hands-on time and the total time of fermentation when incorporating deuterium. Here we outline our approaches to $^{15}N$ labelling and deuterium incorporation, as we developed them for producing the G-domain of the small GTPase KRAS4b. The

Andrew G. Stephen and Dominic Esposito (eds.), *KRAS: Methods and Protocols*, Methods in Molecular Biology, vol. 2797, https://doi.org/10.1007/978-1-0716-3822-4_3,

methods outlined here are compatible with multiple isotope incorporation (e.g., combinations of $^{2}H$, $^{13}C$, $^{15}N$, and site-specific methyl labeling).

### 1.1 $^{15}N$ Incorporation

The standard approach to achieving isotope incorporation is the use of a defined medium (frequently M9 minimal medium) and including the required isotope as the sole source of the target atom component. Cell mass yields (as indirectly measured from optical density measurements of the culture) from minimal medium fermentations are substantially reduced due to the limitation of the carbon source. In our standard (i.e., non-isotopic labeling) fermentations, we applied and expanded upon the ideas put forth by William Studier in his seminal publication on auto-induction to achieve higher cell densities [4]. These approaches are combined in a protocol/medium we refer to as the ZYM-20052 protocol (*see* **Note 1**) [2]. This protocol results in greater cell mass (no change to protein per cell), thus leading to higher protein yields per liter of expression culture (*see* **Note 2**). We combined this approach with $^{15}N$ labeling (in defined medium T-20052) to achieve higher protein yield (mg of protein per volume of expression culture) and observe no loss in terms of the percent incorporation. We routinely produce labeled KRAS4b protein variants in this way for our colleagues to use in NMR experiments (*see* **Note 3** and Fig. 1)

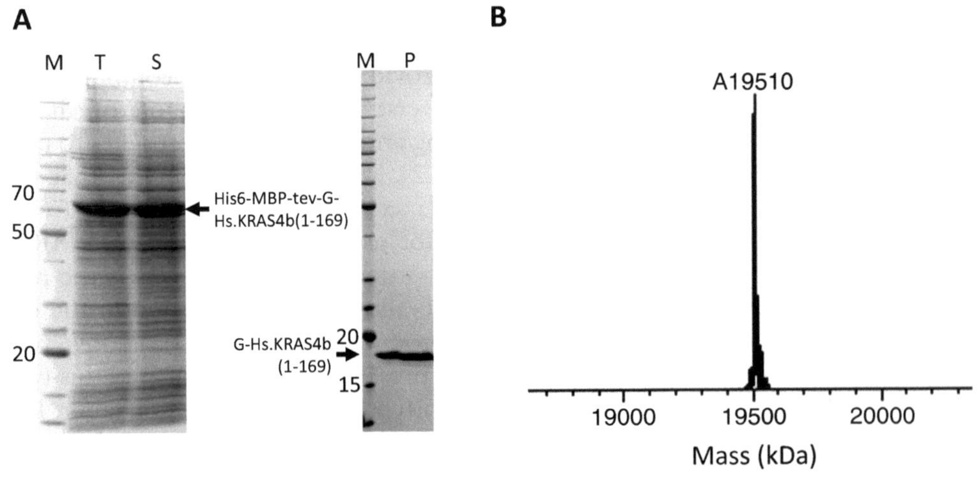


**Fig. 1** Production and intact mass analysis of G-Hs.KRAS4b(1–169) with $^{15}N$ incorporation. (**a**) Coomassie-stained SDS-PAGE analysis of the expression and final protein. M: molecular weight standards (kDa), T: total protein in lysate, S: soluble protein in lysate, P: purified protein. (**b**) Intact mass analysis by electrospray ionization. The figure shows the deconvoluted data derived from raw MS1 data using the Xtract algorithm (ThermoFisher), with the vertical axis showing relative abundance. Only a portion of the deconvoluted data (18,700–20,300 Da) is shown. The predicted mass of the protein with 100% $^{15}N$ incorporation is 19,518 Da, and based on the number of nitrogens present in this protein, we calculate that the protein is ~97% labelled

[5–7]. Here we describe our approach to $^{15}N$ incorporation using $^{15}N$ ammonium chloride and a modified defined auto-induction medium to improve cell yields.

### *1.2 Deuterium Incorporation*

Uniformed labelling of protein with deuterium adds a complication for a fermentation facility, as the standard protocols use extensive weaning processes during which the percentage of deuterium is increased in the culture medium (*see* **Note 4** and Fig. 2). These weaning steps are necessary as immediate shifting to 100% deuterium is detrimental to cell growth [8]. Even a stepped approach results in slowing the growth of the cells as they adjust to the change in levels of deuterium. This results in longer incubation to reach the desired cell density for induction of protein expression (*see* **Note 5**). Additionally, the process typically involves the aseptic harvest and/or washing of the cells and resuspension into the new, higher $D_2O$ content medium over several days (e.g., 0% to 50%, 75%, and 90% $D_2O$). After each of these increases in $D_2O$, the culture enters an extended lag phase as the cells adapt to the higher concentration. Thus, this approach is both lengthy (typically 5 days) and labor intensive. We have incorporated simple alterations in the protocol that in most cases lead to sufficient incorporation of

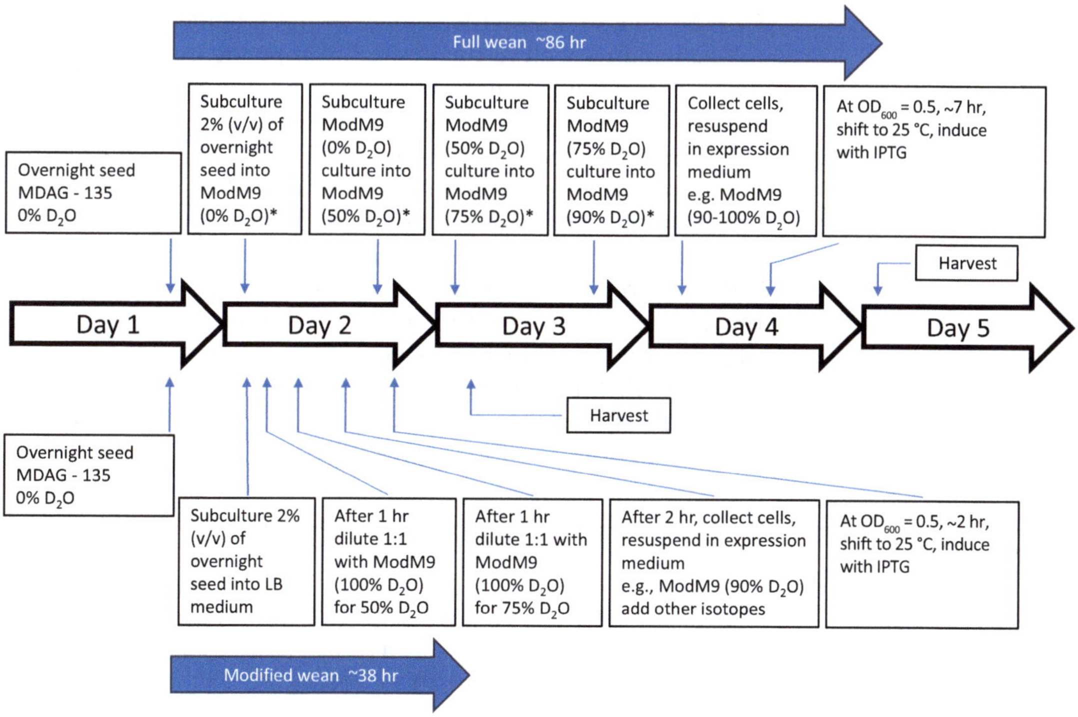


**Fig. 2** Comparison of timelines for deuterium incorporation. The top timeline shows the steps for our full weaning protocol. The bottom timeline shows the steps for our modified weaning protocol. *In the full-wean protocol subculture steps, enough inoculum is used reach $OD_{600}$ > 0.1 for the next culture

$D_2O$ while drastically reducing both the hands-on time and the total length of the fermentation. This allows both faster fermentations and the freeing up of equipment and staff for other purposes. The key difference in our approach is to start the fermentation with a population of cells that are growing at a faster growth rate in complex medium. Cells in this state experience a shorter lag phase, reaching the OD necessary for induction of protein expression faster (perhaps as higher growth rates result in a large increase in the number of ribosomes per cell [9]). Additionally, higher $D_2O$ concentrations are achieved by diluting the growing culture with 100% $D_2O$ medium, rather than through harvesting of cells and resuspension in fresh media, thus streamlining the protocol. The target induction $OD_{600}$ of 0.5 is achieved in high levels of deuterium 48 h earlier in our modified weaning protocol process compared to the standard weaning process (*see* Fig. 2). This allows harvest of induced cells only 38 h after inoculation of a seed culture, compared to 86 h postinoculation using the original weaning approach. This general approach is not novel [10] and was recently improved [11]. However, our modified approach results in shorter fermentations, somewhat reduced costs due to reduced deuterium, and the ability to turn over two weaning projects in 1 week using the same equipment. This modified weaning approach is compatible with incorporation of other isotopes at the induction step (e.g., $^{13}C$, $^{15}N$, and/or methyl labeling), as the modified M9 medium can be altered to achieve the desired isotope incorporation. Additionally, the timing is about the same to bring non-deuterated cultures to the point of induction, allowing for parallel growths and shared laboratory equipment. Thus, the laboratory need not account for any additional fermentation time when considering the production of proteins that are deuterated and labelled with other isotopes. These approaches, while presented here to produce KRAS4b (expressed as His6-MBP-tev-G-Hs.KRAS4b(1–169), *see* **Note 6**), were also used to produce variously labeled additional proteins, including other RAS GTPases, the catalytic domain of SOS1, the RAS-binding domains of RAF1 and BRAF kinases, calmodulin, the GAP domain of neurofibromin, and the RAS-binding domain of PI-3 kinase.

## 2 Materials

### 2.1 High Cell Mass, $^{15}N$ Labelling

1. Expression strain (e.g., R989-X02-566, His6-MBP-tev-Hs. KRAS4b(1–169), Addgene #159539) in strain *E. coli* BL21 (DE3) STAR (*rne131*) TT1 (*see* **Note 7**).
2. LB-agar petri plates: LB (lysogeny broth (Miller): 10 g/L tryptone, 5 g/L yeast extract, 10 g/L NaCl), 2.0% agar with

15 μg/mL chloramphenicol (to select for pRare), and 100 μg/mL ampicillin (to select for expression plasmid)

3. MDAG-135 medium: 25 mM $Na_2HPO_4$, 25 mM $KH_2PO_4$, 50 mM $NH_4Cl$, 5 mM $Na_2SO_4$, 2 mM $MgSO_4$, 50 μM $FeCl_3$, 20 μM $CaCl_2$, 10 μM $MnCl_2$-$4H_2O$, 10 μM $ZnSO_4$-$7H_2O$, 2 μM $CoCl_2$-$6H_2O$, 2 μM $CuCl_2$-$2H_2O$, 2 μM $NiCl_2$-$6H_2O$, 2 μM $Na_2MoO_4$-$2H_2O$, 2 μM $Na_2SeO_3$-$5H_2O$, 2 μM $H_3BO_3$, 19.4 mM glucose, 7.5 mM aspartate, and 200 μg/mL each of 18 amino acids E, D, K, R, H, A, P, G, T, S, Q, N, V, L, I, F, W, and M (*see* **Note 8**).
4. $^{15}N$ $NH_4Cl$: ammonium chloride ($^{15}N$, 99%) Cambridge Isotope Laboratory.
5. T-20052 medium: 25 mM $Na_2HPO_4$, 25 mM $KH_2PO_4$, 50 mM $^{15}N$ $NH_4Cl$, 5 mM $Na_2SO_4$, 2 mM $MgSO_4$, 50 μM $FeCl_3$, 20 μM $CaCl_2$, 10 μM $MnCl_2$-$4H_2O$, 10 μM $ZnSO_4$-$7H_2O$, 2 μM $CoCl_2$-$6H_2O$, 2 μM $CuCl_2$-$2H_2O$, 2 μM $NiCl_2$-$6H_2O$, 2 μM $Na_2MoO_4$-$2H_2O$, 2 μM $Na_2SeO_3$-$5H_2O$, 2 μM $H_3BO_3$, 216 mM glycerol, 2.8 mM glucose, and 5.6 mM α-lactose (*see* **Note 9**).
6. 2 L baffled shake flasks (autoclavable glass or single-use plastic).
7. Shaking incubator with 1″ orbit and temperature control.
8. High-speed centrifuge (e.g., Beckman J-20 XP with JLA8.1 rotor).
9. 15 L bioreactor (e.g., Eppendorf BioFlo IV fermentor, *see* **Note 10**).

### 2.2 Deuteration with Modified Weaning

1. Expression strain (e.g., R989-X02-566, His6-MBP-tev-Hs. KRAS4b(1–169), Addgene #159539) in strain *E. coli* BL21 (DE3) STAR (*rne131*) TT1 (*see* **Note 7**).
2. Lysogeny broth, LB (Miller): 10 g/L tryptone, 5 g/L yeast extract, 10 g/L NaCl, 15 μg/mL chloramphenicol, and 100 μg/mL ampicillin.
3. LB-agar petri plates: LB, 2.0% agar with 15 μg/mL chloramphenicol and 100 μg/mL ampicillin.
4. MDAG-135 medium: 25 mM $Na_2HPO_4$, 25 mM $KH_2PO_4$, 50 mM $NH_4Cl$, 5 mM $Na_2SO_4$, 2 mM $MgSO_4$, 50 μM $FeCl_3$, 20 μM $CaCl_2$, 10 μM $MnCl_2$-$4H_2O$, 10 μM $ZnSO_4$-$7H_2O$, 2 μM $CoCl_2$-$6H_2O$, 2 μM $CuCl_2$-$2H_2O$, 2 μM $NiCl_2$-$6H_2O$, 2 μM $Na_2MoO_4$-$2H_2O$, 2 μM $Na_2SeO_3$-$5H_2O$, 2 μM $H_3BO_3$, 19.4 mM glucose, 7.5 mM aspartate, and 200 μg/mL each of 18 amino acids E, D, K, R, H, A, P, G, T, S, Q, N, V, L, I, F, W, and M (*see* **Note 8**).
5. DLM-4 deuterium oxide (D, 99.9%) Cambridge Isotope Laboratory.

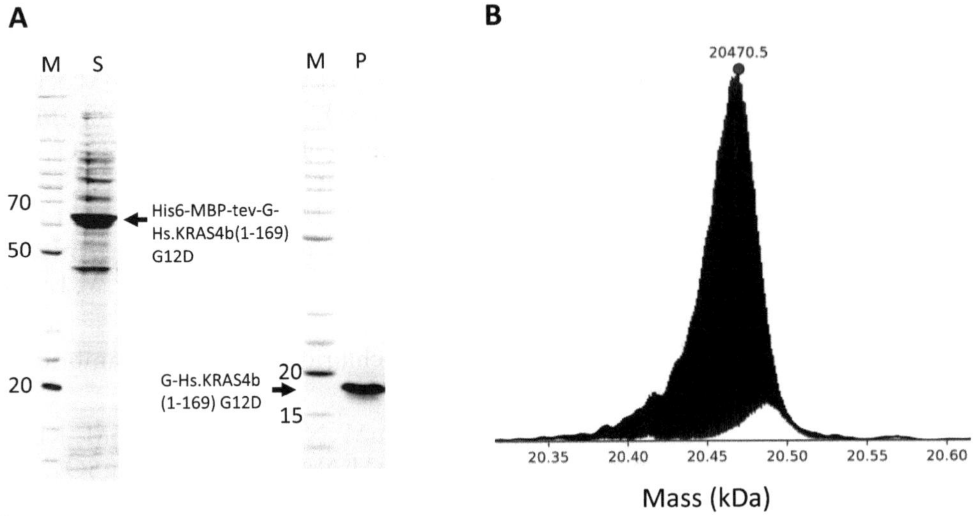


**Fig. 3** Production and intact mass analysis of G-Hs.KRAS4b(1–169) G12D with $^{15}N$, $^{13}C$-methyl-Ile, and $D_2O$ incorporation. (**a**) Coomassie-stained SDS-PAGE analysis of the expression and final protein. M: molecular weight standards (kDa), S: soluble protein. (**b**) Intact mass analysis by electrospray ionization. Shown is deconvoluted data derived from raw MS1 data using the UniDec algorithm (Michael T. Marty, University of Oxford). *See* **Note 17** for comments on determining percent incorporation

6. ModM9 medium: 33.5 mM $Na_2HPO_4$, 44.1 mM $KH_2PO_4$, 17.1 mM NaCl, 11.1 mM glucose, 18.8 mM $NH_4Cl_2$, 2 mM $MgSO_4$, 100 μM $CaCl_2$, 4 μM $ZnSO_4$, 1 μM $MnSO_4$, 4.7μM $H_3BO_3$, and 700 nM $CuSO_4$. Made with 100% deuterium oxide.
7. CDLM-7318 alpha-ketobutyric acid, sodium salt-(methyl-$^{13}C$, 99%; 3,3-D2, 98%). This component and the $^{15}N$ $NH_4Cl$ used in the production shown in Fig. 3 are not needed for this protocol but were included to show the versatility of the protocol.
8. 250 mL baffled shake flasks (autoclavable glass or single-use plastic).
9. 500 mL baffled shake flasks (autoclavable glass or single-use plastic).
10. 1 L baffled shake flasks (autoclavable glass or single-use plastic).
11. 4 L baffled shake flasks (autoclavable glass or single-use plastic).
12. Shaking incubator with 1″ orbit and temperature control.
13. High speed centrifuge (e.g., Beckman J-20 XP with JLA8.1 rotor).

## 3 Methods

### 3.1 High Cell Mass, $^{15}N$ Labelling

Typical $^{15}N$ labeling experiments start with cultures growing in a defined media such as M9. Our approach is to begin with a starter culture grown to log phase (*see* **Note 3**) that serves as the inoculum for a defined medium production culture. For the purposes here, the volumes will be appropriate for a final 15 L expression culture volume at harvest, but the general approach can be scaled up or down as needed.

1. Prepare 300 mL of MDAG-135 medium [2] in a 2 L baffled flask, and inoculate from a glycerol stock (*see* **Note 11**). Grow ~14–16 h overnight at 37 °C, 250 RPM in a shaking incubator.
2. Prepare 15 L of T-20052 medium in a 20 L BioFlo IV bioreactor.
3. Harvest cell pellet from the overnight seed culture by centrifugation 5000 × g for 20 min.
4. Resuspend the pellet in 300 mL of T-20052 medium, and use to inoculate the 15 L of T-20052 medium in the BioFlo IV bioreactor at 37 °C (*see* **Note 12**).
5. Set the airflow to 15 LPM, and the agitation rate to 350 rpm.
6. Grow to $OD_{600}$ of 1.0–1.5 (*see* **Note 13**) and shift the temperature to 20 °C. This normally takes about 6–7 h. Incubate overnight or ~14–16 h.
7. Harvest cells by centrifugation, 5000 × g for 20 min; discard the supernatant; and proceed to lyse the pellet or store the pellet at −80 °C.

### 3.2 Deuteration with Modified Weaning

As shown by others [10, 11], cells at higher growth rates are able to adapt to the presence of deuterium faster than cells growing on defined media such as M9. This shortens the lag phases of the culture during the stepped increases in deuterium levels, thus shortening the adaptation periods and the overall length of the fermentation. Our approach to achieve reproducibly faster growing cells uses an overnight culture grown in MDAG-135 medium (*see* **Note 14**) followed by a subculture into nutrient rich LB the following day (*see* **Note 15**). The protocol described here results in a final culture volume of 1 L in a 4 L shake flask. Scale-up is achieved by using multiple flasks. With incubators common in most labs, this approach puts a 6 L production well within the capabilities of nearly all fermentation laboratories. Additionally, with only 24 h needed to go from a 50 mL overnight culture to the harvest of a 1 L induced culture, a 6 L scale-up occupies the incubator for only those 24 h, thus freeing the incubator for additional work to be scheduled after the harvest of the deuterated cultures.

1. Inoculate an overnight *E. coli* seed culture (50 mL of MDAG-135 in a 250 mL baffled flask) from a glycerol stock (*see* **Note 11**).
2. Grow ~14–16 h overnight at 37 °C, 250 RPM in a shaking incubator.
3. Prepare 150 mL of ModM9 (100% $D_2O$) for weaning purposes and 1 L of ModM9 (100% $D_2O$) production medium (include additional isotopes in the production medium as desired) per liter of desired final expression culture.
4. Inoculate 50 mL of LB in a 250 mL baffled shake flask with 2% (v/v) of the overnight MDAG-135 seed culture, and return to the incubator for 1 h at 37 °C (*see* **Note 14**).
5. Transfer the 50 mL culture to a 500 mL baffled shake flask, and add 50 mL of ModM9 (100% $D_2O$); the culture is now in 50% $D_2O$.
6. Return the flask to the incubator and shake for 1 h at 37 °C.
7. Add 100 mL of ModM9 (100% $D_2O$); the culture is now in 75% $D_2O$, and move it to a 1 L baffled shake flask.
8. Return the flask to the incubator, and shake for 2 h at 37 °C (volume is now 200 mL).
9. Collect the cells by centrifugation, and resuspend in ModM9 with the desired level of deuterium and additional isotopes as needed, and move to a 4 L baffled shake flask. The choice to use ModM9 with 90% vs 100% $D_2O$ should be based on the final application requirements.
10. Grow to an $OD_{600}$ of 0.4–0.6 (*see* **Note 13**); this normally takes about 2 h, then shift the temperature to 25 °C and incubate overnight for 14–16 h.
11. Harvest cells by centrifugation, 5000 × g for 20 min; discard the supernatant; and proceed to lyse the pellet or store the pellet at −80 °C.

### 3.3 Other Isotopic Labeling

We have used our modified deuterium weaning approach in combination with a variety of other incorporations including $^{13}C$, $^{13}C/^{15}N$, and various site-specific methyl labeling (e.g., alpha-ketobutyric acid, sodium salt-(methyl-$^{13}C$, 99%; 3,3-D2, 98%)). The results have been consistent and lead to production yields in the range of 1.0–5.0 mg/L with acceptable incorporation rates (*see* **Note 16**).

## 4 Notes

1. We developed ZYM-20052 for standard protein expression as it is a complex medium.

2. We also frequently use the Dynamite medium/protocol outlined in [2], for routine expression of recombinant proteins (when no isotope incorporation is needed). This protocol requires addition of IPTG to induce protein expression rather than the auto-induction that is used in ZYM-20052. The yields in Dynamite medium are, on average, higher than for ZYM-20052.
3. For $^{15}N$ labelling, we have modified Studier's N-5052 medium by increasing the concentration of glycerol. This new medium, T-20052 [12], can produce three- to fourfold more cells than standard defined medium M9 and two- to threefold more cells than ModM9 with similar levels of protein expression per cell. This is due to the higher carbohydrate content of T-20052. Both ZYM-20052 and T-20052 are auto-induction media, but only T-20052 is a defined medium useful for isotope incorporation.
4. Unfortunately, *E. coli* will not grow directly in media consisting of 100% deuterium oxide [8] and require a weaning period. We typically start by diluting a growing culture to 50% deuterium and then progress to 75% and 100%. We originally performed a week-long weaning process with excellent results but developed the quicker and less expensive weaning method described here.
5. As deuterium levels increase in the culture medium, the lag phase of the culture increases. For example, cells when added directly to ModM9 media containing 75% deuterium failed to multiply, while the same cells, when shifted to the same medium but with only 60% deuterium oxide, eventually resumed dividing (albeit several hours later).
6. The construct of G-Hs.KRAS4b(1–169) is preferred by our structural biology colleagues as the presence of the methionine in native amino acid position one, while not present in the form of KRAS4b that exists in human cells due to N-terminal post-translational processing, provides important crystal contacts. These contacts stabilize the conformation that crystallographers believe is the prevalent form in the cell [13]. We follow standard purification protocols to obtain G-Hs.KRAS4b (1–169) from the expressed fusion protein: His6-MBP-tev-G-Hs.KRAS4b(1–169). Detailed protocols are described in [14].
7. For this expression work, we used the *E. coli* strain BL21 DE3 star from Invitrogen modified to contain a pRare plasmid ($Cm^R$) expressing tRNAs (argU, argW, ileX, glyT, leuW, proL, metT, thrT, tyrU, and thrU). Our strain is an isolate (TT1) from this parental line that is resistant to a phage discovered in our lab.

8. To prepare MDAG-135, we make separate stock solutions for each of these seven components: $Na_2HPO_4$, $KH_2PO_4$, $NH_4Cl$, $Na_2SO_4$, $MgSO_4$, aspartate, and amino acids (all 18 amino acids in single-stock solution). A stock solution of glucose is filter sterilized rather than autoclaved. The remaining components are combined into a single trace metals stock solution. All solutions are then added to $dH_2O$ and brought to final volume. Care should be taken to add the stock solutions to a large volume of $dH_2O$, rather than mixing the stock solutions together initially, as the latter leads to precipitation of some of the components.
9. T-20052, first described in [12], is our modification of Studier's N-5052. Both media include $^{15}N$ as $NH_4Cl$. Considerations in preparing this are the same as for MDAG-135 in **Note 8**.
10. Alternatively, all labeling work can be completed in shake flasks as aeration will be sufficient since the cultures will not exceed $OD_{600}$ of 7.0 due to nutrient limitations of the media.
11. All cultures are inoculated from glycerol stocks that are prepared from a single isolated colony growing on the appropriate antibiotic for plasmid selection, rather than from a freshly picked transformant. We found that this allows a greater level of consistency to the subsequent outgrowths.
12. The temperature of the fresh media should be matched in all cases to the temperature of the inoculum.
13. In this chapter, we often refer to induction timing between a set range. These ranges vary depending on the medium. It is our practice to allow for 2–3 doublings post-induction. For example, we induce by shifting the temperature in T-20052 at an $OD_{600}$ of 1.0–1.5, and the final harvest OD is ~5.0–6.0.
14. The key to this procedure is the stage of growth of the overnight seed in MDAG-135. Our experience is that the reproducibility of the subsequent steps in terms of timing is far greater than in other media. This reproducibility not only allows for a streamlined process during fermentation but also leads to consistent incorporation. It is our practice to start all seed cultures in Studier's MDAG-135 medium. We determined this medium provides healthy cells even at saturation. We routinely store MDAG-135 cultures, grown to saturation, at 4 °C when logistical problems arise in the laboratory. We have not observed issues with cell yield and protein expression per cell and only a minor increase in lag phase. It was these observations that led us to explore the use of MDAG-135 in our deuterium incorporation protocols.
15. This LB culture step allows the cells to be in a rapid growth phase prior to the subsequent exposure to deuterium that we

and others [10, 11] have identified as important for reducing the lag phase in the subsequent step changes in deuterium levels.

16. The level of deuterium in the final induced culture is dependent on the requirement of the intended downstream application. The modified wean protocol described here might not be sufficient for all downstream applications. With 90% $D_2O$, we typically achieve ~75% incorporation of non-exchangeable hydrogens, and with 100% $D_2O$, we attain ~80–83% incorporation.

17. While measuring percent incorporation by electrospray ionization is very precise when incorporating $^{13}C$ and/or $^{15}N$ (*see* Fig. 1b), the incorporation of deuterium produces a distribution of species that does not allow precise assessment of the incorporation of the carbon and/or nitrogen isotopes (*see* Fig. 3b). Nevertheless, the proteins have been uniformly useful for the intended downstream applications.

## Acknowledgments

The authors thank Frank Chao for helpful discussions on many topics involving isotope incorporation, Simon Messing for the review of the manuscript, Jennifer Mehalko, Nitya Ramakrishnan, José A. Sánchez Hernández, and Vanessa Wall for cloning, fermentation, protein purification, and QC, and Matt Drew for comments on electrospray ionization. This work has been funded in whole or in part with federal funds from the National Cancer Institute, National Institutes of Health, under contract 75N91019D00024. The content of this publication does not necessarily reflect the views or policies of the Department of Health and Human Services, nor does mention of trade names, commercial products, or organizations imply endorsement by the US government.

## References

1. Esposito D, Garvey LA, Chakiath CS (2009) Gateway cloning for protein expression. Methods Mol Biol 498:31–54
2. Taylor T, Denson JP, Esposito D (2017) Optimizing expression and solubility of proteins in E. coli using modified media and induction parameters. Methods Mol Biol 1586:65–82
3. Tugarinov V, Kanelis V, Kay LE (2006) Isotope labeling strategies for the study of high-molecular-weight proteins by solution NMR spectroscopy. Nat Protoc 1:749–754
4. Studier FW (2005) Protein production by auto-induction in high density shaking cultures. Protein Expr Purif 41:207–234
5. Chao FA, Dharmaiah S, Taylor T et al (2022) Insights into the cross talk between effector and allosteric lobes of KRAS from methyl conformational dynamics. J Am Chem Soc 144: 4196–4205
6. Morstein J, Shrestha R, Van QN et al (2023) Direct modulators of K-Ras-membrane interactions. ACS Chem Biol 18:2082–2093
7. Travers T, Lopez CA, Van QN et al (2018) Molecular recognition of RAS/RAF complex

at the membrane: role of RAF cysteine-rich domain. Sci Rep 8:8461

8. De Giovanni R (1960) The effects of deuterium oxide on certain microorganisms. Ann N Y Acad Sci 84:644–647
9. Bremer H, Dennis PP (1987) In: Neidhart FC (ed) Eschericia coli and Salmonella typhimurium cellular and molecular biology. American Society of Microbiology, Washington, DC
10. Marley J, Lu M, Bracken C (2001) A method for efficient isotopic labeling of recombinant proteins. J Biomol NMR 20:71–75
11. Li J, Byrd RA (2022) A simple protocol for the production of highly deuterated proteins for biophysical studies. J Biol Chem 298:102253
12. Agamasu C, Ghirlando R, Taylor T et al (2019) KRAS prenylation is required for bivalent binding with calmodulin in a nucleotide-independent manner. Biophys J 116:1049–1063
13. Dharmaiah S, Tran TH, Messing S et al (2019) Structures of N-terminally processed KRAS provide insight into the role of N-acetylation. Sci Rep 9:10512
14. Kopra K, Vuorinen E, Abreu-Blanco M et al (2020) Homogeneous dual-parametric-coupled assay for simultaneous nucleotide exchange and KRAS/RAF-RBD interaction monitoring. Anal Chem 92:4971–4979

# Chapter 4

# Nucleotide Exchange on RAS Proteins Using Hydrolysable and Non-hydrolysable Nucleotides

**Timothy Waybright and Andrew G. Stephen**


## Abstract

Biochemical and biophysical assays using recombinant RAS require the protein to be in either the active or inactive state. Here we describe methods to exchange the nucleotide present in the purified RAS protein with either GDPβS, GppNHp, or GTP depending on the assay requirement. In addition, we also describe the HPLC method used to validate the exchange process and provide information on the efficiency of the nucleotide exchange.


**Key words** Nucleotide exchange, RAS, Ion pairing, Chromatography

## 1 Introduction

The RAS family of small GTPase proteins were discovered over four decades ago. Subsequent work has determined that RAS mutations are present in ~20% of cancer patients, resulting in ~3.4 million new cases each year globally [1]. Of the four RAS isoforms (HRAS, NRAS, KRAS4a, and KRAS4b), KRAS is by far the most frequently identified in three of the most common human tumors (lung, colorectal, and pancreatic cancers) [2]. RAS proteins function as a molecular switch and when bound to guanosine triphosphate (GTP) are in the active or "on" state. Hydrolysis of GTP to guanosine diphosphate (GDP) results in inactivation leaving the protein in the "off" state [3–7]. Mutations at positions G12, G13, and Q61 [8] result in impaired intrinsic and GTPase-activating protein (GAP)-stimulated GTPase activity [9, 10], leaving the proteins in the active state leading to increased cellular proliferation and oncogenic transformation [11–14].

For high-quality biochemical and structural studies, recombinant KRAS protein needs to be homogeneously in the active (GTP) or inactive (GDP) states. The production of KRAS proteins from *E. coli* results in the purified proteins being in either of these states,

Andrew G. Stephen and Dominic Esposito (eds.), *KRAS: Methods and Protocols*, Methods in Molecular Biology, vol. 2797,
https://doi.org/10.1007/978-1-0716-3822-4_4,

depending on the mutation. Removal of the bound nucleotide and replacement with a labeled or non-hydrolysable form is an essential step prior to any further experiments. In addition, careful quantitative analysis of the efficiency of the nucleotide exchange is an important final step required for quality assurance [15–17]. Here we describe a series of protocols for nucleotide exchange and efficiency of nucleotide loading.

## 2 Materials

Prepare all the solutions for the exchange procedures fresh the day of use. They may remain at room temperature unless otherwise noted. For the desalting/nucleotide removal steps, solutions may be prepared ahead of time and kept refrigerated until used. The solvents used for HPLC analysis and exchange efficiency determination may be prepared ahead of time and kept at room temperature. The reagents for HPLC analysis should be mixed for use on the day of analysis. Deionized water ($dH_2O$, >18.2 MΩ $cm^{-1}$ at room temperature) is used in all buffers and is filtered through a 0.22 μm membrane filter.

### 2.1 Non-hydrolysable Nucleotide Exchange (GppNHp, GDPβS, and GTPγS)

1. Required amount of RAS protein, thawed on ice (*see* **Note 1**).
2. Exchange buffer: 40 mM HEPES pH 7.4, 200 mM $(NH_4)_2SO_4$, and 100 μM $ZnCl_2$. Mix 4 mL of 0.5 M HEPES pH 7.4, 5 mL of 2 M $(NH_4)_2SO_4$, and 50 μL of 0.1 M $ZnCl_2$ into a 50 mL conical tube. Bring the final volume to 50 mL by adding 40.95 mL of $dH_2O$ and mix thoroughly. Keep on ice until use.
3. Desalting buffer: 50 mM HEPES pH 7.4, 150 mM NaCl, 1.5 mM $MgCl_2$, and 1 mM TCEP. Mix 100 mL of 0.5 M HEPES pH 7.4, 30 mL of either 5 M NaCl, 1.5 mL of 1 M $MgCl_2$, and 2 mL of TCEP bond breaker into a 1 L glass bottle (*see* **Note 2**). Add 866.5 mL or 746.5 mL of $dH_2O$ to make the final volume 1 L.
4. Centrifuge capable of at least 10,000 rpm.
5. Amicon Ultra 0.5 10 K molecular weight cutoff filter.
6. Amicon Ultra-4 10 K molecular weight cutoff filter.
7. Various pipettes and tips.
8. 1.8–2 mL Eppendorf polypropylene tubes.
9. Alkaline phosphatase acrylic beads from calf intestine.
10. Mini-rotator capable of holding 1.8- to 2 mL Eppendorf tubes.
11. Appropriate nucleotide for the desired exchange (GppNHp, GDPβS, or GTPγS) (*see* **Note 3**).

12. Costar Spin-X centrifuge tube filters, 0.22 μm membrane.
13. Desalting column (*see* **Note 4**).
14. Nanodrop spectrophotometer.

### 2.2 Hydrolysable Nucleotide Exchange (GDP and GTP)

1. Required amount of RAS protein, thawed on ice (*see* **Note 1**).
2. Exchange buffer: 50 mM HEPES pH 7.4, 150 mM NaCl or 150 mM KCl, 200 mM $(NH_4)_2SO_4$, 25 mM EDTA, and 1 mM TCEP. Mix 5 mL of 0.5 M HEPES pH 7.4, 1.5 mL of 5 M NaCl, 5 mL of 2 M $(NH_4)_2SO_4$, 2.5 mL of 0.5 M EDTA, and 100 μL of 0.5 M TCEP into a 50 mL conical tube or another container as appropriate. Bring the final volume to 50 mL by adding 35.9 mL or 29.9 mL of $dH_2O$, and mix thoroughly. Keep on ice until use.
3. Desalting buffer: 50 mM HEPES pH 7.4, 150 mM NaCl or 150 mM KCL, 1.5 mM $MgCl_2$, and 1 mM TCEP. Mix 100 mL of 0.5 M HEPES pH 7.4, 30 mL of either 5 M NaCl, 1.5 mL of 1 M $MgCl_2$, or 2 mL of TCEP bond breaker into a 1 L glass bottle (*see* **Note 2**). Add 866.5 mL or 746.5 mL of $dH_2O$ to make the final volume 1 L.
4. Centrifuge capable of at least 10,000 rpm.
5. Amicon Ultra 0.5 10 K molecular weight cutoff filter.
6. Amicon Ultra-4 10 K molecular weight cutoff filter.
7. Various pipettes and tips.
8. 1.8- to 2 mL Eppendorf polypropylene tubes.
9. Mini-rotator capable of holding 1.8 to 2 mL Eppendorf tubes.
10. Appropriate nucleotide for the desired exchange (GDP or GTP) (*see* **Note 5**).
11. Walk-in cold room or cooler capable of containing the mini rotator at 4 °C.
12. Desalting column (*see* **Note 4**).
13. Nanodrop spectrophotometer.

### 2.3 Ion-Pairing Reversed Phase HPLC

1. Prepare a 1 M solution of $K_2HPO_4$ by weighing 114.1 g of material into a beaker. Add 500 mL of $dH_2O$ and solubilize. Filter through a Corning 500 mL 0.22 filter, and store the solution in a tightly capped bottle at room temperature. Label Di-Basic.
2. Prepare a 1 M solution of $KH_2PO_4$ by weighing 68.1 g of material into a beaker. Add 500 mL $dH_2O$ and solubilize. Filter through a Corning 500 mL 0.22 filter, and store the solution in a tightly capped bottle at room temperature. Label Mono-Basic.

3. Transfer 143.4 mL of $K_2HPO_4$ and 56.6 mL of $KH_2PO_4$ to a 2 L graduated cylinder. Bring the volume up to 2 L with $dH_2O$. This prepares a 100 mM $K_2HPO_4/KH_2PO_4$ solution, pH 7.2. Solution may be stored in the graduated cylinder at room temperature covered by aluminum foil.
4. Prepare a 4 mM solution of tetrabutylammonium hydrogen sulfate by weighing 0.68 g of material into a beaker. Add 500 mL of $dH_2O$ and solubilize. Filter through a Corning 500 mL 0.22 filter, and store the solution in a tightly capped bottle at room temperature for future use.
5. Buffer A: Using a 1 L graduated cylinder, mix 750 mL of the 100 mM $K_2HPO_4/KH_2PO_4$ pH 7.2 solution with 250 mL of the 4 mM tetrabutylammonium hydrogen sulfate solution, and transfer to a glass bottle. Mix well (*see* **Note 6**).
6. Buffer B: Mix 210 mL of Buffer A with 90 mL of acetonitrile.
7. Hichrom Ultrasphere 5 μm C18 ODS HPLC Column, 4.6 × 250 mm (*see* **Note 7**).
8. Macherey-Nagel EC 4/3 Nucleosil 100-5 C18 pre-column with column holder assembly.
9. LCMS-certified clear glass 12 × 32 mm screw neck max recovery vial, with cap and preslit PTFE/silicone septa (*see* **Note 8**).
10. Waters e2695 Alliance System using Empower3 software and Waters 2489 UV/Visible Detector set to 254 nm and column heater (*see* **Note 9**).
11. Nucleotide standards mixture containing appropriate concentrations of nucleotide (*see* **Note 10**).
12. 15 mL conical tube.

## 3 Methods

### 3.1 Non-hydrolysable Nucleotide Exchange

1. All steps may be done at room temperature.
2. Using an Amicon Ultra-0.5 10 K concentrator, add a volume containing approximately 3 mg of protein to two concentrators (6 mg total sample) (*see* **Note 11**).
3. Add exchange buffer to bring total volume in the concentrator to ~500 μL (*see* **Note 12**).
4. Centrifuge at 10,000 rpm for 5 min to bring volume to ~100 μL.
5. Add 350 μL of exchange buffer and centrifuge at 10,000 rpm for 5 min.
6. Add another 350 μL of exchange buffer and centrifuge at 10,000 rpm for 5 min. After this, the protein has been sufficiently exchanged into the exchange buffer.

7. Transfer the remaining supernatant in the concentrator to an Eppendorf tube (~100 μL).
8. Spin the concentrator at 1000 rpm for 2 min.
9. Wash out the concentrator with 350 μL of the exchange buffer, and transfer to the Eppendorf tube.
10. Spin the concentrator at 1000 rpm for 2 min.
11. Wash out the concentrator with 350 μL of the exchange buffer, and transfer to the Eppendorf tube. You should now have ~800 μL of protein in exchange buffer.
12. Solubilize the appropriate nucleotide in $dH_2O$ to prepare a final solution of ~150–200 mM nucleotide (*see* **Note 13**).
13. Add 100 μL of the freshly prepared nucleotide to the protein mixture.
14. Add 48 μL of the alkaline phosphatase acrylic beads (*see* **Note 14**).
15. Place the tubes on the rotator and mix for 3–5 h.
16. Remove the tubes and add an additional 100 μL of the nucleotide to the tubes.
17. Place the tubes on the rotator and mix for 1 h.
18. Remove the tubes, and transfer the entire contents to the Costar Spin-X centrifuge tube filters (*see* **Note 15**).
19. Spin the Costar filters containing the protein and AP beads at 2000 rpm for 3 min.
20. Remove the filter portion and discard. Transfer the filtrate to new Eppendorf tubes.
21. To each tube, add 30 μL of 1 M $MgCl_2$.
22. Place the tubes on the rotator and mix for 15 min.
23. Equilibrate the appropriate desalting column with the desalting buffer.
24. Add the protein solution to the desalting column per manufacturer's instructions or lab protocol.
25. Collect the fractions containing the desalted protein based on the desalting column used.
26. Check the protein concentration on the nanodrop spectrophotometer. Ensure the 260/280 ratio has a value below 1. This ensures that all the excess nucleotide has been removed.
27. If needed, concentrate the protein further by transferring the protein solution to an Amicon Ultra-4 concentrator, and spin at 4700 rpm for 12 min (*see* **Note 16**).
28. Recheck the protein concentration to achieve the desired concentration.

29. Transfer the protein solution to Eppendorf storage tubes, and store at −80 °C until needed.

### 3.2 Hydrolysable Nucleotide (GTP/GDP) Exchange

1. All steps should be done on ice or at 4 °C.
2. Using an Amicon Ultra-0.5 10 K concentrator, add a volume containing approximately 3 mg of protein to two concentrators (6 mg total sample) (*see* **Note 1**).
3. Add exchange buffer to bring total volume in the concentrator to ~500 μL (*see* **Note 12**).
4. Centrifuge at 10,000 rpm for 5 min to bring volume to ~100 μL.
5. Add 350 μL of exchange buffer and centrifuge at 10,000 rpm for 5 min.
6. Add another 350 μL of exchange buffer and centrifuge at 10,000 rpm for 5 min. After this, the protein has been sufficiently exchanged into the exchange buffer.
7. Transfer the remaining supernatant in the concentrator to an Eppendorf tube (~100 μL).
8. Spin the concentrator at 1000 rpm for 2 min.
9. Wash out the concentrator with 350 μL of the exchange buffer, and transfer to the Eppendorf tube.
10. Spin the concentrator at 1000 rpm for 2 min.
11. Wash out the concentrator with 350 μL of the exchange buffer, and transfer to the Eppendorf tube. You should now have ~800 μL of protein in exchange buffer.
12. Solubilize the appropriate nucleotide in $dH_2O$ to prepare a final solution of ~150–200 mM nucleotide (*see* **Note 13**).
13. Add 100 μL of the freshly prepared nucleotide to the protein mixture.
14. Place the tubes on the rotator at 4 °C and mix for 3–4 h.
15. Remove the tubes and add an additional 100 μL of the nucleotide to the tubes.
16. Place the tubes on the rotator and mix for 1 h.
17. Remove the tubes, and add 10 μL of 1 M $MgCl_2$ to each for a final concentration of 10 mM.
18. Place the tubes on the rotator and mix for 15 min.
19. Equilibrate the appropriate desalting column with the desalting buffer.
20. Add the protein solution to the desalting column as per manufacturer's instructions or lab protocol.
21. Collect the fractions containing the desalted protein based on the desalting column used.

22. Check the protein concentration on the spectrophotometer. Ensure the 260/280 ratio is below 1. This ensures that all the excess nucleotide has been removed.
23. If needed, concentrate the protein further by transferring the protein solution to an Amicon Ultra-4 concentrator, and spin at 4700 rpm for 12 min (*see* **Note 16**).
24. Recheck the protein concentration to achieve the desired concentration.
25. Transfer the protein solution to Eppendorf storage tubes, snap-freeze in liquid nitrogen, and store at −80 °C until needed.

### 3.3 Ion-Pairing Reversed Phase HPLC for Nucleotide Exchange Efficiency

1. After preparing the buffers, place them in the desired positions on the HPLC system (*see* **Note 17**).
2. Flush the lines with buffer to prime the system (*see* **Note 18**).
3. Attach the column and the pre-column in the appropriate location in the column heater.
4. Set the column heater to 40 °C (*see* **Note 19**).
5. After the system has been primed, start the buffer flow at 0.6 mL/min, and equilibrate the column using 50% Buffer A and 50% Buffer B for 1 h (*see* **Note 20**).
6. Switch the buffer composition to 98% Buffer A and 2% Buffer B at 0.6 mL/min, and equilibrate for 2 h.
7. After the 2 h column equilibration, set up the sample list on the HPLC system (*see* Tables 1, 2, and 3).
8. Run the samples and collect the data. The order of nucleotide elution with increasing time is GMP, GppNH, GDP, GppNHp, and GTP (*see* Fig. 1).
9. Compare the sample data to the standard data to determine exchange efficiency (*see* **Note 21**).

**Table 1**
**HPLC gradient, flow, and time information used for all routine HPLC nucleotide analyses**

| Time (minutes) | Buffer A | Buffer B | Flow rate (mL/min) |
|---|---|---|---|
| 0 | 98 | 2 | 0.6 |
| 5 | 98 | 2 | 0.6 |
| 33 | 65 | 35 | 0.6 |
| 35 | 98 | 2 | 0.6 |
| 55 | 98 | 2 | 0.6 |

**Table 2**
**HPLC method shutdown profile**

| Time (minutes) | Buffer A | Buffer B | Flow rate (mL/min) |
|---|---|---|---|
| 0 | 2 | 98 | 0.6 |
| 14 | 2 | 98 | 0.6 |
| 15 | 2 | 98 | 0 |

The column is typically stored in these conditions, as it is used quite extensively. For long-term storage, other buffers should be considered

**Table 3**
**A typical sample set used for routine analysis of proteins containing either known or unknown nucleotides**

| Sample # | Sample | Type | Conc. |
|---|---|---|---|
| 1 | Blank | Blank | N/A |
| 2 | GppNHp-GTP-GDP | Standard | 10 μM |
| 3 | GppNHp-GTP-GDP | Standard | 1 μM |
| 4 | Sample 1 | Sample | N/A |
| 5 | Sample 1 | Sample | N/A |
| 6 | Sample 1 | Sample | N/A |
| 7 | GppNHp-GTP-GDP | Standard | 10 μM |
| 8 | Blank | Blank | N/A |
| 9 | End | Blank | N/A |

For a greater quantitative analysis, more dilutions of the standards mixture can be used

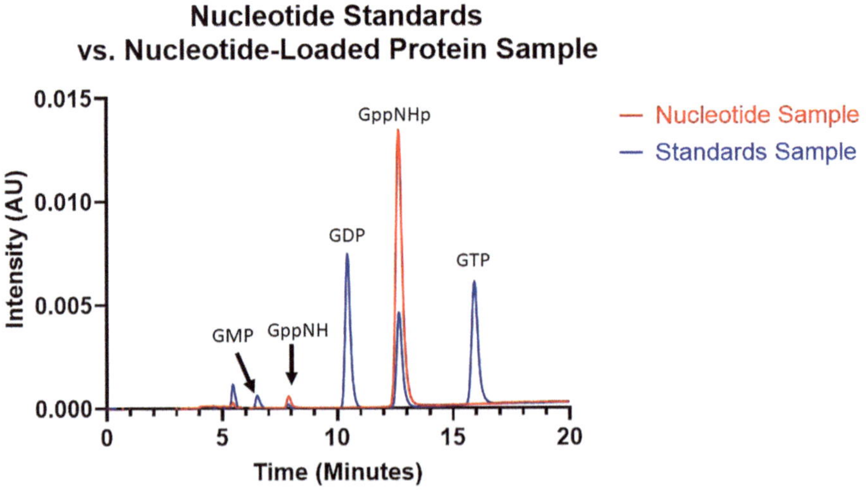


**Fig. 1** HPLC chromatograms of the nucleotide standards mixture and GppNHp exchanged KRAS. The standards mixture ((GDP, GppNHp, and GTP)) is put at the beginning and end of the sample set to ensure retention times remain consistent, or to verify any retention time drift to accurately identify nucleotides in unknown samples. The order of elution, with increasing time, is GMP, GppNH, GDP, GppNHp, and GTP. The KRAS samples are injected between the two sets of nucleotide standards mixture

## 4 Notes

1. The amount of RAS protein exchanged varies based on the requirements of the experimental design. Some experiments require large amounts of exchanged protein (>30 mg), while others only require a small amount.
2. Using wide-mouth mobile phase delivery bottles minimizes contamination of the desalting buffer if it is prepared in a different container. If it is prepared in a different container, it is advisable to filter through a 0.22 μm filter prior to use on any liquid handling system. If using gravity flow columns to desalt the protein, filtering prior to use is not necessary.
3. The GppNHp, GDPβS, or GTPγS nucleotides are non-hydrolysable (or hydrolyze very slowly) since the substitution of a sulfur or nitrogen on the terminal phosphate group slows or prevents hydrolysis and preserves the integrity of the terminal phosphate group.
4. Different methods are available to desalt the exchanged protein. Fast protein liquid chromatography (FPLC) system coupled with a size-exclusion column to desalt and clean up the protein (which is useful for larger amounts of protein) are frequently used; alternatively gravity flow desalting columns can also be used. Based on need, either system, used correctly, can provide a clean protein which is free from excess nucleotide. We use both the FPLC and the Cytiva PD-10 gravity flow columns and obtain equivalent exchange efficiencies.
5. The GTP-bound proteins should be handled either on ice or in a cold room, as the hydrolysis rate is about 0.01 $min^{-1}$, which corresponds to a half-life of wild-type KRAS-GTP at room temperature to be approximately 3–4 h. The GppNHp, GDPβS, and GTPγS nucleotides are significantly more stable and may be kept at room temperature or kept ice.
6. This step is very important, as the separation of the nucleotides is dependent upon thorough mixing of the ion-pairing reagent (in this case, tetrabutylammonium hydrogen sulfate) with the phosphate buffer.
7. There are many commercially available HPLC columns which can perform this separation. The C18 ODS column used in this method utilizes relatively low pressure (approximately 750–900 psi) and delivers very consistent results. U-HPLC or UPLC systems can be used for these separations, as the type of LC system is not the critical factor. Testing of different columns should be performed on any system used to verify that the nucleotides are being separated efficiently and reproducibly.

8. Any autosampler vials may be used here, based on the investigator's preference. The advantage of the screw neck max recovery vials is the bottom of the insert is attached to the bottom of the vial. This can prevent the sampling needle from punching through the bottom of the insert and avoid sample loss if the sampling height of the needle is not set correctly or becomes misaligned.
9. Our lab uses the Waters system, but any HPLC, U-HPLC, or UPLC system may be used, based on availability.
10. Depending upon which nucleotides are being studied, different mixtures are used to identify and determine exchange efficiency. In our lab, we inject 30 μL of a 10 μM and a 1 μM standard solution containing GppNHp, GTP, and GDP. If a non-hydrolysable nucleotide is being assayed, then a 10 μM standard solution of that nucleotide is prepared and 30 μL is injected. These concentrations work well for us as the nucleotide levels in the proteins being studied typically fall within the range of the standards. Prior testing showed a limit of quantitation (LOQ) of approximately 100 nM, with a limit of detection (LOD) between 1 and 10 nM. If higher concentrations of nucleotides are present, the standard curve sample concentrations should be adjusted as well to accurately represent the data.
11. This method may be used with higher amounts of protein, but care should be taken when concentrating large amounts to ensure the protein does not precipitate at the higher concentration. Typically, no more than 10–15 mg of protein in ~3 mL is concentrated using this method. With larger amounts of protein, larger volume concentrators should be used.
12. This step is done to begin the process of buffer exchange. The goal is to remove as much of the buffer that the RAS is stored in and replace it with the exchange buffer. This removes any excess $MgCl_2$ in the buffer, which impedes the nucleotide exchange process. The $MgCl_2$ is added in a later step to essentially lock in the exchanged nucleotide. Poor exchange efficiency as determined by HPLC may be the result of too much $MgCl_2$ present in the exchange buffer.
13. This concentration range works well, as you want the nucleotide:protein exchange ratio to be roughly at a 10:1 molar ratio, but no lower than a 3:1 molar ratio. At lower protein concentrations, the listed nucleotide concentration can be adjusted to save nucleotide amounts, but having an excess of exchange nucleotide present will not adversely affect the exchange.
14. The volume of the alkaline phosphatase beads is calculated based on 250 U/mL of resin. Beads are added at 1 U/mg of protein. Therefore, if you were exchanging 11 mg of protein,

then 11 Units would require 44 μL, but since the beads are in a 50% slurry, you should effectively double the volume to 88 μL.

15. This step is done to remove the alkaline phosphatase beads from the protein solution prior to the addition of $MgCl_2$ to the solution. We use the Costar spin filters, but other methods may be used here to remove the beads, depending upon the investigator's preference.
16. Care should be taken when doing this step to not over-concentrate the protein. Using this concentrator, we have found that 12 min at 4700 rpm typically results in 5 mL being reduced to 0.75–1 mL volume.
17. Based upon the LC system used, place the buffers in positions so that it can produce a binary flow. The separation takes place during a gradient comprised of Buffer A with increasing Buffer B.
18. Typical standard operating procedures for HPLC require the buffer lines be flushed and primed with the buffer system (s) being used in the separation. This is true for nucleotide analysis, as the buffer system contains the ion-pairing reagent, which is essential for separation.
19. With the column described in this method, increasing the temperature to 40 °C helps to lower the column back pressure and decrease the retention time of the nucleotides on the column. We have not observed any hydrolysis of GTP under these conditions as the nucleotide is rapidly eluted from KRAS. The nucleotide is quickly eluted from the protein via the increased organic component of the mobile phase.
20. As with any ion-pairing HPLC method, it is critical that the column has an adequate equilibration time. Since the distribution of an ion-pairing reagent into the organic phase is enhanced by the formation of an ion pair between the two oppositely charged ions (i.e., the ion-pairing reagent and the column stationary phase), failure to adequately equilibrate the column will result in poorly separated, defined chromatography peaks. With our method, we have found that the times and buffer compositions provide a robust and repeatable method for nucleotide analysis.
21. The retention time of the standards is used to confirm the identity of the unknown samples by comparing the two. If multiple standard dilutions are used to generate a standard curve, the concentration of the unknown nucleotide(s) may be calculated.

## Acknowledgments

This project was funded in part with federal funds from the National Cancer Institute, National Institutes of Health Contract 75N91019D00024. The content of this publication does not necessarily reflect the views or policies of the Department of Health and Human Services, and the mention of trade names, commercial products, or organizations does not imply endorsement by the US government.

### References

1. Prior IA, Hood FE, Hartley JL (2020) The frequency of Ras mutations in cancer. Cancer Res 80:2969–2974
2. Moore AR, Rosenberg SC, McCormick F et al (2020) RAS-targeted therapies: is the undruggable drugged? Nat Rev Drug Discov 19:533–552
3. Gibbs JB, Marshall MS, Scolnick EM et al (1990) Modulation of guanine nucleotides bound to Ras in NIH3T3 cells by oncogenes, growth factors, and the GTPase activating protein (GAP). J Biol Chem 265:20437–20442
4. Poulin EJ, Bera AK, Lu J et al (2019) Tissue-specific oncogenic activity of KRAS(A146T). Cancer Discov 9:738–755
5. Rajalingam K, Schreck R, Rapp UR et al (2007) Ras oncogenes and their downstream targets. Biochim Biophys Acta 1773:1177–1195
6. Shutes A, Der CJ (2006) Real-time in vitro measurement of intrinsic and Ras GAP-mediated GTP hydrolysis. Methods Enzymol 407:9–22
7. Simanshu DK, Nissley DV, McCormick F (2017) RAS proteins and their regulators in human disease. Cell 170:17–33
8. Calixto AR, Moreira C, Pabis A et al (2019) GTP hydrolysis without an active site base: a unifying mechanism for Ras and related GTPases. J Am Chem Soc 141:10684–10701
9. Maurer T, Garrenton LS, Oh A et al (2012) Small-molecule ligands bind to a distinct pocket in Ras and inhibit SOS-mediated nucleotide exchange activity. Proc Natl Acad Sci U S A 109:5299–5304
10. Moghadamchargari Z, Shirzadeh M, Liu C et al (2021) Molecular assemblies of the catalytic domain of SOS with KRas and oncogenic mutants. Proc Natl Acad Sci U S A 118: e2022403118
11. Downward J (2003) Targeting RAS signalling pathways in cancer therapy. Nat Rev Cancer 3: 11–22
12. Gysin S, Salt M, Young A et al (2011) Therapeutic strategies for targeting ras proteins. Genes Cancer 2:359–372
13. Khan AQ, Kuttikrishnan S, Siveen KS et al (2019) RAS-mediated oncogenic signaling pathways in human malignancies. Semin Cancer Biol 54:1–13
14. Stephen AG, Esposito D, Bagni RK et al (2014) Dragging ras back in the ring. Cancer Cell 25:272–281
15. Brown PR (1970) The rapid separation of nucleotides in cell extracts using high-pressure liquid chromatography. J Chromatogr 52:257–272
16. Ip CY, Ha D, Morris PW et al (1985) Separation of nucleosides and nucleotides by reversed-phase high-performance liquid chromatography with volatile buffers allowing sample recovery. Anal Biochem 147:180–185
17. Zur Nedden S, Eason R, Doney AS et al (2009) An ion-pair reversed-phase HPLC method for determination of fresh tissue adenine nucleotides avoiding freeze-thaw degradation of ATP. Anal Biochem 388:108–114

# Chapter 5

# Crystallographic Studies of KRAS in Complex with Small Molecules and RAS-Binding Proteins

**Albert H. Chan and Dhirendra K. Simanshu**

## Abstract

RAS proteins play a vital role in regulating downstream signaling and essential cellular processes, positioning them as key players in normal cellular physiology and disease development. Among the various isoforms of RAS, KRAS stands out as one of the most frequently mutated genes in human cancer. The prevalence of RAS mutations in cancer often involves single amino acid substitutions at codons 12, 13, or 61. These mutations disrupt the RAS protein's inherent ability to transition between its active and inactive states, resulting in a constant activation signal and driving uncontrolled cell growth. Crystallization and structural analysis of KRAS with inhibitors and RAS-binding proteins play a pivotal role in unraveling the structural and mechanistic details of KRAS function, aiding in drug discovery efforts, and advancing our understanding of KRAS-driven diseases. Here, we present our experimental methodology for crystallizing KRAS in the presence of covalent or non-covalent small molecules and proteins acting as effectors or regulators of RAS. We detail the techniques for successful crystallization and the subsequent optimization of crystallization conditions. The resulting crystals and their structures will provide valuable insights into the key interactions between KRAS and its partner proteins or potential inhibitors, offering a foundation for developing targeted therapies that are more potent and selective against KRAS-driven cancers.

**Key words** Structural biology, X-ray crystallography, Crystallization, RAS, KRAS, KRAS-inhibitor complex

## 1 Introduction

Mutations in RAS proteins account for approximately 20% of human cancers. Out of the RAS protein family, which includes HRAS, NRAS, and the alternatively spliced isoforms KRAS4a and KRAS4b (collectively referred to as KRAS), mutations in KRAS are specifically responsible for 86% of RAS-driven cancers, such as pancreatic, colorectal, and lung cancers [1]. Despite its prominent role in oncogenesis, the development of targeted therapies against KRAS has been exceedingly challenging, mainly due to its high affinity for GDP/GTP. RAS proteins comprise a GTPase-domain (G-domain; residues 1–166) and a hypervariable region (HVR) of

Andrew G. Stephen and Dominic Esposito (eds.), *KRAS: Methods and Protocols*, Methods in Molecular Biology, vol. 2797, https://doi.org/10.1007/978-1-0716-3822-4_5,

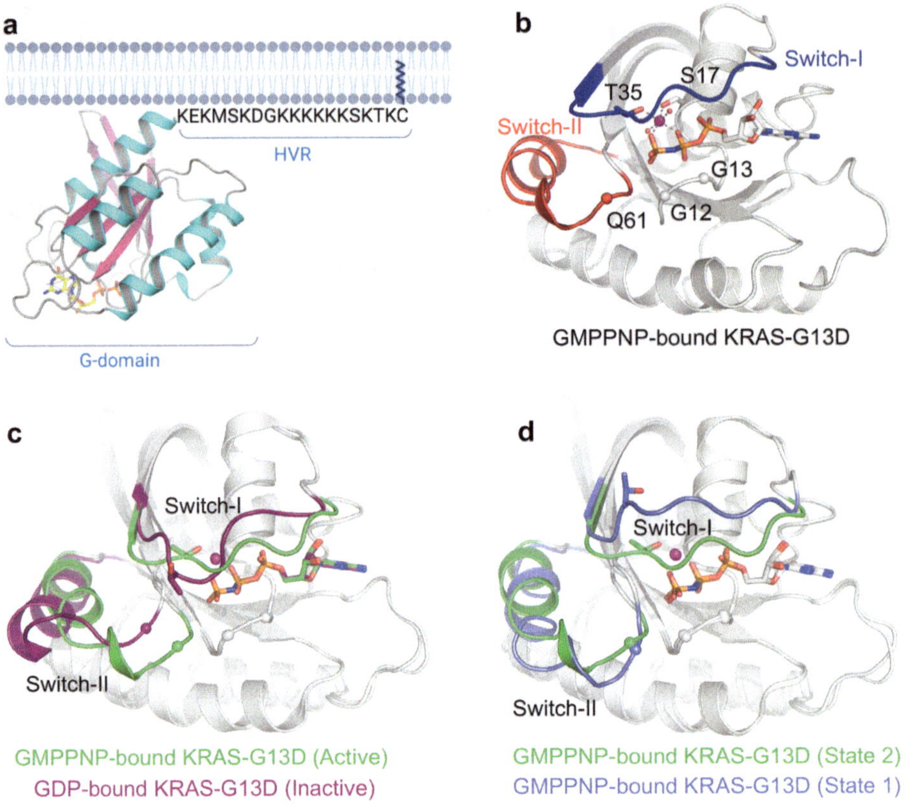


**Fig. 1** Overview of KRAS structure in active and inactive states. (**a**) Full-length KRAS contains a highly conserved catalytic G-domain and a hypervariable region (HVR) that anchors the protein to the cell membrane. (**b**) Details of the G-domain as illustrated by the GMPPNP-bound KRAS-G13D structure (PDB 8EPW) [32]. The flexible switch regions are highlighted (blue: Switch-I and red: Switch-II). The nucleotide analog GMPPNP is shown as sticks, and the magenta sphere represents the $Mg^{2+}$ ion coordinated by T35, S17, GMPPNP, and two water molecules (red sphere). The Cα of the commonly mutated residues G12, G13, and Q61 are shown as spheres as well. (**c**) Comparison of KRAS-G13D in its GMPPNP-bound (active, PDB 8EPW [32]) and GDP-bound (inactive, PDB 4TQA [38]) forms. The flexible switch regions that show the most variation between the two forms are highlighted (green: GMPPNP-bound and magenta: GDP-bound). Note that the side chain of T35 (shown as sticks) in the GDP-bound form does not coordinate with the $Mg^{2+}$ ion. (**d**) Comparison of GMPPNP-bound KRAS-G13D in state 1 (PDB 8EBZ [32]) and state 2 (PDB 8EPW [32]). The flexible switch regions that show the most variation between the two states are highlighted (green: state 2 and blue: state 1). Note that T35 (shown as sticks) in state 1 does not coordinate with the $Mg^{2+}$ ion

20 amino acids (*see* Fig. 1a). The G-domain acts as a molecular switch, toggling between active (GTP-bound) and inactive (GDP-bound) states, regulating RAS activation and signaling. Located at the C-terminus, the flexible HVR with variable amino acid sequence aids in membrane association and localization of RAS proteins. The coordination between the G-domain and HVR allows RAS proteins to govern crucial cellular processes.

In RAS proteins, the transition between active and inactive states is inherently slow. Activation of RAS occurs when guanine nucleotide exchange factors (GEFs) catalyze the exchange of GDP for GTP. Conversely, GTPase-activating proteins (GAPs) facilitate the hydrolysis of GTP to GDP, leading to RAS inactivation. In its active state, RAS engages with various effector proteins, including RAF kinase, PI3K, and RalGDS. This interaction initiates downstream signaling pathways that regulate critical cellular processes. The structural studies on RAS in complex with GAPs, GEFs, and effectors have revealed the presence of two switch regions: switch-I (amino acids 30–38) and switch-II (amino acids 59–76) (*see* Fig. 1b). These regions play a crucial role in mediating protein-protein interactions. Switch-I and switch-II undergo conformational changes between the GDP and GTP states and have been described using a loaded-spring mechanism (*see* Fig. 1c) [2]. Through $^{31}P$ solution-state NMR studies, it has been demonstrated that the switch-I region of RAS exhibits two distinct conformations when bound to GTP or the slowly hydrolyzable analog GMPPNP [3]. These conformations are referred to as state 1 (open state) and state 2 (closed state), and they have been captured crystallographically as well (*see* Fig. 1d). Analysis using $^{31}P$ NMR has revealed that wild-type RAS proteins maintain an equilibrium between states 1 and 2, with state 2 being the predominant conformation [4]. However, oncogenic mutations have been observed to shift the population distribution toward the state 1 conformation. The interaction between RAS and its effectors drives the equilibrium toward the state 2 conformation since this interaction is only compatible with the active state 2 conformation. To impede RAS-effector interactions, a potential therapeutic approach involves stabilizing the state 1 conformation of the switch-I region.

Oncogenic RAS mutations predominantly occur at amino acid positions G12, G13, and Q61. Mutations at these positions impede GAP-mediated GTP hydrolysis, rendering RAS proteins locked in the active GTP-bound state. Consequently, RAS-mediated downstream signaling pathways are continuously activated, contributing to pathogenic cellular outcomes. Over the years, significant progress has been made in determining the structures of KRAS in complex with various effectors and regulators. These structures have provided invaluable insights into the molecular mechanisms underlying KRAS signaling and have paved the way for understanding its interactions with these partner proteins.

Traditionally, small molecule inhibitors have been the cornerstone of targeted cancer therapies. However, the design of KRAS inhibitors has been hindered due to the lack of a well-defined ligand-binding pocket. Recent advances have highlighted the potential of both covalent and non-covalent small molecules in modulating KRAS activity, offering new avenues for therapeutic intervention. Covalent inhibitors have shown promise by covalently modifying a specific cysteine residue in the KRAS active site, while non-covalent small molecules attempt to exploit transient binding sites and allosteric effects. The G12C mutant form of KRAS has been successfully targeted by small molecule covalent inhibitors. These inhibitors form a cross-link with the mutant C12 residue on KRAS, resulting in high selectivity. This is exemplified by sotorasib (AMG 510) and adagrasib (MRTX849)—both have been approved by the FDA to treat non-small-cell lung cancer caused by KRAS G12C mutation. These inhibitors represent a paradigm shift, offering hope for patients with KRAS-driven cancers. Crystal structures show that these two covalent inhibitors reside in a transient pocket (known as the "switch-II pocket") between switch-II and helix α3, covalently and irreversibly cross-linked to C12, and lock KRAS in an inactive GDP-bound state, thereby preventing downstream signaling and tumor growth (*see* Fig. 2a) [5, 6]. Recently, MRTX1133, a non-covalent inhibitor, was discovered to selectively target the KRAS G12D mutant. MRTX1133 was optimized from the adagrasib scaffold to improve affinity for the KRAS switch-II pocket and selectivity for the mutant D12 residue [7]. As expected, the crystal structure shows that MRTX1133 binds in the switch-II pocket and uses its positively charged piperazine to form a salt bridge with the D12 residue (*see* Fig. 2b) [8]. This discovery shows that KRAS mutants without a cross-linkable side chain can be effectively targeted by conventional, non-covalent inhibitors. Recently, Kevan Shokat's laboratory showed the potential of covalent inhibitors in targeting the less common RAS mutants, specifically G12S and G12R [9, 10]. These studies demonstrated the possibility of developing covalent inhibitors that can target non-Cys residues within the RAS protein. Besides the aforementioned switch-II pocket, there is another transient pocket located within the groove between the switch-I and switch-II regions of KRAS where both covalent and non-covalent compounds have been observed to bind. This switch-I/switch-II pocket was initially identified with DCAI from a fragment screening campaign [11], and subsequently several small molecules, including compound 13 [12], Abd-7 [13], and BI-2852 [14], have been demonstrated to bind to this pocket. Figure 2c, d further illustrate the plasticity of this pocket since it can accommodate the KRAS G12C (GDP) covalent hit Cpd3 binding close to the nucleotide [15], as well as the non-covalent hit Ch-3 binding at the posterior end of this cleft in active KRAS [16].

Crystallization provides a unique opportunity to visualize the three-dimensional structure of proteins in complex with ligands, offering invaluable insights into the molecular determinants of protein-ligand interactions. Several factors influence the successful crystallization of protein-ligand complexes. These factors include the affinity between the protein and ligand, as well as the purity, stability, solubility, and concentrations of both components. External factors, such as pH, temperature, ionic strength, and the choice of precipitants and additives in the buffer, also significantly impact

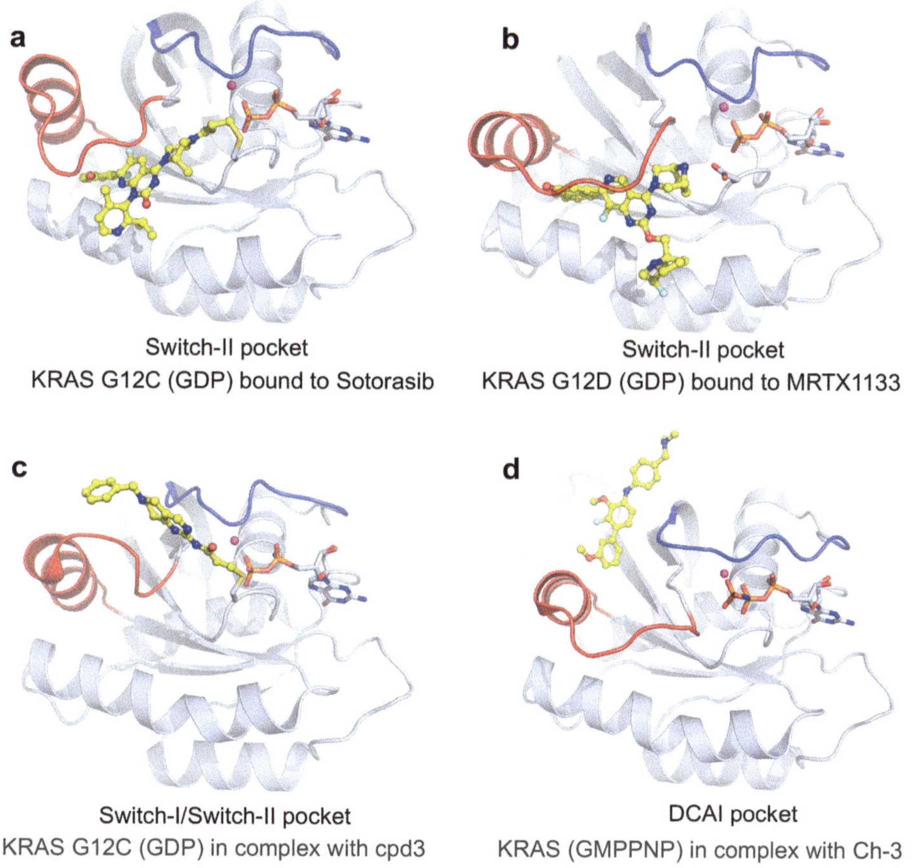


**Fig. 2** Structures of KRAS-inhibitor complexes highlighting binding pockets located on the surface of KRAS. Switch-I and Switch-II are shown in blue and red, respectively, and the small molecule inhibitors are shown in yellow ball-and-stick. (**a**) Sotorasib binds covalently to G12C and resides in the Switch-II pocket (between Switch-II and Helix $\alpha$3) of KRAS-G12C-GDP (PDB 6OIM [5]). (**b**) MRTX1133 binds non-covalently in the Switch-II pocket of KRAS-G12D-GDP (PDB 7RPZ [8]). (**c**) Cpd3 binds covalently to G12C and resides in the Switch-I/II pocket of KRAS-G12C-GDP (PDB 7A1W [15]). (**d**) Ch-3 binds non-covalently to the Switch-I/II pocket of wild-type KRAS-GMPPNP (PDB 6GQY [16])

the crystallization process. Additionally, the flexibility of the protein and the specific nature of crystal lattice contacts play crucial roles in achieving successful crystallization. It is crucial to carefully optimize and fine-tune these factors to obtain high-quality crystals suitable for structural determination. This involves precisely adjusting the experimental conditions to maximize the chances of obtaining well-formed crystals. Addressing these considerations makes it possible to enhance the likelihood of obtaining crystals that provide accurate structural information about protein-ligand complexes.

Crystallizing KRAS with small molecules aids in drug discovery by characterizing their binding modes and interactions with specific amino acids, while co-crystallization with RAS-binding proteins identifies therapeutic targets and reveals new protein-protein interactions for drug discovery. Together, these complexes provide a powerful structural framework for structure-guided drug design, enabling the design of novel compounds with improved binding affinity, selectivity, and pharmacological properties, ultimately leading to more effective KRAS-targeted therapies. Here, we describe our experimental approach to crystallizing KRAS in complex with covalent or non-covalent small molecules, as well as proteins that interact with RAS as effectors or regulators. We outline the methodologies employed for crystallization, as well as the optimization of crystallization conditions. It is important to note that while this chapter describes the crystallization of KRAS, the same methodologies can be extended to crystallizing HRAS and NRAS. The G-domains of these RAS isoforms are highly similar, with complete sequence identity in the first half (residues 1–86) and 82% sequence identity in the second half (residues 87–166). Interestingly, HRAS shows a higher probability of successful crystallization in our experiments than the KRAS protein. Consequently, when faced with challenges in obtaining diffracting crystals using KRAS, HRAS can be used to enable successful crystallization.

## 2 Materials

1. Protein buffer for crystallization: typically 20 mM HEPES pH 7.4, 150 mM NaCl, 5 mM $MgCl_2$, and 1 mM TCEP. Remove TCEP or other reducing agents if a covalent compound will be attached to the protein before setting up crystal trays.
2. Crystallization screening kits available from various commercial sources, such as Hampton Research, Molecular Dimensions, Jena Bioscience, Rigaku, and NeXtal Biotech.
3. Seeding tools, such as Crystal Crusher and Seed Beads from Hampton Research.
4. Crystallization plates, such as 96-well 2- to 3-drop INTELLI-PLATES manufactured by Art Robbins Instruments or MRC plates manufactured by Swissci, available from vendors such as Hampton Research, Molecular Dimensions, MiTeGen, and Jena Bioscience.
5. Plate seals, such as ClearSeal Film from Hampton Research, HD Clear Sealing Tape from MiTeGen, and ClearVue Sheets from Molecular Dimensions.
6. Liquid handling systems for setting up crystal trays. For example, NT8 from Formulatrix, Phoenix from Art Robbins Instruments, Dragonfly, and Mosquito from SPT Labtech.

7. Plates incubator with an automated visualization system, such as Rock Imager from Formulatrix.
8. Stereo microscope equipped with a cold light source and a large stage that can fit a crystal tray, available from various vendors such as MiTeGen and Molecular Dimensions.
9. Loops and bases or the preassembled systems (Hampton Research, MiTeGen, Molecular Dimensions) and pucks (MiTeGen, Molecular Dimensions, Crystal Positioning Systems) compatible with the synchrotron beamline.
10. Cryogenic handling tools including magnetic wands, tongs, and foam dewars (Hampton Research, MiTeGen, Molecular Dimensions).
11. CX100 cryo express dry shipper (MiTeGen, Crystal Positioning Systems) for puck storage and shipment.
12. X-ray source from in-house diffractometer (Rigaku, Bruker) or synchrotron radiation (e.g., APS, NSLS-II, SSRL, ALS, CHESS, and DLS).
13. Data processing and refinement programs such as XDS [17], HKL2000 [18], iMosflm [19], AIMLESS [20], POINTLESS [20], Xtriage [21], Phaser [22], Refmac [23], Phenix.refine [24], and COOT [25], many of them available in the CCP4 [26] or Phenix [24] programs suite and can be downloaded individually or provided by the SBGrid consortium [27].

## 3 Methods

### *3.1 Crystallization of KRAS*

Different diffusion-based methods, including hanging drop vapor diffusion, sitting drop vapor diffusion, and microbatch-under-oil, are commonly employed for the crystallization of protein, protein-ligand, and protein-protein complexes [28]. Here, we will focus on the sitting drop vapor diffusion method while noting that the other techniques operate similarly. Vapor diffusion methods rely on the controlled diffusion of solutes and vapor to facilitate the formation of protein crystals. By establishing a concentration gradient through precipitants in the reservoir solution, these methods promote the nucleation and growth of crystals for detailed structural analysis.

This section describes a general crystallization scheme for crystallizing the KRAS-nucleotide-$Mg^{2+}$ complex, the native form of KRAS in cells. The nucleotide can be GDP or any slowly or non-hydrolyzable GTP analogs such as GTPγS, GMPPNP, or GMPPCP (*see* **Note 1**). KRAS can be any mutant (*see* **Note 2**) and can be full-length or mainly the G-domain (residues 1–169) (*see* **Note 3**). For methods to produce different KRAS constructs in different nucleotide-bound forms, *see* Chapters 3 and 4.

1. Add 65 μL of crystallization solution from commercially available crystallization screening kits to the reservoir of a 96-well sitting-drop plate (*see* **Note 4**).
2. Combine 0.2 μL of 10–20 mg/mL protein sample with 0.2 μL of crystallization solution in the wells (*see* **Note 5**).
3. Seal the plate and incubate at 20 °C. Check the plate for crystal growth periodically for up to 2 months.
4. If no hits are observed after 2–3 weeks, a new screen might need to be set up, with changes to protein concentration, crystallization condition, or incubation temperature.
5. If only low-quality (tiny, clusters, no sharp edges, etc.) crystals are observed, the crystallization condition might need to be optimized by performing a grid screen (varying the pH and/or concentration of the crystallization ingredients) or an additive screen (adding 10% additive to the original crystallization condition). Seeding might be a good alternative (*see* **Note 6**).
6. If good crystals are observed, depending on the crystallization condition, 20–30% cryoprotectant such as glycerol or ethylene glycol might need to be added before crystal harvesting. Crystals should be flash-frozen and stored in a liquid nitrogen dewar until data collection time (*see* **Note 7**).

### *3.2 Crystallization of KRAS with a Small Molecule Compound*

To target KRAS-driven cancers, both covalent and non-covalent inhibitors are being developed, each with their advantages and limitations. Covalent inhibitors provide prolonged target engagement, but careful design is required to ensure specificity. Non-covalent inhibitors offer reversibility, allowing for fine-tuning of the target interaction, but may require higher potency for sustained inhibition. The choice between covalent and non-covalent inhibitors depends on the specific therapeutic goals, the target protein's characteristics, and considerations for safety and selectivity.

The best way to obtain crystals of KRAS with a small molecule compound bound to it depends on the nature of the compound—covalent or non-covalent, and whether it is a strong binder ($K_D$ < 10 μM) or a weak binder ($K_D$ > 100 μM) if non-covalent. Figure 3 outlines the decision tree and refers readers to different sections below. If co-crystallization with KRAS does not produce crystals and soaking does not work either, HRAS with the proper "KRAS-like" mutation (such as Q95H) in the expected small molecule binding pocket can be used as a surrogate. This suggestion is based on our observation that HRAS exhibits a higher tendency to crystallize compared to KRAS. Compound-binding poses in KRAS and HRAS are often very similar, if not identical.

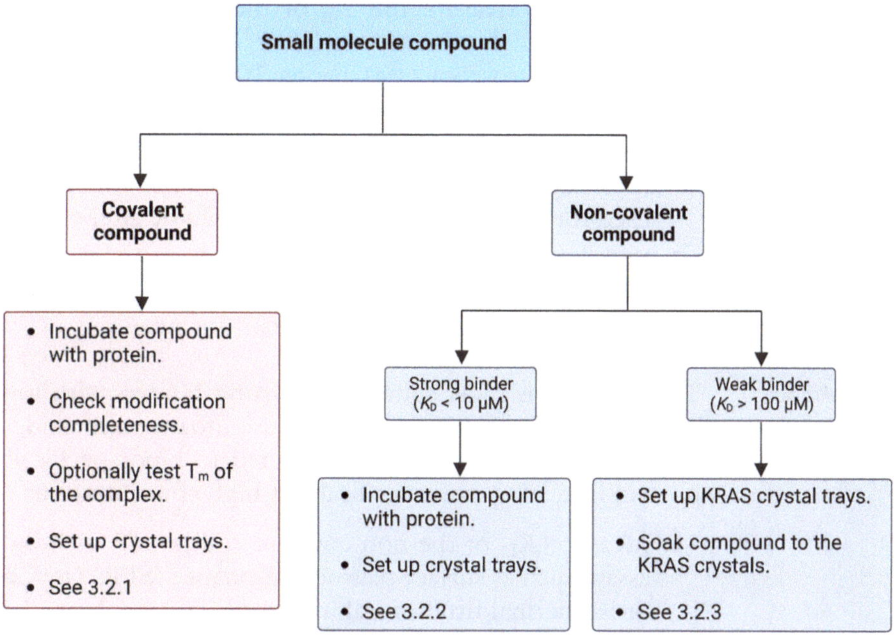


**Fig. 3** Flowchart showing crystallization strategy for obtaining crystal structures of KRAS in complex with a small molecule compound

*3.2.1 Co-crystallizing with a Covalent Inhibitor*

When working with a new covalent inhibitor, it is often helpful to perform a small volume test to optimize the compound-to-protein ratio and incubation time before producing a large batch of sample for crystallization. Moreover, when there are multiple covalent inhibitor candidates, it might be helpful to compare the melting temperature ($T_m$) of the different complexes. The more stable (higher $T_m$) complex will have a higher chance of being crystallized.

1. Prepare KRAS sample (*see* **Note 8**) suitable for covalent modification. If the covalent compound is expected to form a disulfide bond with the protein, remove any reducing agent such as TCEP or DTT by dialysis or buffer exchange.
2. Add an excess of covalent compound to KRAS. A compound-to-KRAS ratio of 20:1 is a good starting point for compounds in the early development phase with low non-covalent binding affinity. Do not exceed 10% DMSO final concentration in the reaction mix.
3. Incubate at 4 °C with regular mixing. Take time points periodically and monitor modification completeness by mass spectrometry (*see* Chapter 11 and **Note 9**).
4. When >90% KRAS has been modified by the compound, separate the excess compound from the protein by passing the mixture through a desalting column. Alternatively, the excess compound can be removed by dialysis or buffer exchange.

5. Optional: Measure the $T_m$ of the complex with differential scanning fluorimetry (DSF) or circular dichroism (CD) spectroscopy (*see* Chapter 10 and **Note 10**).
6. Concentrate the complex to 10–20 mg/mL prior to setting up crystal trays. As the compound might absorb UV, concentration might have to be estimated from the amount of protein used for modification and the final volume after concentration.
7. Follow the steps outlined in Subheading **3.1** to set up crystal trays (*see* **Note 11**). Be wary of crystals that appear in conditions containing reducing agents.

#### *3.2.2 Co-crystallization with a Non-covalent Inhibitor*

This method is most suitable for strong binders with low-micromolar $K_D$ or below. It may also be used for weaker binders if they have good solubility and are predicted to bind specifically to the protein instead of non-specifically at high concentrations.

1. Measure $K_D$ of the non-covalent compound with biophysical assays such as surface plasmon resonance (SPR) (*see* Chapter 8) or isothermal titration calorimetry (ITC) (*see* **Note 12**).
2. Add the non-covalent compound to KRAS. To ensure >90% bound complex, compound concentration should be >10 times the $K_D$ and at least 10% more than the protein concentration (*see* **Note 13**).
3. Incubate at 4 °C with regular mixing for 1 h.
4. Follow Subheading 3.1, **steps 1–5** to set up crystal trays (*see* **Note 14**).
5. If crystals have to be cryoprotected before freezing, either add the cryoprotectant (e.g., glycerol at 20–30% final concentration) directly to the drop containing the crystals or transfer the crystals to a drop containing the crystallization solution with 20–30% cryoprotectant and the same amount of compound used during crystal tray setup (*see* **Note 15**).

#### *3.2.3 Soaking Compounds to KRAS Crystals*

This method is popular for studying the binding poses of early hits with relatively low binding affinity, as co-crystallization might not be practical due to solubility limit and/or potential nonspecific binding at high concentrations. However, strong binders or even covalent compounds can be studied as well if there is a compatible crystal form for soaking. The compounds that benefit most from soaking are usually of low molecular weight, such that they can migrate through the solvent channels to their binding pockets more easily and may require less rearrangement of the protein once they bind. These low-molecular-weight compounds tend to have higher solubility, too. The main obstacle of performing the soaking experiment is finding a suitable KRAS crystal form. Below we outline the steps from determining if a KRAS crystal is suitable for soaking to incubating with compounds and harvesting.

1. Follow Subheading 3.1 to produce KRAS crystals. Crystals should be very reproducible and diffract to 2.5 Å or better, and ideally do not require the addition of cryoprotectant for freezing.
2. Examine the protein structure and crystal packing. The putative binding pocket should not be obstructed by nearby protomers and should be open to solvent channel(s) (*see* **Note 16**).
3. Test DMSO tolerance of the crystal. Transfer the crystals to drops containing crystallization solution and an increasing amount of DMSO (5–30%). Observe for crystal cracking/melting over time, and harvest to test diffraction at several time points. Ideally, the crystal should tolerate 10% DMSO over 1–3 h of soaking without losing diffraction quality (*see* **Note 17**).
4. Reproduce the KRAS crystals if they are determined to be suitable for the soaking experiment. Otherwise, identify a different crystal form and repeat **steps 1–3**. Examples from the literature are given in **Note 18**.
5. Prepare 25–50 mM compound soaking solution by adding a highly concentrated (250–500 mM) compound stock to the crystallization solution (*see* **Note 19**). Final DMSO concentration should be within the range that the crystal can tolerate.
6. Test the pH of the soaking solution by spotting the solution on a pH strip. Adjust pH with acid or base if the pH is more than 2 units off.
7. Transfer a crystal to 1–2 μL of soaking solution (*see* **Note 20**). Incubate for at least 5 min, more commonly 1–3 h, up to 24 h.
8. If the crystal has to be cryoprotected before freezing, either add the cryoprotectant (e.g., glycerol at 20–30% final concentration) directly to the drop or prepare the cryo-solution separately by adding the cryoprotectant to the soaking solution and then transferring the crystal into it.
9. Loop the crystal and store it in a liquid nitrogen dewar until data collection time.

### *3.3 Crystallization of KRAS with Other Proteins*

The main challenge of crystallizing KRAS with some of its partner proteins (effectors, GEFs, and GAPs) is the relatively low binding affinity, often in the single-digit micromolar range. Careful construct designs and exploring different KRAS mutants might be needed for successful co-crystallization.

1. Measure the binding affinity of the partner protein to KRAS with methods such as SPR or ITC.

2. If $K_D$ is in the nanomolar range, mix the partner protein with KRAS and purify the complex with a size exclusion column. Depending on the stability of the partner protein and where the complex elutes relative to the monomers, one of the proteins might need to be in slight excess relative to the other (*see* **Note 21**).
3. If $K_D$ is in the single-digit micromolar range and the complex is at risk of falling apart during size-exclusion column purification, simply mix the partner protein with KRAS in a 1:1 molar ratio, without further purifying the complex with the size-exclusion column (*see* **Note 22**).
4. If $K_D > 10\ \mu M$, redesign the protein construct by varying the length to include or exclude certain domain or loop region or by introducing mutations to increase the affinity between the partner proteins and KRAS. Certain mutants of KRAS might have a higher affinity as well (*see* **Note 23**).
5. Follow the steps in Subheading 3.1 to set up crystal trays.

### 3.4 Data Collection and Structure Determination

With over 250 structures of KRAS or KRAS complexes in the PDB, structure determination should be rather straightforward with good-quality data. Typically, the main challenge of structure determination lies in data quality; it is not uncommon to have pseudo-translation or twinning problems or both in the same dataset. Since the majority of the KRAS backbone remains the same in different structures, with most of the variation localized in the switch regions and the C-terminus, excellent initial models can often be obtained by molecular replacement without the need for experimental phasing. Many structures of KRAS-binding proteins or domains such as GAP, GEF, and RBD have been solved as well, so molecular replacement is applicable to many KRAS complexes, too. Once the structure is solved, it is important to be careful while interpreting the conformation of switch regions due to potential crystal contacts. Crystal structures provide valuable insights into the three-dimensional arrangement of proteins, but they capture only a single snapshot. Crystal packing forces and interactions can influence the conformation of flexible regions, such as the switch regions in RAS proteins. These switch regions in RAS proteins are highly dynamic and undergo conformational changes to fulfill their functional roles. However, crystal packing forces can limit the conformational flexibility of switch regions, resulting in a single snapshot that may not fully represent the protein's behavior in solution. To better understand the conformations of the switch regions, it is helpful to analyze structures in multiple crystal forms with different packing interactions.

1. Collect data from an in-house X-ray diffractometer or a synchrotron beamline. Native data is usually sufficient.

2. Perform data reduction and assess the quality of the processed data with Xtriage. Data might need to be reindexed if twinning and/or pseudo-translation, the most common pathologies of KRAS crystals, especially in the P622 point group, is suspected.
3. Perform molecular replacement with Phaser. It is recommended to use an initial model with the switch regions (residues 30–38 and 59–76) removed, as they tend to adopt different conformations across different structures.
4. Iteratively build the model with Coot and refine the model with Phenix.refine or Refmac (*see* **Note 24**). It is not unusual to have no density for the switch regions unless they are interacting with a partner protein, small molecule, or neighboring protomer in symmetry.

## 4 Notes

1. The GDP-bound form usually crystallizes more readily, whereas the GTP-analog-bound form is usually more challenging to obtain good diffracting crystals or any crystals at all. Most popular slowly or non-hydrolyzable GTP analogs used are GMPPNP, GMPPCP, and GTP$\gamma$S. Among these analogs, GTP$\gamma$S demonstrates the highest affinity, followed by GMPPNP and then GMPPCP [29]. However, in terms of stability, the rank order is reversed. Therefore, GMPPNP is the preferred GTP analog for conducting structural studies on RAS proteins. GTP can be used too for KRAS mutants with a very slow intrinsic hydrolysis rate, and it has been successfully used in crystallizing KRAS-Q61H (PDB 6MNX) [30].
2. Most of the mutations found in the KRAS structures in the PDB are those that promote oncogenicity (e.g., G12C, G12V, G12D, G13D, Q61H, Q61L, Q61R, A146T, etc.), but there are examples of mutation designed to improve crystallization, such as C118S, which is especially helpful when crystallizing the GTP-analog-bound form of KRAS.
3. While full-length KRAS has been used for crystallization, when crystallized without a partner protein, the hypervariable region outside of the G-domain often has poor electron density and cannot be modelled. Therefore, most of the constructs used in crystallography consist of the G-domain only.
4. KRAS crystallizes in a wide range of conditions. For example, wild-type KRAS4b(1–169)-GDP crystallized in 19% PEG 3350, 0.2 M calcium chloride (PDB 4LPK) [31]; KRAS4b (1–169)G12D/C118S-GDP crystallized in 0.1 M sodium phosphate pH 7.5, 30% PEG 4000, 0.2 M lithium sulfate

(PDB 4EPR) [12]; KRAS4b(1–169)G13D-GMPPNP crystallized in 0.1 Tris–HCl pH 8.5, 20% PEG 400, 20% PEG 8000, 0.05 M magnesium chloride (PDB 8EBZ) [32]; KRAS4b (1–169)A59G-GMPPCP crystallized in 0.1 M citric acid pH 3.5, 20% PEG 1500, 4% MPD (PDB 6BP1) [33]; and wild-type KRAS4b(2–188)-GMPPNP crystallized in 0.1 M Tris–HCl pH 8.0, 31.4% PEG 4000, 0.2 M sodium acetate (PDB 6GOD) [16]. To improve the chance of success, for every new protein sample, we typically test 10–12 different crystallization screening kits. Difficult targets such as the GMPPNP-bound protein sometimes only form crystals in one condition out of over a thousand conditions tested.

5. Generally, 15–20 mg/mL of protein concentration will yield crystals. However, it is not unusual that crystallization may occasionally require a much higher protein concentration 40–75 mg/mL [16]. Typically, the ideal protein concentration is achieved when approximately 10–25% of the crystallization drops exhibit noticeable signs of precipitation. The Pre-Crystallization Test kit from Hampton Research can be used to determine the appropriate protein concentration for crystallization screening.

6. There are many different seeding methods, such as streaking and microseeding. We recommend using Seed Beads (Hampton Research) and following the manufacturer's protocol to generate seeds.

7. If the nucleotide is GTP or GTPγS, crystals should be harvested as soon as they have grown to decent sizes to avoid nucleotide hydrolysis. On the contrary, GDP, GMPPNP, and GMPPCP are very stable and will not be hydrolyzed for months.

8. To ensure homogeneity of the compound-modified sample, it might be necessary to mutate one or all the native cysteines in KRAS prior to modification. C118 is the most solvent-exposed cysteine and is commonly mutated to a serine to prevent compound mistargeting. In addition, it might be necessary to mutate C51 to serine and C80 to leucine as well if the compound has a very reactive covalent warhead. The KRAS G12C/C51S/C80L/C118S construct was first described by Kevan Shokat's group [31] and has since been used to study the structures of many different G12C-targeting covalent inhibitors, including sotorasib [5] and adagrasib [6].

9. For covalent compounds with relatively weak tethering efficiency, it might take up to 24 h to achieve >90% modification. If 90% modification is not achieved, or if multiple modification is observed, changing the compound:protein ratio and/or incubation temperature is recommended. If none of these

strategies work, it might be necessary to swap the covalent warhead for something more reactive (e.g., chloroacetamide, if modification is inefficient and incomplete) or milder (e.g., 2-F substituted acrylamide like adagrasib, if the original compound is too reactive and modifies the protein in different places).

10. For KRAS G12C-targeting compounds that bind in the Switch-II pocket, $T_m$ of the modified protein can increase by as much as 20 °C over unmodified protein [34].
11. KRAS that has been modified by covalent compounds can crystallize in a variety of conditions. For example, vinyl sulfonamide compound 8 tethered to KRAS4b(1–169)G12C/C51S/C80L/C118S-GDP crystallized in 0.1 M sodium citrate pH 5.6, 31% PEG 4000, 0.2 M ammonium acetate (PDB 4LYF) [31]; sotorasib tethered to KRAS4b(1–169)G12C/C51S/C80L/C118S-GDP crystallized in 0.1 M MES pH 6.5, 30% PEG 4000, 1 mM magnesium chloride (PDB 6OIM) [5]; and adagrasib tethered to KRAS4b(1–169)G12C/C51S/C80L/C118S-GDP crystallized in 32% PEG 4000, 0.1 M sodium acetate, 8% 2-propanol (PDB 6UT0) [6].
12. If data from direct-binding assays are not available, $IC_{50}$ obtained from protein-protein interaction (PPI) or nucleotide exchange assays (*see* Chapters 12 and 18) can be used to estimate the binding affinity of the compound.
13. If compound solubility is a problem, for a strong binder, the compound and KRAS can be incubated at a lower concentration (e.g., 50 μM compound and 20 μM protein) before concentrating to proper final concentration and setting up crystal trays. Alternatively, the protein may be incubated with the compound suspension for an extended period of time (2–3 h up to overnight) at 4 °C, and the mixture should be filtered to remove particulates before setting up crystal trays.
14. KRAS has been co-crystallized with non-covalent compounds many times and in a variety of conditions. For example, compound 13 was co-crystallized with KRAS4b(1–169)G12V/C118S-GDP in 0.1 M MMT pH 4, 28% PEG 4000, 0.2 M ammonium chloride (PDB 4EPY) [12]; BI2852 was co-crystallized with KRAS4b(1–169)G12D-GMPPCP in 0.1 M bis-Tris pH 6, 25% PEG 3350, 0.2 M ammonium sulfate (PDB 6GJ8) [14]; and MRTX1133 was co-crystallized with KRAS4b(1–169)G12D/C51S/C80L/C118S-GDP in 0.1 M sodium citrate pH 4.2, 29% PEG 8000, 0.2 M ammonium acetate (PDB 7RPZ) [8].
15. If the compound is a very strong binder with $K_D$ in the nanomolar range, it might not be necessary to include the compound in the cryo-solution, as the tight binder is not likely

to diffuse out of the binding pocket during the brief handling time.

16. The pocket does not necessarily need to be open or fully formed, as the compound binding can induce conformational changes to form the pocket. However, we need to make sure there is no nearby protomer(s) that could prevent the conformational changes from happening.
17. While DMSO is the most popular solvent for small molecules, if the crystals cannot tolerate DMSO, there are other solvents one can test, such as methanol or ethylene glycol.
18. A popular system for soaking is the His-tagged version of KRAS Q61H (MHHHHHHSSGRENLYFQG-KRAS4b (1–169)Q61H-GMPPNP). Crystals can be obtained by mixing 75 mg/mL of protein with an equal volume of crystallization solution consists of 0.2 M lithium citrate pH 5.5, 8–15% PEG 3350, and incubating at 4 °C (PDB 6GOG) [13, 16]. The second example is GS-KRAS4b(2–188)G12D-GMPPNP (His tag cleaved off by TEV protease). Crystals can be obtained by mixing 45 mg/mL of protein with an equal volume of crystallization solution consists of 0.1 M Tris–HCl pH 8, 0.2 M sodium acetate, and 32% PEG 4000 (PDB 6GOF) [13, 16]. The third example is G-KRAS4b(1–169) G12C-GDP (His tag cleaved off by TEV protease). Crystals can be obtained by mixing 44 mg/mL of protein with an equal volume of crystallization solution consists of 0.1 M Tris–HCl pH 8.5, 23% PEG 4000, and 0.1 M sodium acetate (PDB 7A1X) [15].
19. Alternatively, a cocktail of multiple compounds can be prepared by mixing multiple compounds with the crystallization solution. However, this will lower the final concentration of individual compound as we do not recommend exceeding final DMSO concentration more than 10–20%.
20. Alternatively, instead of transferring the crystal to a soaking solution, compound can be added directly to the drop containing crystals. It is possible to use ECHO (Beckman Coulter), a liquid handler that employs acoustic energy for precise nanoliter ejections, to deliver a very small volume (20–40 nL) of compound to the drop. This method is convenient if one wishes to soak many different compounds. However, it might not be suitable if the compounds are expected to alter the pH of the drop a lot (e.g., if they are in HCl or TFA salt form).
21. An example is complexing the RAS-binding domain and cysteine-rich domain of Raf1 (Raf1(RBD-CRD)) with KRAS. Raf1(RBD-CRD) has a nanomolar binding affinity ($K_D$ ~ 152 nM as measured by SPR) with KRAS. However, since Raf1(RBD-CRD) is less stable and tends to form

aggregates, we incubated it with KRAS in a 1.2:1 molar ratio and purified the 1:1 complex from the size exclusion column. Ultimately, we were able to obtain crystals of Raf1(RBD-CRD) with wild-type or different oncogenic mutants of KRAS in crystallization conditions such as 0.1 M sodium cacodylate pH 6.5, 0.7 M sodium acetate, and 0.35 mM D-myo-phosphatidylinositol 3,4,5-triphosphate (PDB 6XGV) [35]. These crystals diffracted to a wide range of resolution limits (1.95–2.87 Å).

22. An example is complexing PDEδ with the full-length, farnesylated-methylated KRAS4b. According to ITC, the binding affinity is 2.3 μM. Moreover, the lipidated KRAS tends to bind to the column resin. To avoid the complex falling apart and the lipidated KRAS component preferentially sticking to the column resin during size-exclusion column purification, we mixed the two proteins in a 1:1 molar ratio and set up trays directly. We were able to obtain crystals in two different crystallization conditions: Condition 1 (diffracted to 2.0 Å): 0.1 M citrate pH 5.0 and 20% PEG 6000 (PDB 5TB5) and Condition 2 (diffracted to 1.9 Å): 0.1 M HEPES pH 7.5 and 2 M ammonium sulfate (PDB 5TAR) [36].
23. An example is complexing RBD and Pleckstrin homology (PH) domain of Sin1 with KRAS. While Sin1(RBD) has $K_D$ ~ 5.4 μM with KRAS, Sin1(RBD-PH) has a significantly higher $K_D$ ~ 65 μM. Fortunately, we were able to obtain the structure of Sin1(RBD) in complex with KRAS, and from there we identified a Q25A mutation in KRAS that resulted in a modest but significant decrease in $K_D$ (~3.6 μM). This KRAS-Q25A also has improved binding affinity to Sin1 (RBD-PH) ($K_D$ ~ 10.5 μM), and we were able to obtain a KRAS-Q25A/Sin1(RBD-PH) complex crystal (diffracted to 2.35 Å) from the crystallization condition containing 0.1 M HEPES pH 7.5, 25% PEG 3350, and 0.2 M ammonium sulfate (PDB 7LC1) [37].
24. If the crystal is twinned with multiple twinning operators available, Refmac might be a better refinement program to use than Phenix.refine, since Refmac supports refinement with multiple twinning operators, while Phenix.refine supports only one. We have come across several KRAS-mutant crystals with almost equal ratio of twinning across four twinning operators, and refinement with Refmac gave far superior results than Phenix.refine.

## Acknowledgments

This project has been funded in full or in part with Federal funds from the National Cancer Institute, National Institutes of Health, under Contract No. 75N91019D00024. The content of this publication does not necessarily reflect the views or policies of the Department of Health and Human Services nor does mention of trade names, commercial products, or organizations imply endorsement by the US government.

### References

1. Prior IA, Hood FE, Hartley JL (2020) The frequency of Ras mutations in cancer. Cancer Res 80:2969–2974
2. Vetter IR, Wittinghofer A (2001) The guanine nucleotide-binding switch in three dimensions. Science 294:1299–1304
3. Spoerner M, Herrmann C, Vetter IR et al (2001) Dynamic properties of the Ras switch I region and its importance for binding to effectors. Proc Natl Acad Sci U S A 98:4944–4949
4. Spoerner M, Nuehs A, Ganser P et al (2005) Conformational states of Ras complexed with the GTP analogue GppNHp or GppCH2p: implications for the interaction with effector proteins. Biochemistry 44:2225–2236
5. Canon J, Rex K, Saiki AY et al (2019) The clinical KRAS(G12C) inhibitor AMG 510 drives anti-tumour immunity. Nature 575:217–223
6. Fell JB, Fischer JP, Baer BR et al (2020) Identification of the clinical development candidate MRTX849, a covalent KRAS(G12C) inhibitor for the treatment of cancer. J Med Chem 63: 6679–6693
7. Hallin J, Bowcut V, Calinisan A et al (2022) Anti-tumor efficacy of a potent and selective non-covalent KRAS(G12D) inhibitor. Nat Med 28:2171–2182
8. Wang X, Allen S, Blake JF et al (2022) Identification of MRTX1133, a noncovalent, potent, and selective KRAS(G12D) inhibitor. J Med Chem 65:3123–3133
9. Zhang Z, Guiley KZ, Shokat KM (2022) Chemical acylation of an acquired serine suppresses oncogenic signaling of K-Ras(G12S). Nat Chem Biol 18:1177–1183
10. Zhang Z, Morstein J, Ecker AK et al (2022) Chemoselective covalent modification of K-Ras (G12R) with a small molecule electrophile. J Am Chem Soc 144:15916–15921
11. Maurer T, Garrenton LS, Oh A et al (2012) Small-molecule ligands bind to a distinct pocket in Ras and inhibit SOS-mediated nucleotide exchange activity. Proc Natl Acad Sci U S A 109:5299–5304
12. Sun Q, Burke JP, Phan J et al (2012) Discovery of small molecules that bind to K-Ras and inhibit Sos-mediated activation. Angew Chem Int Ed Engl 51:6140–6143
13. Quevedo CE, Cruz-Migoni A, Bery N et al (2018) Small molecule inhibitors of RAS-effector protein interactions derived using an intracellular antibody fragment. Nat Commun 9:3169
14. Kessler D, Gmachl M, Mantoulidis A et al (2019) Drugging an undruggable pocket on KRAS. Proc Natl Acad Sci U S A 116:15823–15829
15. Mathieu M, Steier V, Fassy F et al (2022) KRAS G12C fragment screening renders new binding pockets. Small GTPases 13:225–238
16. Cruz-Migoni A, Canning P, Quevedo CE et al (2019) Structure-based development of new RAS-effector inhibitors from a combination of active and inactive RAS-binding compounds. Proc Natl Acad Sci U S A 116:2545–2550
17. Kabsch W (2010) Integration, scaling, space-group assignment and post-refinement. Acta Crystallogr D Biol Crystallogr 66:133–144
18. Otwinowski Z, Minor W (1997) Processing of X-ray diffraction data collected in oscillation mode. Methods Enzymol 276:307–326
19. Battye TG, Kontogiannis L, Johnson O et al (2011) iMOSFLM: a new graphical interface for diffraction-image processing with MOSFLM. Acta Crystallogr D Biol Crystallogr 67:271–281
20. Evans PR, Murshudov GN (2013) How good are my data and what is the resolution? Acta Crystallogr D Biol Crystallogr 69:1204–1214
21. Zwart PH, Grosse-Kunstleve RW, Adams PD (2005) Xtriage and Fest: automatic assessment

of X-ray data and substructure structure factor estimation. CCP4 Newsl 43:contribution 7

22. McCoy AJ, Grosse-Kunstleve RW, Adams PD et al (2007) Phaser crystallographic software. J Appl Crystallogr 40:658–674
23. Murshudov GN, Skubak P, Lebedev AA et al (2011) REFMAC5 for the refinement of macromolecular crystal structures. Acta Crystallogr D Biol Crystallogr 67:355–367
24. Adams PD, Afonine PV, Bunkoczi G et al (2010) PHENIX: a comprehensive Python-based system for macromolecular structure solution. Acta Crystallogr D Biol Crystallogr 66:213–221
25. Emsley P, Lohkamp B, Scott WG et al (2010) Features and development of Coot. Acta Crystallogr D Biol Crystallogr 66:486–501
26. Agirre J, Atanasova M, Bagdonas H et al (2023) The CCP4 suite: integrative software for macromolecular crystallography. Acta Crystallogr D Struct Biol 79:449–461
27. Morin A, Eisenbraun B, Key J et al (2013) Collaboration gets the most out of software. elife 2:e01456
28. McPherson A, Gavira JA (2014) Introduction to protein crystallization. Acta Crystallogr F Struct Biol Commun 70:2–20
29. Rensland H, John J, Linke R et al (1995) Substrate and product structural requirements for binding of nucleotides to H-ras p21: the mechanism of discrimination between guanosine and adenosine nucleotides. Biochemistry 34: 593–599
30. Zhou ZW, Ambrogio C, Bera AK et al (2020) KRAS(Q61H) preferentially signals through MAPK in a RAF dimer-dependent manner in non-small cell lung cancer. Cancer Res 80: 3719–3731
31. Ostrem JM, Peters U, Sos ML et al (2013) K-Ras(G12C) inhibitors allosterically control GTP affinity and effector interactions. Nature 503:548–551
32. Chao FA, Chan AH, Dharmaiah S et al (2023) Reduced dynamic complexity allows structure elucidation of an excited state of KRAS (G13D). Commun Biol 6:594
33. Lu J, Bera AK, Gondi S et al (2018) KRAS switch mutants D33E and A59G crystallize in the state 1 conformation. Biochemistry 57: 324–333
34. McGregor LM, Jenkins ML, Kerwin C et al (2017) Expanding the scope of electrophiles capable of targeting K-Ras oncogenes. Biochemistry 56:3178–3183
35. Tran TH, Chan AH, Young LC et al (2021) KRAS interaction with RAF1 RAS-binding domain and cysteine-rich domain provides insights into RAS-mediated RAF activation. Nat Commun 12:1176
36. Dharmaiah S, Bindu L, Tran TH et al (2016) Structural basis of recognition of farnesylated and methylated KRAS4b by PDEdelta. Proc Natl Acad Sci U S A 113:E6766–E6775
37. Castel P, Dharmaiah S, Sale MJ et al (2021) RAS interaction with Sin1 is dispensable for mTORC2 assembly and activity. Proc Natl Acad Sci U S A 118:e2103261118
38. Hunter JC, Manandhar A, Carrasco MA et al (2015) Biochemical and structural analysis of common cancer-associated KRAS mutations. Mol Cancer Res 13:1325–1335

# Chapter 6

# Considerations Around Structure-Based Drug Discovery for KRAS Using DOCK

**Mayukh Chakrabarti, Y. Stanley Tan, and Trent E. Balius**


## Abstract

Molecular docking is a popular computational tool in drug discovery. Leveraging structural information, docking software predicts binding poses of small molecules to cavities on the surfaces of proteins. Virtual screening for ligand discovery is a useful application of docking software. In this chapter, using the enigmatic KRAS protein as an example system, we endeavor to teach the reader about best practices for performing molecular docking with UCSF DOCK. We discuss methods for virtual screening and docking molecules on KRAS. We present the following six points to optimize our docking setup for prosecuting a virtual screen: protein structure choice, pocket selection, optimization of the scoring function, modification of sampling spheres and sampling procedures, choosing an appropriate portion of chemical space to dock, and the choice of which top scoring molecules to pick for purchase.


**Key words** Molecular docking, DOCK, Virtual screening, Druggable pockets, Computational drug discovery

## 1 Introduction

The objective of molecular docking is to generate and score poses of small molecules to cavities on biological macromolecules, typically proteins, to enable drug discovery. We focus on UCSF DOCK, the first docking software [1]. UCSF DOCK deploys a receptor-based approach with physics-based scoring functions. Currently, there are two actively developed versions of DOCK: DOCK 6 and DOCK 3. These programs differ in their ligand sampling algorithms: *Anchor-and-grow* (DOCK 6), and *hierarchical database (HDB) search* (DOCK 3). We have implemented *HDB search* into DOCK 6 [2]. DOCK 3 can perform covalent docking by using a modified *HDB search* method [3]. We direct the reader

Mayukh Chakrabarti and Y. Stanley Tan contributed equally.

Andrew G. Stephen and Dominic Esposito (eds.), *KRAS: Methods and Protocols*, Methods in Molecular Biology, vol. 2797, https://doi.org/10.1007/978-1-0716-3822-4_6,

to a protocol paper [4] for a discussion of large-scale docking with DOCK 3.

There are numerous molecular docking software programs, including Autodock, Autodock Vina, Glide, FRED, EnzyDock, Rosettaligand, Gold, Cdocker, Surflex, PLANTS, DockThor, FlexX, ICM, LigandFit, MCDock, MOE-Dock, LeDock, rDOCK, RxDock, DiffDOCK, DPL, GNINA, and others [5–8]. Recently, great excitement has emerged around machine learning, particularly deep neural networks, and seems to be impacting computational drug discovery and design [9, 10]. Docking programs deploy an array of sampling and scoring methods. Scoring functions classify into physics-based, knowledge-based, and empirical [11–16]. Recently, scoring functions have also been derived from machine learning, e.g., GNINA [17], MedusaNet [18], and DeepVS [19].

KRAS, a member of the superfamily of guanine nucleotide-binding proteins, is a GTPase, hydrolyzing GTP to GDP by releasing a phosphate group. The first 165 residues of RAS, referred to as the G-domain, adopts a topology consisting of 5 α-helices surrounding a 6-stranded β-sheet, arranged in a 3-layer α/β/α sandwich architecture known as a Rossmann fold. The secondary structural elements of KRAS are arranged as follows: β-sheet 1, phosphate-binding loop (p-loop), α-helix 1, Switch I, β-sheet 2, β-sheet 3, Switch II (which contains α-helix 2), β-sheet 4, α-helix 3, β-sheet 5, α-helix 4, β-sheet 6, α-helix 5, and the hypervariable region (HVR) (Fig. 1). The G-domain has two highly flexible regions: Switch I (residues 30–38) and Switch II (residues 60–76). The C-terminal HVR of RAS (residues 167–188/189) can be subdivided into a disordered linker region and a membrane interacting region that is the locus of posttranslational modifications (palmitoylation and prenylation), facilitating RAS association with the cell membrane. In all RAS isoforms, a farnesyl group present on the hypervariable region (residue 185/186) acts as a lipid membrane anchor. GTP and GDP bind under Switch I and interact with the p-loop, Switch I, α-helix 1, and Switch II.

In the Protein Data Bank (PDB), there are two main pockets on KRAS that bind small molecules (*see* Fig. 1b, c): the Switch II pocket formed by molecules wedged between Switch II and α-helix 3 (*see* Fig. 1d, e), and the Switch I/Switch II pocket (also called the SOS pocket, *see* Fig. 1f). Other pockets have also been observed, including a pocket under Switch I that is less well-represented in the PDB [20]. Switch I and Switch II are dynamic and play a role in effector binding [21–30].

Here, we focus on considerations for drug discovery against KRAS. Misregulation of RAS is implicated in several human disease conditions. RAS has historically been a difficult molecular target for several reasons, including the highly potent binding affinity of GTP to its orthosteric site ($K_d$ of $10^{-11}$ M), precluding the development

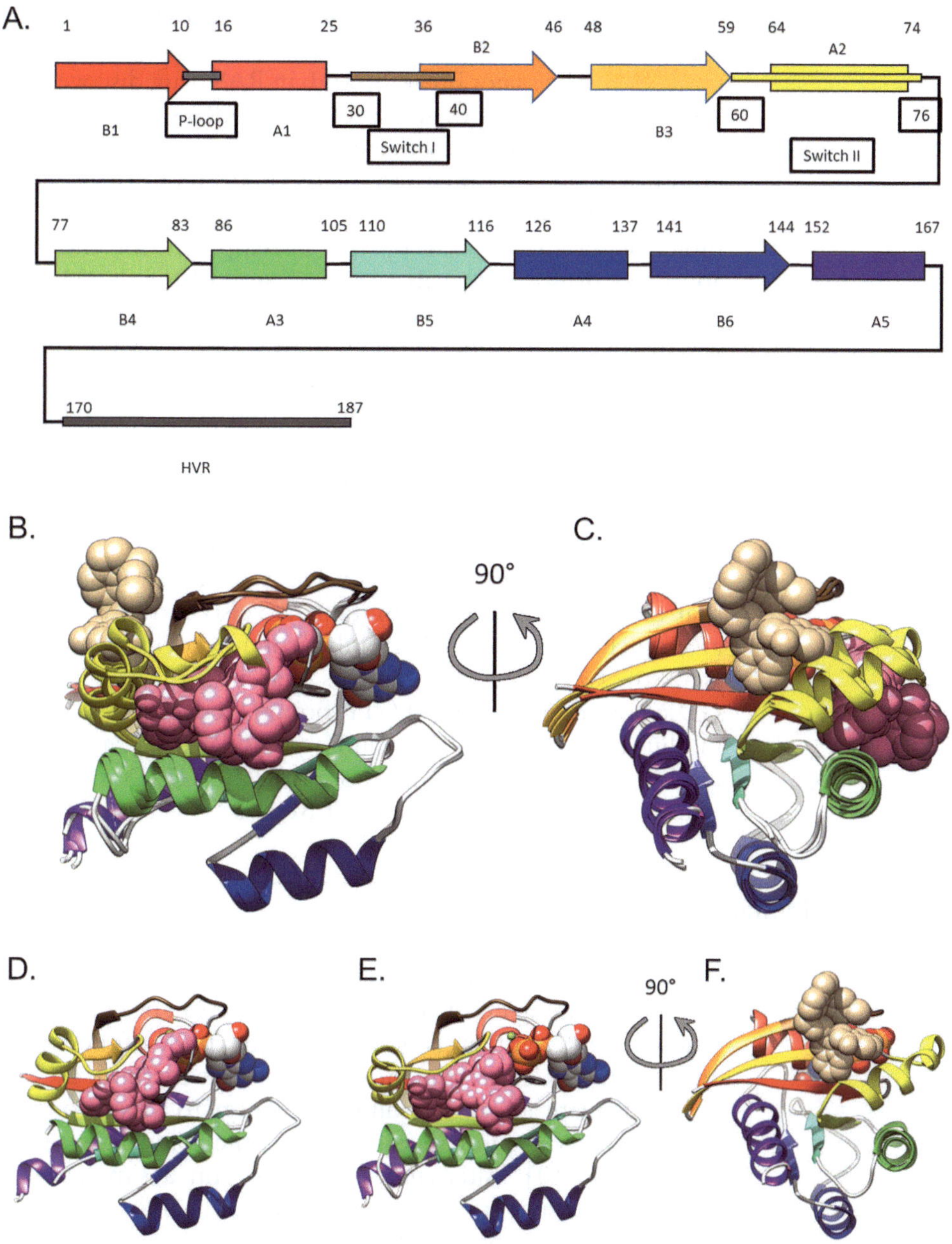


**Fig. 1** KRAS secondary and tertiary structure, with two prominent binding pockets indicated in tertiary depictions. (**a**) Secondary elements are shown in a cartoon (not to scale). Rainbow colors applied where β-sheet 1, p-loop, α-helix 1, Switch I, β-sheet 2, β-sheet 3, Switch II (which includes α-helix 2), β-sheet 4, α-helix 3, β-sheet 5, α-helix 4, β-sheet 6, α-helix 5, and the hypervariable region are colored red, grey (p-loop), red-orange, brown (Switch I), orange, yellow-orange (amber), yellow (Switch II), yellow-green (lime), green, blue-green, blue, indigo, violet, and dark gray, respectively. (**b**) Overlay of three RAS structures with three small molecules in two distinct pockets. (**c**) Rotation of B by 90°. (**d**) PDB Code 6OIM with covalent ligand sotorasib (AMG 510) in the Switch II pocket. (**e**) PDB Code 7RPZ with ligand MRTX-1133 in the Switch II pocket. (**f**) PDB Code 6GJ8 SOS pocket (Switch I/Switch II pocket) shown, indicated by small molecule BI-2852. (**b**)–(**f**) are made with UCSF Chimera (orthographic mode)

of competitive inhibitors, and its relatively featureless surface topology, making it challenging to identify suitable pockets for small molecules. The major breakthrough in RAS drug discovery came in 2013, when Ostrem et al. targeted the KRAS G12C mutant in a covalent screen and discovered the galvanizing small molecule inhibitor termed Compound 12 [31], although three decades of effort to discover inhibitors of RAS proteins preceded it. This molecule eventually led to the development of the first two approved drugs: sotorasib (Amgen) and adagrasib (Mirati). The lead molecule, which binds to KRAS in the GDP-bound state, was discovered in a pocket that is located under Switch II (between Switch II and α-helix 3), termed the "Switch II pocket." The inhibitor was aided by two factors: the relatively high intrinsic hydrolysis rate of the G12C mutant relative to wild-type KRAS, allowing RAS to be trapped in the inactive state during the process of nucleotide exchange, and the catalytic interaction of residue K16 with the covalent warhead [32]. Other molecules have been identified that bind this site non-covalently [33, 34]. Other inhibitors that target multiple KRAS mutants are also currently being evaluated, including a pan-KRAS inhibitor by Boehringer Ingelheim, BI-2865 [35]. Several inhibitors that target KRAS in both the GDP-bound (inactive) and GTP-bound (active) states are presently in clinical trials [36–38], including MRTX-1133, a non-covalent inhibitor that binds to GDP-bound KRAS G12D with picomolar affinity. Docking to sotorasib and BI-2852 is explored below.

## 2 Theory and Methods

Before we can understand molecular docking, we first must understand the binding event (Fig. 2). This event can be computationally studied at different levels of theory, with a trade off between speed and accuracy. Usually, the more accurate the method, the slower it is. To calculate a ligand binding free energy to a receptor to form a complex, we can simulate each species: the complex, receptor, and ligand [39, 40]. Alternatively, we can run a simulation of the complex and split it into a receptor and a ligand piece, as is done in end-state free energy calculations, like MM-GBSA [41], with a binding energy for each snapshot. These snapshot energies can be averaged over the simulation, resulting in a binding energy comparable to simulating each species. In docking, the receptor is typically held rigid, while different orientations and conformations of the ligand are sampled, generating a large number of complexes [42]. Each complex is evaluated with a score that typically attempts to calculate the binding event. The binding event is an ensemble of conformations, while the docking score is generated by a single complex. Thus, docking is an approximation from the start, because of this one complex per score.

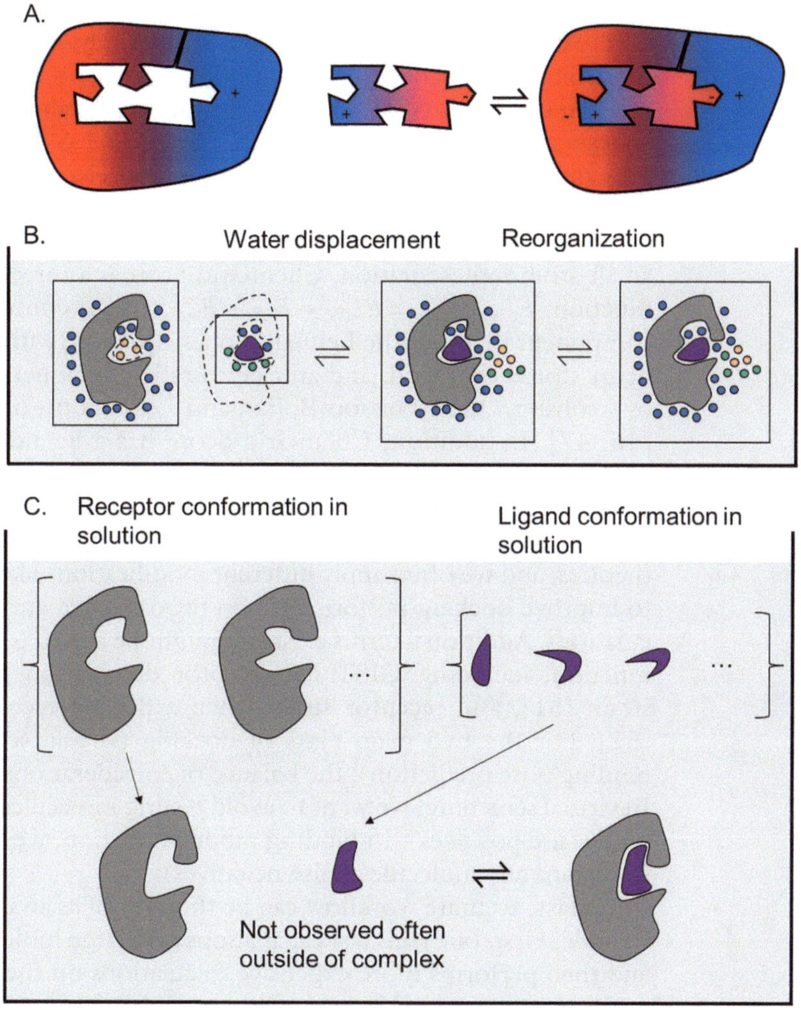


**Fig. 2** Cartoons that describe the binding event. (**a**) Cartoon that shows the jigsaw analogy to the binding event. The receptor has a missing region that is perfectly filled by the ligand. The ligand and receptor have complementary charges (represented by the red and blue colors). This stresses the importance of shape (van der Waals (vdW)) and charge (electrostatics (ES)) complementarity. (**b**) Cartoon that illustrates the importance of water displacement and consideration of waters that remain. (**c**) Cartoon that shows the presence of multiple states of the receptor and ligand, and the possibility that the complex stabilizes a state that is not highly populated in the unbound ensemble. This figure illustrates how the receptor and ligand undergo conformational changes upon binding

Docking methods seek to approximate the binding event while remaining fast. Most molecular docking scoring functions crudely predict the small molecule binding to a protein. Because docking methods make approximations, we test the software to verify the speed advantage, and the deleterious impacts on success are manageable using retrospective tests before performing screens. Despite the approximations made, experimental approaches have validated docking predictions [43–45].

The two main scoring functions of DOCK 6 and DOCK 3 are Grid Score and ChemGrid Score, respectively. Grid Score is a two-component scoring function: $E_{Grid} = E_{vdW} + E_{ES}$. It consists of a van der Waals (vdW) component modeled using the Lennard-Jones equation, typically with 6–9 or 6–12 exponents pre-computed and stored on a grid, and an electrostatic component, calculated using Coulomb's law and typically with a distance-dependent dielectric, also stored on a grid. Grid Score typically uses an all-atom representation. ChemGrid Score is a three-component function: $E_{ChemGrid} = E_{vdW} + E_{ES} + E_{lig,\ desolv}$. It consists of a vdW component that uses the Lennard-Jones equation, with 6–12 exponents stored on a grid, and an electrostatic component, calculated by solving the Poisson-Boltzmann equation using *Qnifft* [46, 47]. In addition, ChemGrid Score has a ligand desolvation component [48, 49]. ChemGrid score uses a united atom representation, where only polar hydrogens are explicitly modeled. This DOCK 3.7 scoring function is cobbled together from different theories, and we often apply different modifications like *thin spheres* to improve docking performance on retrospective enrichment calculations. Additional terms exist and might be added to this scoring function, including GIST for receptor desolvation [50], ligand strain [51], and receptor strain when a flexible receptor is used [52]. For the two main tasks of docking (virtual screening and binding pose prediction), the balance of considerations is different. In virtual screening, we want to avoid testing molecules that do not bind (false positives). In binding mode prediction, we do not want to discard any molecules (false negatives).

A fast, accurate workflow can be thought of as an upside-down triangle. First, one runs fast calculations on a large molecular library and then performs more expensive calculations on the top scoring molecules, performed in layers. For example, (1) a fast docking can be performed with DOCK on hundreds of millions (or billons) of molecules. (2) The top 100,000 ranked molecules can be redocked with increased sampling in DOCK, and more poses can be saved. (3) These poses can then be minimized and rescored with other methods like MM-GBSA calculations or molecular footprint score [53]. (4) The top thousand or so could then be passed to Amber-Score [54, 55] or Amber for molecular simulations, to conduct stability analysis and MM-GBSA binding energy calculations. (5) For these top scoring molecules, poses, properties, and energies are then visualized, and molecules are selected for purchase and experimental testing. (6) For experimentally verified hits, MM-GBSA, thermodynamic integration, or free energy perturbation calculations could be used for analog selection.

In docking, sampling approximations are made for speed and include the use of spheres for ligand placement, a rigid receptor, and an under-sampling of ligand internal degrees of freedom. Although there are many different sampling methods in other docking software, here we discuss two methods implemented in DOCK. First, in *anchor-and-grow*, the ligand internal degrees of freedom are explored on-the-fly using an incremental construction procedure. The *anchor-and-grow* algorithm is a breadth-first search method. Second, in *Hierarchical database search*, the ligand internal degrees of freedom are sampled outside of docking. The *hierarchical database search* algorithm is a depth-first search method that traverses and scores pre-computed ligand conformations. This method is 16× faster than anchor-and-grow [2].

In standard docking, a ligand is placed into a pocket on a receptor by exploring the three translational and three rotational degrees of freedom, and the ligand's internal degrees of freedom. A pose can be thought of as the internal conformation and orientation (relative to the protein) of a molecule. *Anchor-and-grow* and *hierarchical database search* are both standard docking algorithms. Matching spheres are generated by converting crystallographic ligand atoms to spheres, or by filling a cavity on the protein with spheres, as is done with the auxiliary program *sphgen*. These *sphgen* spheres can be thought of as occupying the inverse image of the molecular surface of the protein. Spheres are used for orienting the ligand into the binding site.

Covalent molecules react with the protein (or biomolecule) to form an adduct. Most often, these molecules react with a cysteine residue, as is the case for sotorasib, which targets the KRAS C12 mutation. For covalent docking, we only explore orientations that are possible considering the covalent bond. Thus, the relative degrees of freedom between the ligand and receptor are greatly reduced compared to standard docking. The degrees that are explored are the dihedral angles, distances, and an angle around the bond. However, to be effective, the fineness of sampling is higher. In DOCK 3.6, 3.7, and 3.8, a covalent method has been implemented that attaches the ligand to the residue by placing a dummy atom onto the gamma sulfur and adjusting two dihedrals, the bond angle and a distance about the sulfur [3]. The ligand internal degrees of freedom are explored by traversing through the hierarchical database (*db2* file) [3].

We recommend, if structural and ligand data exists, that pose reproduction and enrichment control calculations be performed before any screen. In pose reproduction, our docking method is used to reproduce an experimentally determined binding pose. We define three outcomes of this type of calculation, guided by Mukherjee et al. [56]: (1) *Docking success* is when the experimental pose is reproduced by our docking method as the top scoring pose. (2) *Scoring failure* is when the experimental pose is reproduced, but not as the top scoring pose. (3) *Sampling failure* is when the experimental pose is not reproduced at all. To calculate these outcomes, we consider the number of poses kept during docking, and the method used to determine similarity to the experimental pose.

Here, we use a symmetry-corrected RMSD (Hungarian algorithm) [57] with a threshold of 2.0 Å. That is, any pose less than 2.0 Å RMSD is considered as reproducing the experimental pose. Alternatively, *docking success* can be defined by considering the top scoring pose, or the top five scoring poses. In enrichment calculations, a set of known-binding active molecules and a set of presumed nonbinding decoy molecules are docked to a protein, and the ability of our docking method to distinguish between the actives and decoys is measured. There are several metrics used to quantify enrichment. One metric is the area under the curve (AUC) of the receiver operator characteristic (ROC) curve [58]. The ROC curve AUC, plotted with a logarithmic scale on the $x$-axis (% of decoys found, or false positive rate), can be computed to weight early enrichment more (logAUC). Different docking parameter options can be examined in a semiautomated format to explore the enrichment space [59]. Other computational considerations include choice of ligands, decoy background, and receptor preparation. Focusing on decoy background, we can choose property-matched decoys, extrema decoys, a set of known non-binders, or other backgrounds. The background choice greatly impacts enrichment quantification.

Retrospective calculations are meant to be used as a control to ensure the setup is behaving as intended or that a code change is working as expected, before performing virtual screens. Despite their utility in virtual screening, there are still notable limitations. These include appropriately weighting early enrichment more than later enrichment [60], which can be addressed with log-adjusted AUCs; however, the lower bound then becomes important to consider [58]. In addition to retrospective tests, one can dock prospectively to simplified sites, i.e., host-guest systems and model cavities [50, 61–63].

The main application of molecular docking is virtual screening in pursuit of discovering small molecules that bind to a protein. These molecules can be developed further to become drug leads or biological tool compounds. We examine pitfalls in selecting molecules from a virtual screen (*see* **Note 1**) and rules of thumb for prioritizing molecules from a virtual screen (*see* **Note 2**). As a community, we aspire to dock all accessible chemical space. Libraries are growing ever larger. We examine the impact of chemical space growth on our calculations in a thought experiment (*see* **Note 3**) and discuss strategies for docking fewer molecules while maintaining the ability to accurately identify top-scoring hits (*see* **Note 4**). Before running a screen, we advise optimizing the docking setup with retrospective calculations and having a binding assay that is clear and high throughput, where 20–50 molecules and analogs can be tested without burden.

To prepare the receptor for ChemGrid Score, run the *blastermaster.py* script distributed with DOCK 3, which performs the following: (1) Uses UCSF Chimera's Dock Prep function to remove solvent, and remove alternative conformations [64]. (2) Uses the program *reduce* [65] to add hydrogens to the protein. (3) Calculates hydrogens and partial charges for the cofactor as described below. These partial charges are stored, and vdW parameters are looked up based on the atom type and stored in text files for use in the next steps. (4) Uses *ChemGrid* to calculate the vdW grid. (5) Uses *Qnifft* to calculate the electrostatic grid. (6) Uses *Solvmap* to calculate the desolvation grid. (7) Creates matching spheres, using the experimental ligand atoms converted to spheres, supplemented with *sphgen* spheres. Run *blastermaster.py* once with the protein and the ion, but without the nucleotide cofactor (performing **steps 2**, **4–7**), and then a second time using a protonated protein and cofactor (obtained from **steps 2** and **3**, respectively), where *blastermaster.py* does not add hydrogens but uses existing ones (*see* **steps 4–7**). For both the ES grid and desolvation grid, thin spheres can be added to perturb the balance of energetic terms (use *blastermaster* defaults with desolvation spheres from *sphgen*). For more information about *blastermaster.py*, *see* refs. 4 and 47.

To prepare the receptor for *Grid Score* (the default DOCK 6 scoring method), perform the following: (1) Use UCSF Chimera's Dock Prep function to remove solvent, remove alternative conformations, and add charges to the protein and ion [64]. (2) Calculate hydrogens and partial charges for the cofactor, as described below. Chimera is used to combine the cofactor and protein. (3) Use the auxiliary program *sphgen* to calculate spheres that fill pockets on the protein and select spheres within 3 Å of the experimental ligand as matching spheres. (4) Use the program *Grid* to calculate the vdW and ES grids (see above). The grid is defined and bounded by a box placed about the spheres, padded with a margin of 10 Å.

Cofactors bound to the protein need to be included in docking calculations and should be prepared with the receptor. For RAS, both the nucleotide and a magnesium ion are treated as cofactors. There are four possibilities for the nucleotide cofactor: not present, GDP, GTP, or a GTP analog. There are three main analogs: GMPP[NH]P, GMPP[CH2]P, and GMPPP[γS], all of which differ from GTP by a single heavy atom. This atom can be mutated back to GTP, and either the GTP or GTP analog can be used. UCSF Chimera is used to protonate the cofactor and visually check the protonation state. Partial charges are computed using AM1-BCC charges calculated with Antechamber using the SQM program, as distributed with AmberTools [66]. Partial charges may also be obtained with Chimera's Dock Prep module, which for GDP and GTP can be looked up.

When starting from a chemical structure (SMILES), prepare a molecule for docking by performing the following steps (*see* Balius et al. [2] for details): (1) tautomerization, (2) protonation, (3) conformation generation (going from a chemical structure to a 3D molecular conformation), (4) partial charge assignment, (5) desolvation decomposition, (6) conformational expansion (sampling the internal degrees of freedom of the molecule), and (7) *db2* file generation. To use *anchor-and-grow* and Grid Score, only **steps 1–4** are needed to generate *mol2* files with a single conformation. To use ChemGrid Score, **step 5** is needed, and desolvation per atom values need to be added to the mol2 file in a solvation section. For *HDB search*, **steps 1–7** should be performed. These steps are deployed by the ZINC database [67] to build molecules. If the molecule you are interested in docking exists in the ZINC database, simply download the molecule. There is no need to build the molecule yourself.

## 3 Examples

### 3.1 Preparing RAS for Molecular Docking

#### 3.1.1 Protein Structure Choice

There are over 119 KRAS protein-ligand PDB structures. How can you pick the best system for docking? Perform retrospective calculations on each, including both cognate and cross-docking pose reproduction experiments, and enrichment calculations.

#### 3.1.2 Pocket Selection

How can you pick the right pocket for docking? The procedure of blind docking, where the whole surface of the protein is explored instead of selecting a pocket, is not practical with UCSF DOCK. We recommend that the user pick a pocket. The most obvious way is to pick a pocket with a small molecule present. The nucleotide binding site on RAS is not considered targetable. When a small molecule is not present, it is less obvious how to pick a good pocket. Cryptic pockets are difficult to target unless experimental structures or long molecular simulations reveal them. You should use a holo-structure with a bound ligand. If this is not possible, then you may consider other factors, e.g., pocket volume, how enclosed it is, how hydrophilic, how hydrophobic, or pocket shape. All can help in making the choice of what pocket to target. The *sphgen* module can be used to identify pockets. Deep, narrow sites that are mostly hydrophobic, but with polar patches, are often the most accessible to target with molecular docking.

#### 3.1.3 Optimization of the Scoring Function

There are a few ways to adjust the scoring to change the energetics of the pocket. (1) Adjust the protein structure in the following ways: (i) Modify the protonation state on titratable residues: the most obvious residue is histidine, which can be charged, or protonated on the ε- or δ-nitrogen. (ii) Flip residues: glutamines and asparagines are good candidates because the amide nitrogen (H-bond donor) vs carbonyl oxygen (h-bond acceptor) offer up a different environment. (iii) Rotate hydroxyls on serines and threonines, or flip the hydroxyl on tyrosines. (iv) Explore alternative sidechain conformations: sometimes there are alternative conformations in the PDB, or a rotamer library can be used. (2) Alter the balance between electrostatics and ligand desolvation terms. (i) In DOCK 3.7 and 3.8, this is primarily accomplished with *thin spheres*, which will adjust the molecular surface used in the electrostatics and desolvation grid calculations. The thicker the layer added to the surface, the stronger the electrostatics contribution; likewise, the thicker the layer of desolvation spheres, the more the poses are desolvated. Because the geometric placement of this layer is extending the surface, this strengthening of the contribution is dependent on the poses. The electrostatics *thin spheres* will encourage the placement of complementary charges close to the surface to be inside the boundary, so that it experiences a dielectric of 2 rather than 80. You can use the DOCK 3.7 and 3.8 scoring function in DOCK 6 and thus take advantage of these procedures. (ii) You can also increase the polar nature of groups by subtracting partial charge from one atom and adding it to another. (3) Add structural waters to the pocket. A structural water may be determined by looking at preserved water across multiple structures, or, alternatively, with inhomogeneous solvation theory. If a structural water is unfavorable to displace, then including the water as part of the receptor is advisable. It might be clear how the water hydrogens are oriented based on the receptor; if it is unclear, alternative water orientations might impact the calculation and could be considered.

#### 3.1.4 Modification of Matching Spheres and Sampling Procedure

The matching spheres control where to place a small molecule. Only the anchor of the small molecule is placed on the spheres in anchor-and-grow, and likewise, its rigid segment is placed in *HDB search*. If the spheres are altered, the sampling will be perturbed. To alter the sampling behavior, you can perform the following: (1) Change the spheres. (i) Delete one or more spheres. For example, if the molecule that is being docked is being placed outside the pocket, then remove all spheres near where the molecule is being undesirably placed, and re-dock. (ii) Move one or more spheres. For example, if you notice that the binding mode is not exactly to your liking, you can move spheres to better position the ligand. There are automated scripts in *teb_scripts_programs* [68] that will randomly perturb the sphere coordinates. This is to help understand the sensitivity of the results on sphere locations. (2) In DOCK 6, there are parameters that can be applied, like *critical points*. This allows the user to specify that a specific sphere be used for all placements (orients). Thus, instead of deleting the spheres, as was discussed above, you can insist that DOCK match the ligand with specific spheres.

#### 3.1.5 Chemical Space

The ZINC database is a publicly available resource that provides accessible chemical space, aspiring to store all vendor catalogs in standard file formats as well as DOCK-ready formats. The choice of the slice of the database to dock depends on your objectives. Docking *fragments* are useful for initial hits. They are often more soluble but are also lower in affinity. *Lead-like* is the most general portion to dock. *Drug-like* are large (and thus more flexible) and more expensive to dock (we often approach *drug-like* for selecting for analogs). In a typical workflow, you can dock *fragments* and *lead-like* to generate hits for experimental testing. *See* **Note 4** for strategies for docking fewer molecules without detection-accuracy loss for top-scoring hits. Library bias is a concern for screens on the order of $10^6$ molecules. Perhaps, as libraries continue to grow, this will become less of an issue. However, there are still clear biases that exist in libraries [69]. For specific targets, it is possible to design bespoke libraries to bias toward the region of chemical space that likely is richer in binding molecules [70].

#### 3.1.6 Hit Picking

**Notes 2** and **1**, respectively, provide guidance on which top-scoring molecules to purchase and which to discard because of artifacts (i.e., broken molecules).

### 3.2 Two KRAS Examples

Here, we will walk through retrospective tests of docking to two KRAS systems (Table 1). (1) KRAS bound to inhibitor BI-2852, with GTP analog GMPPCP (PDB ID: 6GJ8). This ligand induces a dimer, and we dock to both the dimer and the monomer. (2) KRAS bound to covalent inhibitor sotorasib, the first approved drug that binds to KRAS, also called AMG 510, with cofactor GDP (PDB ID: 6OIM). We created online tutorials, available at https://github.com/tbalius/teb_docking_test_sets/wiki, where we go through these examples in step-by-step detail. In addition, we are developing a RAS docking test set, which includes these systems, that we plan to publish. These tutorials will be updated over time as DOCK develops and the RAS test set grows.

For both systems, start by preparing the receptor for docking by generating spheres and computing energetic grids (see above). Select the spheres within a certain distance from the experimentally determined ligand, and convert the experimental ligand atoms to spheres. To prepare the receptor for grid generation, complete missing side chains, add hydrogens to the protein, add hydrogens to the cofactor, and calculate the partial charges on the cofactors (*see* Subheading 2). Calculate the energetic grids. Next, prepare the crystallographic ligand for docking, and perform pose reproduction.

For BI-2852, dock to both the monomer and the dimer (*see* Table 2 and Fig. 3). When docking to the monomer, notice that DOCK correctly positions the portion of the ligand in the pocket to match the crystallographic ligand (*see* Fig. 3). However, DOCK

**Table 1**
**Ligand and structures used for docking**

| Molecule | Chemical structure | SMILES | Notes |
|---|---|---|---|
| BI-2852 | 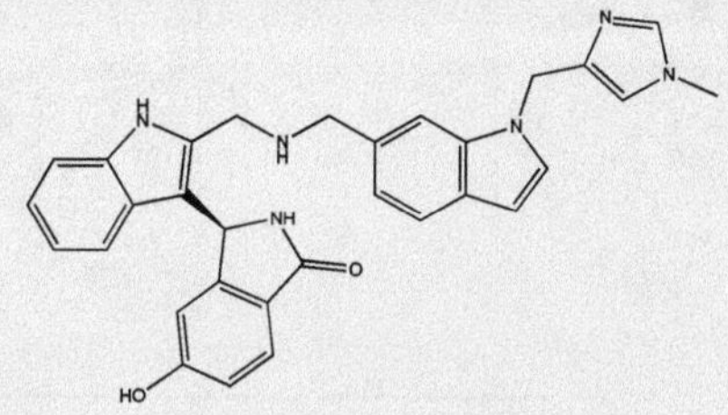 | OC(C=C1)=CC([C@@H](C2=C(CNCC3=CC4=C(C=©CC5=CN(C)C=N5)C=C3)NC6=CC=CC=C62)N7)=C1C7=O | Binds to the SI/SII pocket. Appears to stabilize a nonfunctional dimer of RAS<br>PDB ligand name: F0K<br>PDB entries: 6GJ8, 6ZL5 |
| Sotorasib | 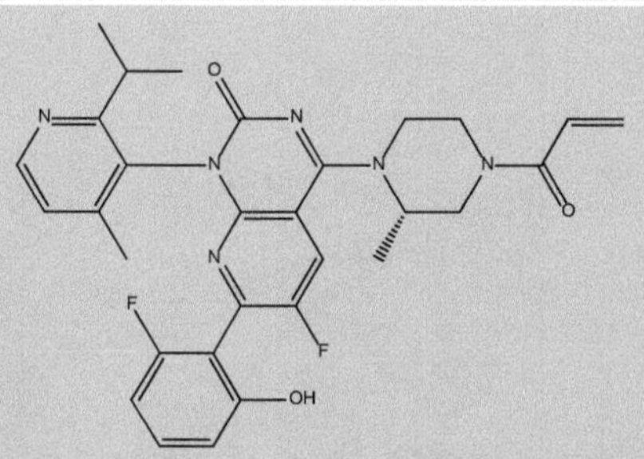 | O=C(C=C)N1C[C@H]©N(C2=NC(N(C©(C)C=©C3C(C)C)C4=C2C=C(F)C(C5=C(O)C=CC=C5F)=N4)=O)CC1 | First FDA-approved drug. Has two trade names: Lumakras and Lumykras<br>Covalent molecule targeting G12C. Binds to the SII pocket<br>PDB ligand name: MOV<br>PDB entry: 6OIM |

Underlined PDB entries used for docking in examples

**Table 2**
**Pose reproduction results for non-covalent docking of BI-2852**

| Molecule | RMSD[a] (BE)[b] | Energy (BE) | RMSD (BR)[c] |
|---|---|---|---|
| BI-2852 monomer | 3.43 | −76.71 | 2.32 |
| BI-2852 dimer | 0.61 | −116.33 | Same |

[a]RMSD calculated in DOCK using the Hungarian algorithm to correct for symmetry. Units are in Å

[b]BE stands for best energy pose

[c]BR stands for best RMSD pose from 25 poses kept

has difficulty orienting the methylimidazole ring. Despite a portion of the ligand matching the experimental pose closely, the RMSD is poor (RMSD = 3.43, *see* Table 2) because the pose diverges. If the ligand were to adopt the experimental pose in the context of the monomer, the methylimidazole ring would be solvent-exposed. Instead, a pose that lies along the surface is chosen. When docking to the dimer, additional interactions from the second protein allow DOCK to reproduce the experimental pose (RMSD = 0.61 Å).

For sotorasib, perform both non-covalent and covalent docking (*see* Table 3 and Fig. 4). Of the two, sotorasib is the most

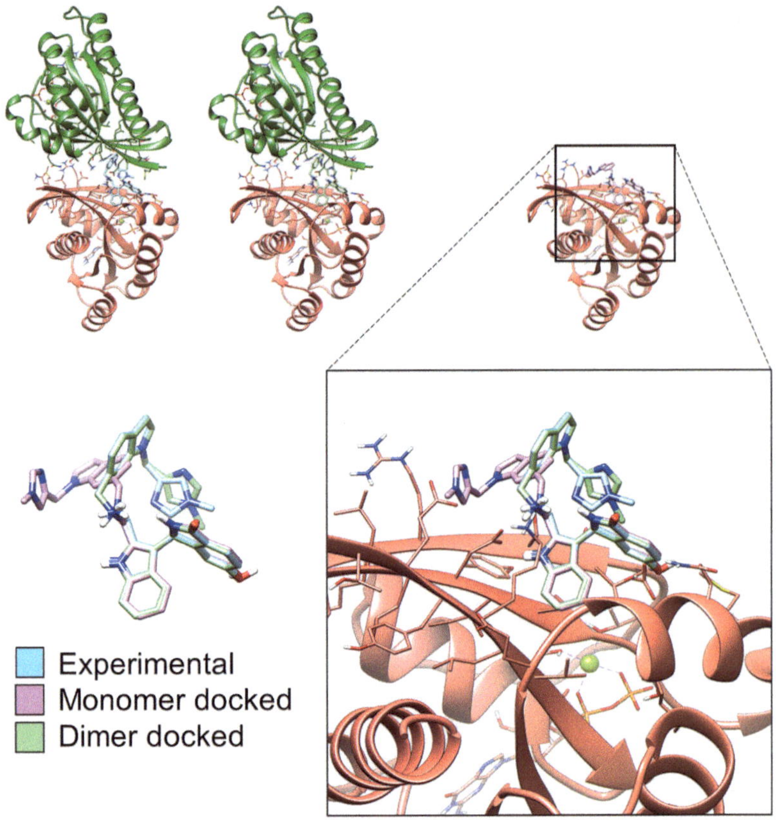


**Fig. 3** Comparing poses of BI-2852 docked to the monomer or the dimer. Clockwise, starting from the top left: the dimer of KRAS is shown with BI-2852, with experimental pose in cyan, the pose docked to the dimer in light green, and the pose docked to the monomer in pink. In the expanded inset is the zoom-in of the pocket with all three poses overlaid. Finally, on the bottom left, the overlay of just the three poses is shown. (Images made with UCSF Chimera (orthographic mode))

challenging to successfully dock, due to considerations for how to deal with the covalent adduct, and issues arising from ligand strain of the experimental structure.

For non-covalent docking of sotorasib, alter the ligand by removing the atom that participates in the covalent bond, and computationally mutate the cysteine C12 to an alanine. The internal energy for the experimental pose is high. To reproduce the experimental pose, increase the internal energy parameter to 1,000,000.0 kcal/mol. This internal strain is not as large for the ligand built from SMILES, and the normal value of 100 can be used. This highlights that, for some ligands, particularly structurally congested molecules, the internal conformations may have large internal strain, and thus it is important to consider the baseline or minimum strain for that molecule (conformation). In pose

**Table 3**
**Pose reproduction results for non-covalent docking and covalent docking of sotorasib**

| Molecule | RMSD[a] (BE)[b] | Energy (BE) | RMSD (BR)[c] | Notes |
|---|---|---|---|---|
| Sotorasib non-covalent from experimental structure | 5.27 | −57.17 | 4.29 | Because of ligand strain of the experimental structure, you will need to adjust the parameter *internal_energy_cutoff* (1,000,000.0) |
| Sotorasib non-covalent from experimental structure (LAS subset[d]) | 0.40 | −79.52 | Same | DOCK reproduces the experimental pose |
| Sotorasib non-covalent from SMILES | 4.56 | −59.24 | 1.19 | Conformation has less ligand strain energy (internal energy) |
| Sotorasib non-covalent from SMILES (LAS subset[d]) | 1.40 | −62.25 | Same | The pose looks good, but ring pucker is different and the 3-chlorophenol ring is flipped |
| Sotorasib covalent (SMILES)[e] | 9.03 | −11.97[f] | 8.69 | DOCK fails to reproduce the experimental pose |
| Sotorasib covalent (experimental structure)[e] | 0.91 | −43.93[f] | 0.86 | DOCK mostly reproduces the pose (but the pyridine ring is flipped) |

[a]RMSD calculated in DOCK using the Hungarian algorithm to correct for symmetry. Units are in Å
[b]BE stands for best energy pose
[c]BR stands for best RMSD pose from 25 poses kept
[d]The *l*igand *a*toms are converted to *s*pheres. A solvent accessible group was removed to focus sampling in the cavity
[e]Result generated with DOCK 3.7
[f]ChemGrid Score (all others are Grid Score)

reproduction for sotorasib, the spheres make a difference. If you use spheres generated with *sphgen*, you will not find a good pose, with the best RMSD of 4.29 Å, and 1.19 Å when using SMILES (*see* Table 3). However, if you use the ligand atoms as spheres (*see* Fig. 4a, b), you will get better results: the RMSD is 0.40 Å for the ligand built from mol2 and 1.40 Å from SMILES (*see* Table 3). When you prepare the ligand database from SMILES, the pucker for the piperazine ring will not match the experimental structure unless additional, higher-energy ring conformations are generated.

For covalent docking, use DOCK 3.7, the DOCKovalent method [3]. DOCK 3 uses ChemGrid Score. Before generating energy grids, remove the gamma sulfur atom and its attached hydrogen atom from covalent residue CYS12. If you use the molecule generated from SMILES, you will not be able to reproduce the experimental pose (the best RMSD is 8.69 Å, *see* Table 3, Fig. 4c). However, when you use the molecule generated from the experimental structure (bond length and angles are preserved, but dihedral information is forgotten), you will be able to reproduce the

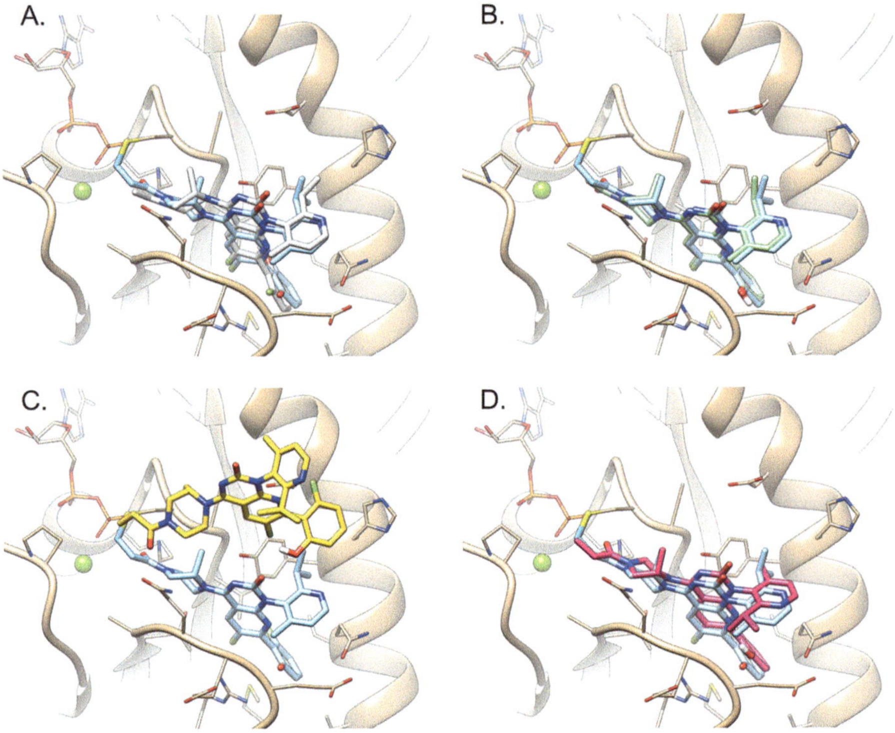


**Fig. 4** Sotorasib docking pose reproduction experiments. (**a**) Non-covalent docking pose (white) generated from SMILES docked with DOCK6 using the anchor-and-grow algorithm and compared to experimental pose (sky blue). (**b**) Non-covalent docking pose (light green) generated from the experimental structure (i.e., bonds and angles are preserved from the experimental structure) docked with DOCK6 using the anchor-and-grow algorithm. (**c**) Covalent docking pose (yellow) from SMILES docked with DOCK3 using the DOCKovalent algorithm. (**d**) Covalent docking pose (magenta) from experimental structure docked with DOCK3 using DOCKovalent algorithm

pose of the experimental structure: 0.91 Å (best score) and 0.86 Å (best RMSD), as shown in Table 3 and Fig. 4d.

To perform enrichment, begin by preparing a database of ligand and property-matched decoys for docking. The database should include five ligands from the PDB that bind to the Switch I/Switch II (SOS) pocket, including the BI-2852 molecule (see tutorial instructions). Property-matched decoys are property-matched to the ligands by formal charge, molecular weight, calculated water-octanol partition coefficient, rotatable bonds, number of hydrogen-bond donors, and number of hydrogen-bond acceptors. Include potential decoys for each ligand protomer, discarding decoys too similar to ligands, and clustering decoys that are similar to each other. Aim to assign 50 decoys to each ligand protomer. Online resources for generating property-matched decoys from a set of ligand SMILES are available at https://dude.docking.org and https://tldr.docking.org [71, 72]. DUDE decoys can be used for this calculation (see the tutorial for more information). Using the

**Table 4**
**Enrichment for the SOS pocket**

| | AUC | logAUC |
|---|---|---|
| Grid Score | 77.60 | 19.45 |
| Grid Score, keep 10 | 77.60 | 17.37 |
| Grid Score, keep 10, torsion min | 75.93 | 18.13 |
| ChemGrid Score | 63.60 | 17.01 |
| ChemGrid Score, keep 10 | 63.67 | 19.99 |
| ChemGrid Score, keep 10, torsion min | 64.33 | 18.64 |

DOCK 3.8 pipeline, prepare ligand and decoy molecules for docking in the *db2* file format, and then perform docking with DOCK 6's *HDB search*.

Upon performing the calculation, you will see that Grid Score seems to outperform ChemGrid Score for AUC (77.6 compared to 63.60), but that they perform more similarly for logAUC (19.45 compared to 17.01) (*see* Table 4). For ChemGrid Score, if you conduct additional sampling, it seems to help (17.01 compared to 19.99), but minimization does not seem to help (*see* Table 4).

### 3.3 Illustrative Virtual Screening Results

We performed a large-scale virtual screen of ~158.7 million molecules with DOCK 3.8 on the SOS pocket defined by the BI-2852 molecule (PDB: 6GJ8) on RAS as a monomer. A similar docking setup was used to that described above, but the GTP analog was kept as GMPPCP (not converted to GTP). We visualized the top 1000 poses. Although many of the top-scoring molecules look promising, here we showcase four molecules, which we would *discard* based on the docked pose, to illustrate docking *issues*. We discard molecules that (1) have multidentate interactions (an atom of one residue should not accept many interactions from the small molecule pose) (*see* Fig. 5a); (2) look like known binders (*see* Fig. 5a); (3) are occupying a tautomeric or protonation state that is not well occupied at the relevant experimental pH, most often pH 7 (*see* Fig. 5b); (4) float above the pocket and do not fill the pocket or make good contact with the pocket (*see* Fig. 5c); and (5) have stranded interactions (*see* Fig. 5d). *See* **Note 2** for more details of rules of thumb. To go into more detail on the example from point 3 (Fig. 5b), the molecule is incorrectly protonated. The 2-aminoquinoline ring has a predicted pKa of 6.16, according to ChemAxon, and will be 91.64% in the neutral state at pH 7.2. When chlorine is in the ring, the predicted pKa is lower (5.41), resulting in 98.41% of the species being in a neutral state at pH 7.2. The conformation is strained because the 2-aminoquinoline ring group should be planar.

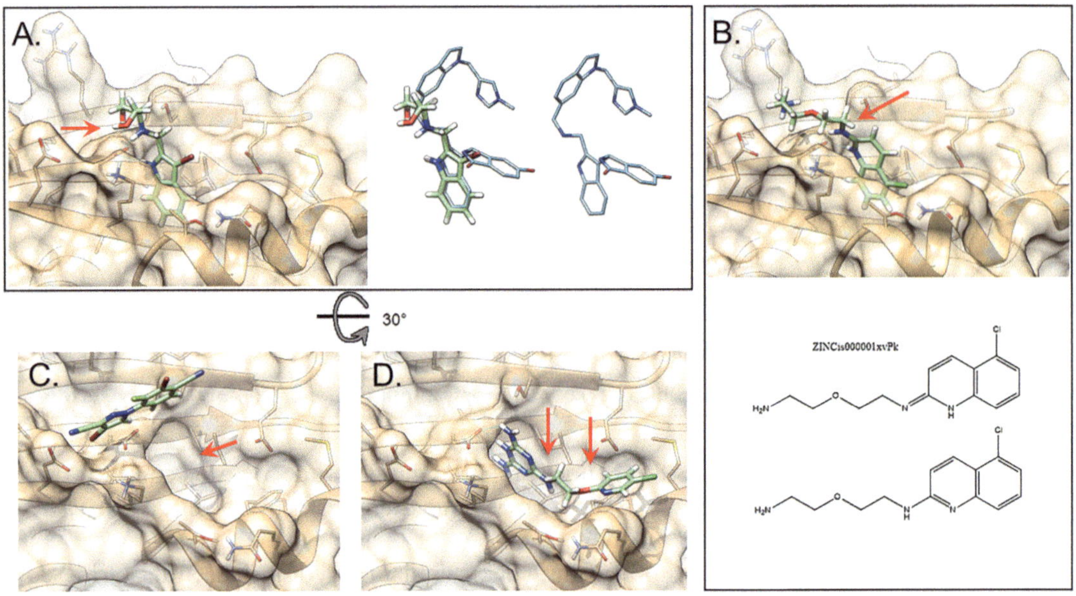


**Fig. 5** Molecules that exhibit unwanted behavior or properties from an example screen. (**a**) A molecule that resembles the experimental molecule and that has a congested tri- or quadridentate interaction. Red arrow points to congested interaction. (**b**) A molecule that is an incorrect protomer and tautomer and is in a strained conformation. (**c**) A molecule that does not fill the site. (**d**) A molecule that has stranded h-bond donors and stranded acceptors

## 4 Concluding Thoughts

Here, we hope that you have learned about the following: (1) The molecular binding event, and how docking uses approximations to model binding at scale. (2) Molecular docking of covalent and non-covalent ligands using DOCK. (3) KRAS, and issues with docking to KRAS. Potential issues in these docking calculations associated with the small molecules, including ring puckers and internal strain, have been explored. (4) Rules of thumb for picking molecules from virtual screens. (5) How to perform retrospective calculations, including pose reproduction and enrichment calculations to evaluate a docking setup. (6) How to choose a database of small molecules for virtual screening. (7) Potential pitfalls using molecular docking for ligand discovery, and how to avoid them.

## 5 Notes

1. Pitfalls in molecular docking for virtual screening.

   Broken molecules habitually score well: These molecules exploit issues with the scoring function. Examples of broken molecules include (1) incorrect protonation states,

(2) unphysical tautomerization, (3) strained ligand conformation, (4) incorrect partial charge assignment, and (5) incorrect desolvation per atom assignment.

2. Rules of thumb for prioritizing molecules from a virtual screen.

   Discard the docked poses of molecules if they possess the following: (1) Unsatisfied h-bond acceptors and donors. Pay attention to internal h-bonds. One strained donor and two strained acceptors are permitted. (2) Poor surface contacts, large gaps between receptor and ligand surfaces. (3) Internal issues that indicate that it is a broken molecule (*see* **Note 1**). (4) Many internal degrees of freedom (entropic cost). (5) Extreme hydrophobicity. If it is too polar, it will pay a cost to come out of water. If it is too hydrophobic, it might not be able to be tested because it falls out of solution. (6) Extreme size. If it is too big, it will not fit in the pocket. Conversely, if it is too small, it might not fill the site, or it might be nonspecific. (7) Reactive groups. We might permit weak reactive groups, but will discard very reactive molecules. (8) Pan-assay interference compounds (PAINS). We will often permit molecules with these vulnerabilities, since these molecules are meant to be leads. (9) Similarity to known aggregators. One of the biggest nonspecific binding artifacts is colloidal aggregation, where a compound will form colloids and will absorb the proteins on their surface and interfere with assays. (10) Molecules that resemble undesirable molecules, for example, known binders (we are looking for the new, not the old), or bad actors (points 7–9). Molecules selected for experimental testing should satisfy critical interactions: e.g., they should make a key salt-bridge or a key hydrogen bond. These rules of thumb are not hard and uncompromising. If you believe in a molecule, test it.

3. A thought experiment: Chemical space tractability.

   Assume a docking rate of 1 molecule per second, the average time for DOCK [73]. Docking 0.5 billion molecules takes 500 million seconds on one CPU core. If we run on 1000 CPU cores, it will take 500,000 s, which is ~5.8 days. If we dock 5 billion molecules on the same number of machines, it will take about 10 times longer: ~58 days, or ~2 months. This is too long, but there are ways to make it faster: (1) Dock with more CPUs: one way to do this is to use the cloud [74]; (2) make docking faster through algorithm improvements; and (3) focus on the productive portion of chemical space.

   Another consideration is disk space. 0.5 billion molecules use about 50 TB of disk space. Five billion would use ten times that, about 500 TB or 0.5 PB. This is used to store the precomputed conformations of the ligands. At scale, the overhead incurred by filesystem operations begins to play a role in the throughput of molecular docking algorithms [75]. Parallelization of docking necessitates consideration of the costs incurred by file writing, reading, and memory usage, and algorithms developed for docking billions of molecules must consider their use of filesystem resources to ensure reasonable throughput and stability. Do we have any real advantage to docking more molecules? How big is too big? In Lyu et al. [76], the authors state that docking larger and larger databases in the foreseeable future is worth the additional investment in computational time and resources.

4. Approaches to more efficiently sample rapidly expanding chemical space.

   To address expanding chemical libraries, ideas for improving the efficiency and throughput of docking have been developed. For example, chemical space can be reduced. One can dock *synthons*, which are fragments based on reactants used to generate a chemical library, like Enamine Real. The molecules from the chemical library associated with the top scoring synthons can then be docked [77, 78]. These *synthons* scale on the order of the second or third root of the library size. Another method is to train a machine learning model, using chemical structures and docking screens, to predict docking scores to prefilter libraries to help prioritize molecules for docking. This has an effective speed up of 10×, with the potential for speedups of 100–1000× given improvements in model accuracy, and does not significantly impact the quality of hits [79]. This type of method would be trained on a specific docking setup (protein, pocket, setup parameters used), and this training needs to happen before each unique docking campaign. Related to this are active learning approaches, which have also been used effectively to increase throughput and reduce computing costs by more than an order of magnitude. In this process, a small subset of molecules is randomly selected from a library, docked, and used to train a machine learning model to predict the docking scores of the remaining library, followed by iterative rounds of selecting the top-ranked machine learning predictions for further docking to refine the model [80–82]. A similar method is Adaptive Target-Guided Virtual Screens [83]. Here, a sparse docking is performed, and then the algorithm focuses on the perceived richest part of chemical space. Alternatively, the user could pick the space based on their own knowledge of the pocket. Other approaches have been hardware-based, e.g., adapting docking algorithms to run on GPUs [84], and in combination with software that permits the usage of high-performance computing or cloud-based resources [74, 85].

## Acknowledgments

We thank Alyssa Klein, Muyan Zhou, Dillon Chu, Maryam Abdur-Rahman, Yilin Liang, and Yunpeng Wang for contributions to the tutorials. We thank colleagues at the NCI RAS Initiative, including Dwight Nissley and Frank McCormick, and the Frederick Research Computing Environment (FRCE). This project was funded in whole or in part with federal funds from the National Cancer Institute, NIH Contract 75N91019D00024. The content of this publication does not necessarily reflect the views or policies of the Department of Health and Human Services, and the mention of trade names, commercial products, or organizations does not imply endorsement by the US government.

### References


1. Kuntz ID, Blaney JM, Oatley SJ et al (1982) A geometric approach to macromolecule-ligand interactions. J Mol Biol 161:269–288
2. Balius TE, Tan YS, Chakrabarti M (2024) DOCK 6: incorporating hierarchical traversal through precomputed ligand conformations to enable large-scale docking. J Comput Chem 45:47–63. https://doi.org/10.1002/jcc.27218
3. London N, Miller RM, Krishnan S et al (2014) Covalent docking of large libraries for the discovery of chemical probes. Nat Chem Biol 10: 1066–1072
4. Bender BJ, Gahbauer S, Luttens A et al (2021) A practical guide to large-scale docking. Nat Protoc 16:4799–4832
5. Kamal IM, Chakrabarti S (2023) MetaDOCK: a combinatorial molecular docking approach. ACS Omega 8:5850–5860
6. Pagadala NS, Syed K, Tuszynski J (2017) Software for molecular docking: a review. Biophys Rev 9:91–102
7. Zev S, Raz K, Schwartz R et al (2021) Benchmarking the ability of common docking programs to correctly reproduce and score binding modes in SARS-CoV-2 protease Mpro. J Chem Inf Model 61:2957–2966
8. Shuya N, Yoshiharu M, Shigenori T (2022) End-to-end protein-ligand complex structure generation with diffusion-based generative models. bioRxiv. https://doi.org/10.1101/2022.12.20.521309
9. Dara S, Dhamercherla S, Jadav SS et al (2022) Machine learning in drug discovery: a review. Artif Intell Rev 55:1947–1999
10. Vamathevan J, Clark D, Czodrowski P et al (2019) Applications of machine learning in drug discovery and development. Nat Rev Drug Discov 18:463–477
11. Kitchen DB, Decornez H, Furr JR et al (2004) Docking and scoring in virtual screening for drug discovery: methods and applications. Nat Rev Drug Discov 3:935–949
12. Chen P, Ke Y, Lu Y et al (2019) DLIGAND2: an improved knowledge-based energy function for protein-ligand interactions using the distance-scaled, finite, ideal-gas reference state. J Cheminform 11:52
13. Guedes IA, Pereira FSS, Dardenne LE (2018) Empirical scoring functions for structure-based virtual screening: applications, critical aspects, and challenges. Front Pharmacol 9:1089
14. Liu J, Wang R (2015) Classification of current scoring functions. J Chem Inf Model 55:475–482
15. Meli R, Morris GM, Biggin PC (2022) Scoring functions for protein-ligand binding affinity prediction using structure-based deep learning: a review. Front Bioinform 2:885983
16. Torres PHM, Sodero ACR, Jofily P et al (2019) Key topics in molecular docking for drug design. Int J Mol Sci 20:4574
17. McNutt AT, Francoeur P, Aggarwal R et al (2021) GNINA 1.0: molecular docking with deep learning. J Cheminform 13:43
18. Jiang H, Fan M, Wang J et al (2020) Guiding conventional protein-ligand docking software with convolutional neural networks. J Chem Inf Model 60:4594–4602
19. Pereira JC, Caffarena ER, Dos Santos CN (2016) Boosting docking-based virtual screening with deep learning. J Chem Inf Model 56: 2495–2506

20. Mathieu et al., KRAS G12C fragment screening renders new binding pockets, Small GTPases, 13:1, 225–238, https://doi.org/10.1080/21541248.2021.1979360
21. Liao J, Shima F, Araki M et al (2008) Two conformational states of Ras GTPase exhibit differential GTP-binding kinetics. Biochem Biophys Res Commun 369:327–332
22. Spoerner M, Hozsa C, Poetzl JA et al (2010) Conformational states of human rat sarcoma (Ras) protein complexed with its natural ligand GTP and their role for effector interaction and GTP hydrolysis. J Biol Chem 285:39768–39778
23. Ye M, Shima F, Muraoka S et al (2005) Crystal structure of M-Ras reveals a GTP-bound "off" state conformation of Ras family small GTPases. J Biol Chem 280:31267–31275
24. Shima F, Ijiri Y, Muraoka S et al (2010) Structural basis for conformational dynamics of GTP-bound Ras protein. J Biol Chem 285:22696–22705
25. Kalbitzer HR, Spoerner M, Ganser P et al (2009) Fundamental link between folding states and functional states of proteins. J Am Chem Soc 131:16714–16719
26. Chao F-A, Chan AH, Dharmaiah S et al (2023) Reduced dynamic complexity allows structure elucidation of an excited state of KRASG13D. Commun Biol 6:594
27. Shima F, Yoshikawa Y, Ye M et al (2013) In silico discovery of small-molecule Ras inhibitors that display antitumor activity by blocking the Ras-effector interaction. Proc Natl Acad Sci USA 110:8182–8187
28. Rosnizeck IC, Filchtinski D, Lopes RP et al (2014) Elucidating the mode of action of a typical Ras state 1(T) inhibitor. Biochemistry 53:3867–3878
29. Rosnizeck IC, Graf T, Spoerner M et al (2010) Stabilizing a weak binding state for effectors in the human ras protein by cyclen complexes. Angew Chem Int Ed Engl 49:3830–3833
30. Rosnizeck IC, Spoerner M, Harsch T et al (2012) Metal-bis(2-picolyl)amine complexes as state 1(T) inhibitors of activated Ras protein. Angew Chem Int Ed Engl 51:10647–10651
31. Ostrem JM, Peters U, Sos ML et al (2013) K-Ras(G12C) inhibitors allosterically control GTP affinity and effector interactions. Nature 503:548–551
32. McCormick F (2020) Sticking it to KRAS: covalent inhibitors enter the clinic. Cancer Cell 37:3–4
33. Broker J, Waterson AG, Smethurst C et al (2022) Fragment optimization of reversible binding to the switch II pocket on KRAS leads to a potent, in vivo active KRAS(G12C) inhibitor. J Med Chem 65:14614–14629
34. Wang X, Allen S, Blake JF et al (2022) Identification of MRTX1133, a noncovalent, potent, and selective KRAS(G12D) inhibitor. J Med Chem 65:3123–3133
35. Kim D, Herdeis L, Rudolph D et al (2023) Pan-KRAS inhibitor disables oncogenic signalling and tumour growth. Nature 619:160
36. Kwan AK, Piazza GA, Keeton AB et al (2022) The path to the clinic: a comprehensive review on direct KRAS(G12C) inhibitors. J Exp Clin Cancer Res 41:27
37. Punekar SR, Velcheti V, Neel BG et al (2022) The current state of the art and future trends in RAS-targeted cancer therapies. Nat Rev Clin Oncol 19:637–655
38. Rosen JC, Sacher A, Tsao MS (2023) Direct GDP-KRAS(G12C) inhibitors and mechanisms of resistance: the tip of the iceberg. Ther Adv Med Oncol 15:17588359231160141
39. Dong L, Qu X, Zhao Y et al (2021) Prediction of binding free energy of protein-ligand complexes with a hybrid molecular mechanics/generalized born surface area and machine learning method. ACS Omega 6:32938–32947
40. Kollman PA, Massova I, Reyes C et al (2000) Calculating structures and free energies of complex molecules: combining molecular mechanics and continuum models. Acc Chem Res 33:889–897
41. Genheden S, Ryde U (2015) The MM/PBSA and MM/GBSA methods to estimate ligand-binding affinities. Expert Opin Drug Discov 10:449–461
42. Lorber DM, Shoichet BK (1998) Flexible ligand docking using conformational ensembles. Protein Sci 7:938–950
43. Wang Z, Sun H, Yao X et al (2016) Comprehensive evaluation of ten docking programs on a diverse set of protein-ligand complexes: the prediction accuracy of sampling power and scoring power. Phys Chem Chem Phys 18:12964–12975
44. Warren GL, Andrews CW, Capelli AM et al (2006) A critical assessment of docking programs and scoring functions. J Med Chem 49:5912–5931
45. Xu M, Shen C, Yang J et al (2022) Systematic investigation of docking failures in large-scale structure-based virtual screening. ACS Omega 7:39417–39428
46. Meng EC, Shoichet BK, Kuntz ID (1992) Automated docking with grid-based energy evaluation. J Comput Chem 13:505–524

47. Coleman RG, Carchia M, Sterling T et al (2013) Ligand pose and orientational sampling in molecular docking. PLoS One 8:e75992
48. Mysinger MM, Shoichet BK (2010) Rapid context-dependent ligand desolvation in molecular docking. J Chem Inf Model 50: 1561–1573
49. Shoichet BK, Leach AR, Kuntz ID (1999) Ligand solvation in molecular docking. Proteins 34:4–16
50. Balius TE, Fischer M, Stein RM et al (2017) Testing inhomogeneous solvation theory in structure-based ligand discovery. Proc Natl Acad Sci USA 114:E6839–E6846
51. Gu S, Smith MS, Yang Y et al (2021) Ligand strain energy in large library docking. J Chem Inf Model 61:4331–4341
52. Fischer M, Coleman RG, Fraser JS et al (2014) Incorporation of protein flexibility and conformational energy penalties in docking screens to improve ligand discovery. Nat Chem 6:575–583
53. Balius TE, Mukherjee S, Rizzo RC (2011) Implementation and evaluation of a docking-rescoring method using molecular footprint comparisons. J Comput Chem 32:2273–2289
54. Lang PT, Brozell SR, Mukherjee S et al (2009) DOCK 6: combining techniques to model RNA-small molecule complexes. RNA 15: 1219–1230
55. Allen WJ, Balius T, Bickel J et al (2023) DOCK 6.10 users manual. [cited May 3, 2023]. Available from https://dock.compbio.ucsf.edu/DOCK_6/dock6_manual.htm
56. Mukherjee S, Balius TE, Rizzo RC (2010) Docking validation resources: protein family and ligand flexibility experiments. J Chem Inf Model 50:1986–2000
57. Allen WJ, Rizzo RC (2014) Implementation of the Hungarian algorithm to account for ligand symmetry and similarity in structure-based design. J Chem Inf Model 54:518–529
58. Knight IS, Naprienko S, Irwin JJ (2022) Enrichment Score: a better quantitative metric for evaluating the enrichment capacity of molecular docking models. arXiv:2210.10905. https://doi.org/10.48550/arXiv.2210.10905
59. Knight IS, Mailhot O, Tang KG, Irwin JJ, (2024) DockOpt: A tool for automatic optimization of docking models. J Chem Inf Model Article ASAP. https://doi.org/10.1021/acs.jcim.3c01406
60. Hawkins PC, Warren GL, Skillman AG et al (2008) How to do an evaluation: pitfalls and traps. J Comput Aided Mol Des 22:179–190
61. Amezcua M, Setiadi J, Ge Y et al (2022) An overview of the SAMPL8 host-guest binding challenge. J Comput Aided Mol Des 36:707–734
62. Graves AP, Shivakumar DM, Boyce SE et al (2008) Rescoring docking hit lists for model cavity sites: predictions and experimental testing. J Mol Biol 377:914–934
63. Kamenik AS, Singh I, Lak P et al (2021) Energy penalties enhance flexible receptor docking in a model cavity. Proc Natl Acad Sci USA 118:e2106195118
64. Pettersen EF, Goddard TD, Huang CC et al (2004) UCSF Chimera--a visualization system for exploratory research and analysis. J Comput Chem 25:1605–1612
65. Word JM, Lovell SC, Richardson JS et al (1999) Asparagine and glutamine: using hydrogen atom contacts in the choice of side-chain amide orientation. J Mol Biol 285:1735–1747
66. Case DA, Ben-Shalom IY, Brozell SR et al (2019) AMBERTools 2019. University of California, San Francisco
67. Tingle BI, Tang KG, Castanon M et al (2023) ZINC-22 – a free multi-billion-scale database of tangible compounds for ligand discovery. J Chem Inf Model 63:1166–1176
68. Balius TE (2023) teb_scripts_programs. Available from https://github.com/tbalius/teb_scripts_programs/tree/master/zzz.scripts/
69. Hert J, Irwin JJ, Laggner C et al (2009) Quantifying biogenic bias in screening libraries. Nat Chem Biol 5:479–483
70. Kaplan AL, Confair DN, Kim K et al (2022) Bespoke library docking for 5-HT(2A) receptor agonists with antidepressant activity. Nature 610:582–591
71. Irwin JJ, Shoichet BK, Mysinger MM et al (2009) Automated docking screens: a feasibility study. J Med Chem 52:5712–5720
72. Mysinger MM, Carchia M, Irwin JJ et al (2012) Directory of useful decoys, enhanced (DUD-E): better ligands and decoys for better benchmarking. J Med Chem 55:6582–6594
73. Beroza P, Crawford JJ, Ganichkin O et al (2022) Chemical space docking enables large-scale structure-based virtual screening to discover ROCK1 kinase inhibitors. Nat Commun 13:6447
74. Tingle BI, Irwin JJ (2023) Large-scale docking in the cloud. J Chem Inf Model 63:2735–2741
75. Lu H, Wei Z, Wang C et al (2021) Redesigning Vina@QNLM for ultra-large-scale molecular docking and screening on a Sunway supercomputer. Front Chem 9:750325

76. Lyu J, Irwin JJ, Shoichet BK (2023) Modeling the expansion of virtual screening libraries. Nat Chem Biol 19:712–718
77. Sadybekov AA, Sadybekov AV, Liu Y et al (2022) Synthon-based ligand discovery in virtual libraries of over 11 billion compounds. Nature 601:452–459
78. Sadybekov AV, Katritch V (2023) Computational approaches streamlining drug discovery. Nature 616:673–685
79. Clyde A, Liu X, Brettin T et al (2023) AI-accelerated protein-ligand docking for SARS-CoV-2 is 100-fold faster with no significant change in detection. Sci Rep 13:2105
80. Gentile F, Yaacoub JC, Gleave J et al (2022) Artificial intelligence-enabled virtual screening of ultra-large chemical libraries with deep docking. Nat Protoc 17:672–697
81. Graff DE, Shakhnovich EI, Coley CW (2021) Accelerating high-throughput virtual screening through molecular pool-based active learning. Chem Sci 12:7866–7881
82. Yang Y, Yao K, Repasky MP et al (2021) Efficient exploration of chemical space with docking and deep learning. J Chem Theory Comput 17:7106–7119
83. Christoph G, AkshatKumar N, Matt K et al (2023) VirtualFlow 2.0 – the next generation drug discovery platform enabling adaptive screens of 69 billion molecules. bioRxiv. https://doi.org/10.1101/2023.04.25.537981
84. Yu Y, Cai C, Wang J et al (2023) Uni-Dock: GPU-accelerated docking enables ultralarge virtual screening. J Chem Theory Comput 19:3336–3345
85. Zhang B, Li H, Yu K et al (2022) Molecular docking-based computational platform for high-throughput virtual screening. CCF Trans High Perform Comput 4:63–74

# Chapter 7

# Measurement of KRAS-GTPase Activity


## Dana Rabara and Andrew G. Stephen



## Abstract

Oncogenic mutations in KRAS typically impact the GAP-mediated and intrinsic GTP hydrolysis activity resulting in elevated levels of cellular KRAS-GTP. The development of biochemical assays for GTPase activity provides an opportunity to quantitatively measure the impact of these mutations on GTP hydrolysis. Here we describe a biochemical assay that measures the release of free phosphate upon hydrolysis of the GTP nucleotide and allows the measurement of intrinsic or GAP-stimulated GTP hydrolysis by KRAS. This assay can be used to measure GTPase activity under single turnover conditions.



**Key words** RAS, GTPase activity, Phosphate sensor, GAP, NF1, RASA1, Kinetic measurement, Free phosphate


## 1 Introduction

Historically the GTPase activity of RAS proteins was measured using [$\gamma$-$^{32}$P] GTP [1–3]. This method is sensitive and allows for the direct measurement of GTP hydrolysis by detecting the release of $^{32}$P-labeled inorganic phosphate. However, the challenges of working with radioactivity spurred the development of absorbance and fluorescence read-outs for the detection of hydrolysis activity. The preparation of N-methylanthraniloyl ester derivatives of nucleotides [4] including GTP (mant-GTP) was used to monitor intrinsic and GTPase-activating protein (GAP)-stimulated GTP hydrolysis [5–7]. The environmentally sensitive mant-fluorophore changes fluorescence as the switch regions in RAS rearrange during the hydrolysis of GTP to GDP. In addition, changes in RAS conformation upon GTP hydrolysis can be monitored using NMR and can report on the GTPase activity [8]. However, this approach requires the RAS protein to be labeled with $^{15}$N during protein production.

Other measurements detect the release of the inorganic phosphate from the hydrolysis reaction, such as malachite green [9] or the use of purine nucleoside phosphorylase (PNP) that transfers


Andrew G. Stephen and Dominic Esposito (eds.), *KRAS: Methods and Protocols*, Methods in Molecular Biology, vol. 2797, https://doi.org/10.1007/978-1-0716-3822-4_7,

inorganic phosphate to 2-amino-6-mercapto-7-methylpurine ribonucleoside (MESG, a guanosine analogue) as a substrate [10, 11]. The MESG/phosphorylase-coupled assay measures the absorbance change of MESG and is available commercially as the EnzChek Phosphate Assay. Modification of the bacteria phosphate-binding protein with the N-[2-(1-maleimidyl)ethyl]-7-(diethylamino)coumarin-3-carboxamide (MDCC) fluorophore results in a highly sensitive phosphate detection system [12]. This "phosphate sensor" assay can be used to effectively measure real-time intrinsic and GAP-stimulated GTP hydrolysis [4, 5, 13].

Finally, the detection of remaining GTP levels at the end of the assay can be used to quantitate the extent of the hydrolysis reaction. Coupled assays, such as "GTPase Glo," convert the remaining GTP into ATP which is then used as a substrate for luciferase and the luminescence is measured to quantify the level of GTP [14]. Other approaches are available that detect free GDP or GTP using antibodies that selectively detect these nucleotides and are coupled with fluorescent tracers [15–17].

GTP hydrolysis can be followed as a single turnover reaction, where no additional substrate (GTP) is added to the reaction and only the hydrolysis of GTP bound to KRAS is monitored. This approach allows quantitation of the rate of GTP hydrolysis at the active site of KRAS and requires careful preparation of KRAS-GTP such that all excess GTP is removed during sample preparation [18]. Alternatively, the reaction can be analyzed in a multi-turnover format where excess substrate or GTP is available to bind to the active site once the GDP product dissociates from the active site. In this format, the measured rate is a mixture of the hydrolysis, GDP dissociation, and GTP association rates.

The phosphate sensor assay allows real-time kinetic analysis of the phosphate released during KRAS-catalyzed GTP hydrolysis. We have successfully measured the intrinsic and GAP-mediated GTP hydrolysis of wild-type (WT) KRAS and other mutants using this approach. In this chapter, we present a protocol that describes using the phosphate sensor assay to measure the single turnover reactions of WT, G12D, and G13D KRAS.

## 2 Materials

### 2.1 Lab Equipment

1. Plate reader equipped with fluorescence detection (the BioTek Synergy Neo2 was used for these experiments).
2. 1 to 30 μL E1-ClipTip electronic pipette.
3. Light box for microplates (*optional*).
4. Centrifuge suitable for centrifuging plates.

### 2.2 Phosphate Sensor Materials

1. Phosphate Sensor kit.
2. Phosphate standard: $KH_2PO_4$ (50 mM) (*see* **Note 1**).
3. Phosphate sensor assay buffer: 50 mM HEPES, pH 7.3; 150 mM KCl, 1.5 mM $MgCl_2$, and 5 mM DTT.
4. Recombinant KRAS proteins: KRAS (1–169) preloaded with GTP (*see* Nucleotide exchange on RAS proteins using hydrolysable and non-hydrolysable nucleotides [18]).
5. Recombinant GAPs: RASA1 (714–1046) (p120) and NF1 (1198–1530) (*see* **Note 2**).
6. Assay plate: Corning 3540 Low Volume 384-well Black/Clear Flat Bottom Polystyrene Not Treated Microplate.
7. Reagent plate: Corning 3762 384-well Black/Clear Flat Bottom Polystyrene Not Treated Microplate (*see* **Note 3**).
8. Dilution plate: Greiner Bio-One 650201 microplate, 96-well, polypropylene, U-bottom.
9. Pipetting reservoir: 12-well basin, polypropylene, 3 mL.
10. ClipTip™ Filtered 384-format pipette tips (30 μL).
11. Black plate seal: TopSeal-A black.

### 2.3 Phosphate Sensor Assay Buffer (PSAB) (500 mL)

1. Measure out 399.25 mL of water to a graduated beaker (*see* **Note 4**).
2. Add 25 mL of 1 M Hepes, pH 7.3.
3. Add 75 mL of 1 M KCl.
4. Add 750 μL of 1 M $MgCl_2$.
5. Mix reagents and then filter using a 0.2 μM filter.
6. Check pH of buffer using a pH meter or pH strips.
7. Store at room temperature (*see* **Note 5**).
8. Final buffer composition will be 50 mM HEPES, pH 7.3; 150 mM KCl, 1.5 mM $MgCl_2$, and 5 mM DTT once the DTT is added before use.

## 3 Methods

The method outlined below is for running the assay with two KRAS proteins and one GAP titration. However, the method can be adjusted to run the assay you desire. Reagents were kept on ice during assay preparation. Preparation of plates were kept at room temperature.

### 3.1 Single Turnover Phosphate Sensor Assay

1. Thaw 1 M DTT, 50 mM $KH_2PO_4$ stock, phosphate sensor, and GAP on ice (*see* **Note 6**).
2. Set up plate reader with the following settings:

> **BioTek Neo Synergy 2 Plate Reader Settings:**
> **Temperature:** Setpoint 25 °C
> **Kinetic Read:**
> *Run:* 8:00:00 (8 h)
> *Interval:* 0:02:20 (2 min, 20 s)
> **Fluorescence Read:**
> *Wells:* A1-G9
> *Excitation:* 430/5
> *Emission:* 450/5
> *Optic Position:* Bottom
> *Gain:* Automatic Gain Adjustment, Scale to High Wells
> *Scale Wells:* A1
> *Scale Percent:* 40%
> *Read Speed:* Normal

3. Take 6 mL of the premade phosphate sensor assay buffer (PSAB), and add 30 μL of 1 M DTT to a final concentration of 1 mM DTT.
4. Dilute the 50 mM $KH_2PO_4$ stock to 1 mM in PSAB by adding 2 μL of the stock into 98 μL of PSAB in a 500 μL microcentrifuge tube and vortex.
5. Dilute the 1 mM $KH_2PO_4$ stock to 6 μM in PSAB by adding 1.2 μL of the 1 mM stock into 198.8 μL of PSAB in a 500 μL microcentrifuge tube and vortex.
6. Using the 6 μM stock of $KH_2PO_4$, create a twofold titration in the 96-well dilution plate.
7. Place 100 μL of PSAB in wells A2–A8, and place 200 μL of water in A9 (*see* Fig. 1).
8. Add 100 μL of the 6 μM stock of $KH_2PO_4$ to A2, then add the remaining 100 μL of the 6 μM stock of $KH_2PO_4$ to A1.
9. Pipette 100 μL up and down in well A2 and then transfer 100 μL to well A3, discard tip, and repeat this process until you transfer 100 μL to well A7. Mix the final dilution made in well A7 by pipetting 100 μL up and down. A7 should have a final volume of 200 μL.
10. Make a 9 μM stock of phosphate sensor (PS) diluted in 2400 μL of PSAB by adding 43.2 μL of 500 μM PS (*see* **Note 7**).

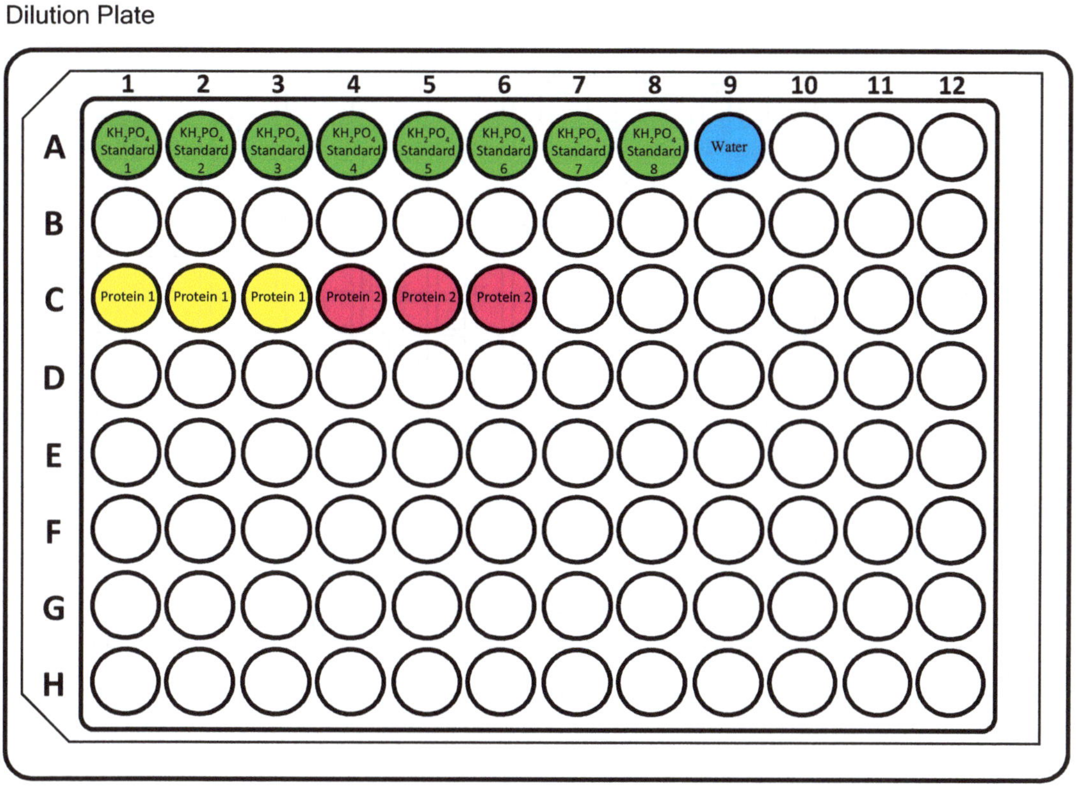


**Fig. 1** Dilution plate: Greiner 650201: this plate is used to create the $KH_2PO_4$ standard curve and to dispense the two KRAS-GTP proteins

11. Thaw the two KRAS-GTP proteins on ice while prepping the GAP titration.
12. Prepare a 6 μM stock of your selected GAP (*see* **Note 8**).
13. Create an eight-point (last point is 9 μM PS only), twofold dilution series of your selected GAP starting at 6 μM in the 9 μM PS using the pipetting reservoir: 12-well basin (*see* **Note 9**).
14. Using the electronic pipette, set to repetitive and dispense 25 μL of the 9 μM PS to wells A1–A8 of the 384-well reagent plate.
15. Place 25 μL of water in A9 of the reagent plate (*see* Fig. 2).
16. Using the electronic pipette, set to repetitive and dispense 25 μL the eight-point GAP titration from the 12-well basin to wells B1–B8, repeating this for rows C through G of the reagent plate.
17. Dilute the thawed KRAS-GTP proteins from **step 11** to 6 μM in PSAB at a final volume of 600 μL (*see* **Note 10**).

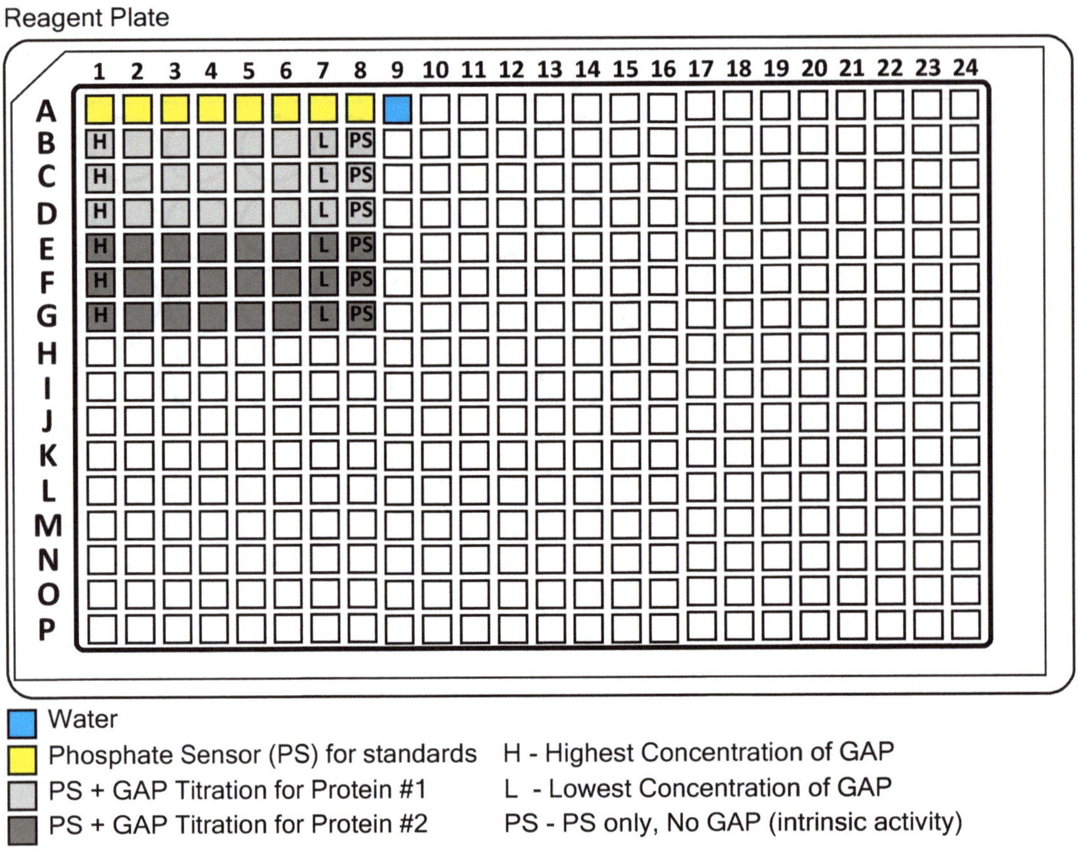


**Fig. 2** Reagent plate: Corning 3762: this plate is used to dispense the PS, the PS + GAP, and the water well

18. Using the electronic pipette, set to repetitive and dispense add 20 μL of standards and water control into wells A1–A9 of the assay plate (*see* Fig. 3) from wells A1 to A9 of the dilution plate (*see* Fig. 1).
19. Pipette 190 μL/well of the first 6 μM KRAS-GTP into C1–C3 of the dilution plate (*see* Fig. 1), and then pipette 190 μL/well of the second 6 μM KRAS-GTP into well C4–C6.
20. Using the electronic pipette, set to repetitive and dispense 20 μL of 6 μM KRAS from wells C1–C6 of the dilution plate (*see* Fig. 1) to the assay plate wells B1–G1, repeating the addition to columns 2 through 8 of the assay plate (*see* Fig. 3).
21. Spin the assay plate and the reagent plate at 931 × *g* for 1 min at room temperature in the centrifuge.
22. Using the electronic pipette, set to matrix (*see* **Note 11**), add 20 μL of PS (wells A1–A8) and water from A9 from the reagent plate to wells A1–A9 of the assay plate.

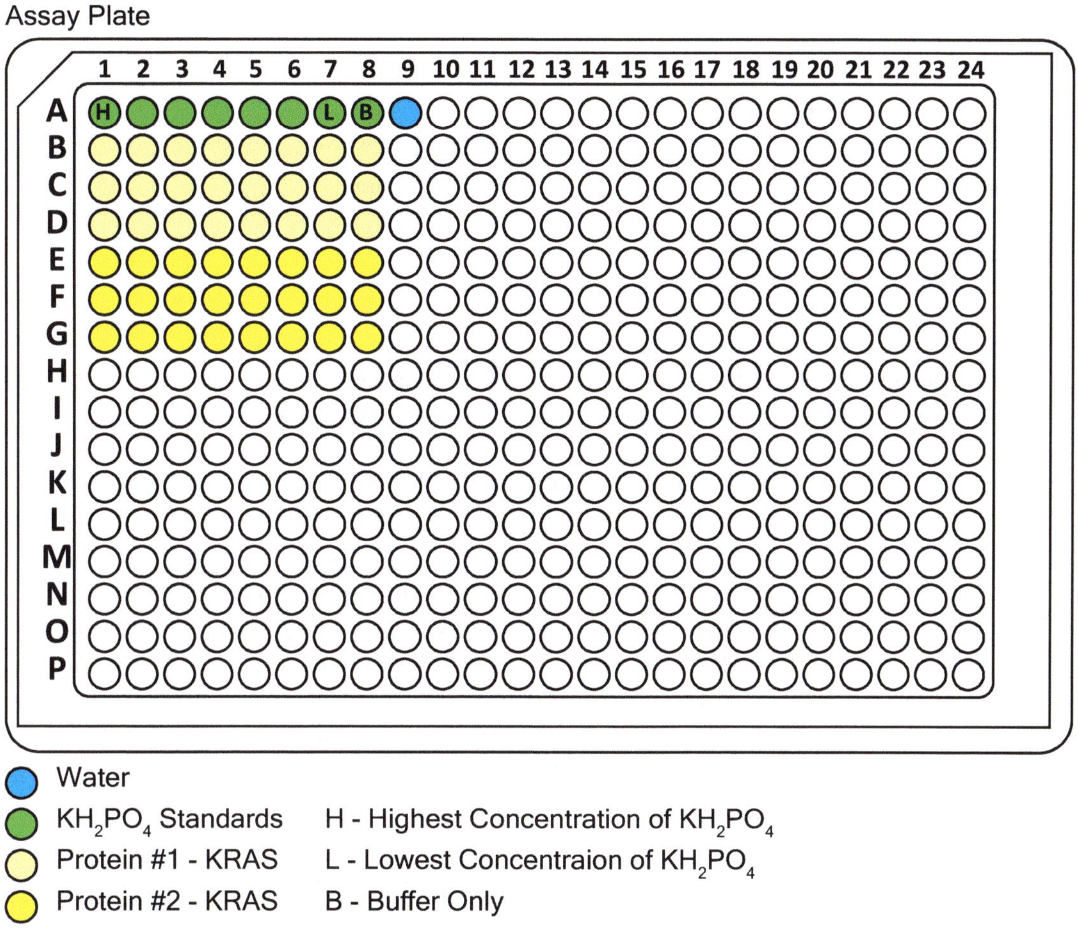


**Fig. 3** Assay plate: Corning 3540: row A is for the standards and water control; row B to D, for KRAS Protein 1; and Row E to G for KRAS Protein 2

23. Using the electronic pipette, set to matrix with the same settings, and add 20 μL of the PS + GAP titration B1–B8 from the 384-well reagent plate to B1–B8 of the 384-well assay plate.
24. Repeat above step for rows C through G of reagent and assay plate.
25. Seal the Corning assay plate with the black seal.
26. Immediately place the plate in the plate reader, and start reading the plate for 8 h (*see* **Note 12**).

### *3.2 Phosphate Sensor Data Analysis*

1. Plot the standards in GraphPad Prism (*see* Fig. 4).
2. Take the 30 min and 20-s time point read for the $KH_2PO_4$ standards (*y*-axis), and plot it against the $KH_2PO_4$ concentrations (*x*-axis) (*see* Fig. 4).
3. Perform a linear analysis using GraphPad Prism.

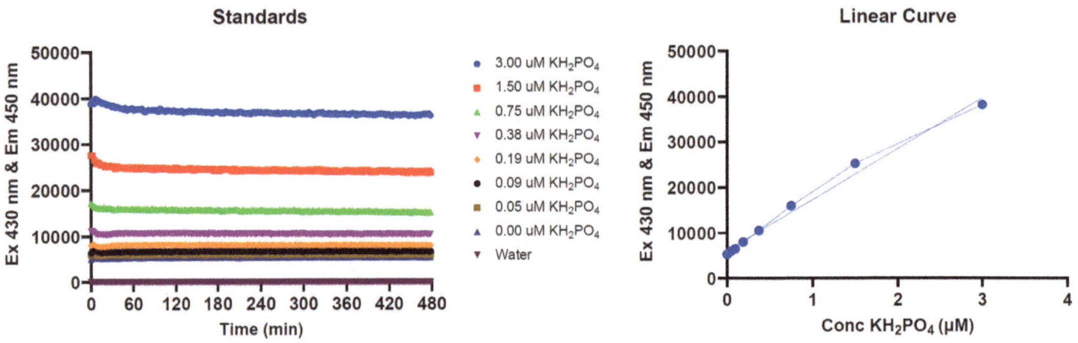


**Fig. 4** The signal intensity of $KH_2PO_4$ standards is constant as measured over the 8 h assay. Quantitation of the fluorescence signal at the 30-min time point shows a linear relationship with the $KH_2PO_4$ standards

4. Use the slope and intercept from the linear analysis to calculate the amount of GTP hydrolyzed in the sample over time (*see* **Note 13**).
5. Plot analyzed data (*y*-axis) over time (*x*-axis) (*see* Fig. 5).
6. Using GraphPad Prism use the one phase association analysis to determine the hydrolysis rate for each GAP concentration with KRAS (*see* **Note 14**).
7. To compare the impact of GAP sensitivity of the mutant KRAS proteins, plot the GTPase rate against GAP concentration (*see* Fig. 6).

## 4 Notes

1. The 50 mM stock of $KH_2PO_4$ can also be made up by diluting 68 mg of $KH_2PO_4$ into 10 mL of water and can be used instead of the stock provided in the EnzChek kit.
2. Protein expression constructs of His6-MBP-tev-protein of interest (POI) for all proteins were created using the protocols outlined in Wall et al. [19]. All proteins were expressed using the Dynamite media protocol with overnight induction at 16 °C as outlined in Taylor et al. [20]. Proteins of the form His6-MBP-tev-POI were purified using the protocols described in Kopra et al. [21].
3. Plates must be "Not Treated" since treated or surface coated plates contain a significant amount of phosphate and therefore can cause phosphate contamination and interfere with the assay.
4. Water in these methods refers to deionized water ($dH_2O$, >18.2 MΩ·cm at room temperature) which is used making all buffers and is filtered through a 0.22 μm membrane filter.

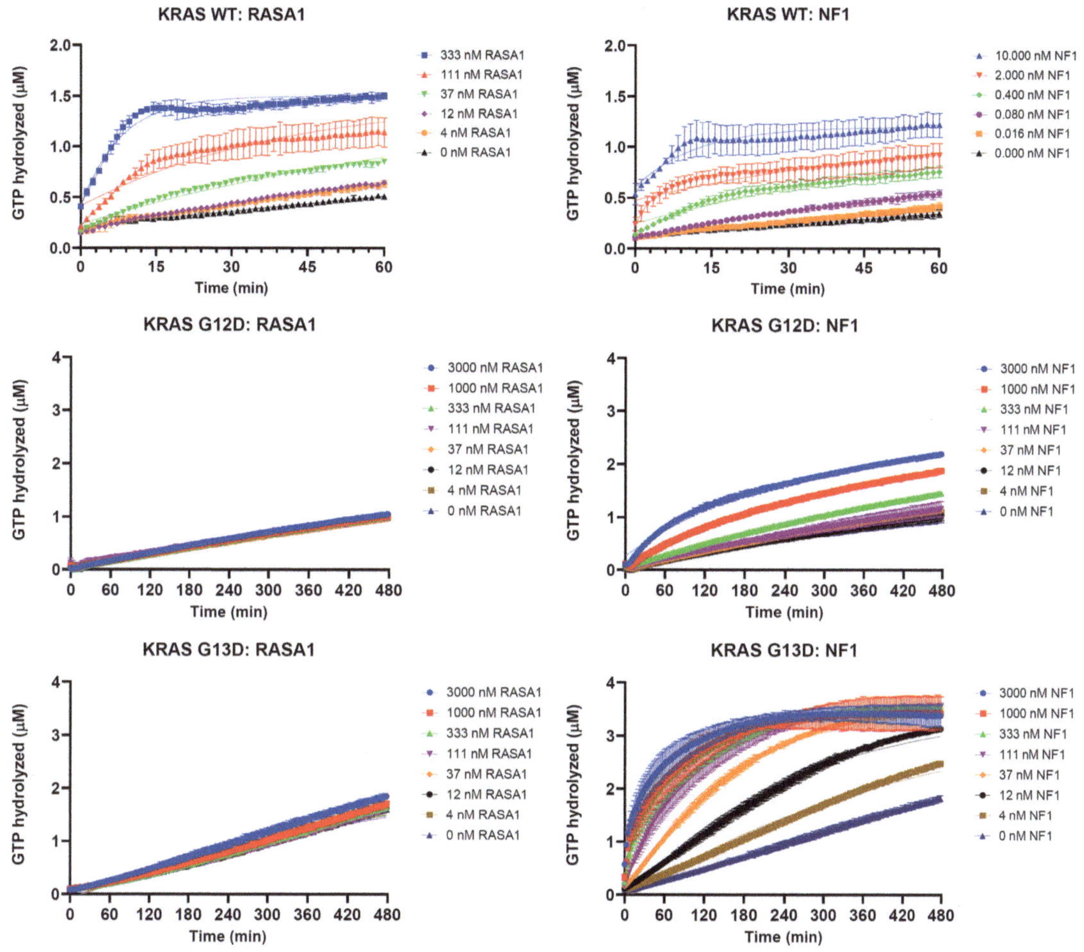


**Fig. 5** All reactions consisted of 3 μM KRAS-GTP with 4.5 μM phosphate sensor. The GTP hydrolysis of KRAS WT, G12D, and G13D was measured with increasing concentrations of RASA1 and NF1 GAP

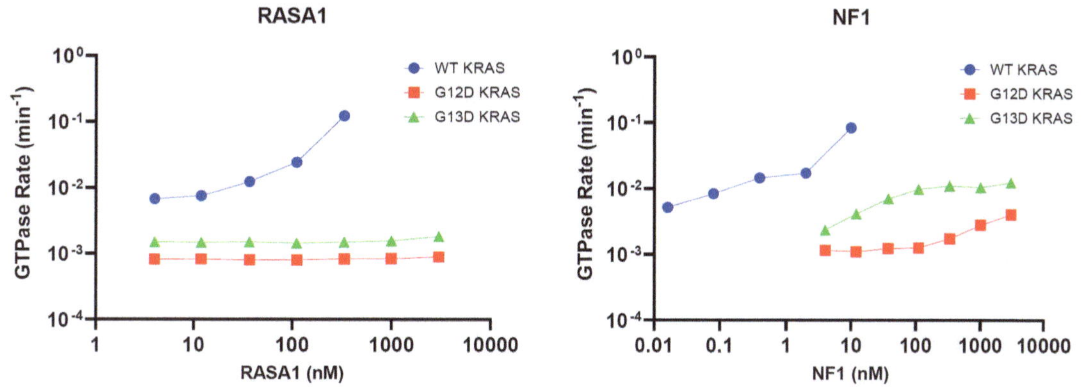


**Fig. 6** The GTPase rates were calculated by fitting the curves in Fig. 5 to a one phase association using GraphPad Prism and then plotted against the respective GAP concentrations

5. Since DTT at room temperature will degrade and oxidize, it is added fresh to an aliquot of PSAB prior to assay.
6. Leave the aliquoted stocks of GTP-loaded KRAS proteins in the −80 °C until just before use. Do not vortex GAP proteins when thawing or after thawed. Vortexing can cause the protein to unfold making the protein solution cloudy. Instead, invert tubes to mix or pipette up and down to mix prior to use.
7. Be sure to prepare enough of the 9 μM phosphate sensor diluted in PSAB for the assay you design. For the assay described here, the amount of 2400 μL works well.
8. For the assay described here, a 6 μM stock of GAP was used for the titration, but the top end concentration can be adjusted for the assay you design. If starting with very high amounts of GAP, say over 6 μM, create your top concentration in PSAB, and add in PS to 9 μM. Otherwise your 6 μM stock of GAP can be diluted in the 9 μM PS stock created in **step 9**. This is done to ensure the final concentration of PS remains at 9 μM prior to being added to the well and 4.5 μM final in the well.
9. The dilution can also be performed in the 96-well dilution plate if you are doing two different GAP proteins or if you are using a different dilution series than the one described here.
10. For the assay described here, 600 μL of 6 μM of KRAS-GTP is enough; however, make as much as you need for the assay you design. For instance, if the assay is run with two GAP titrations and one KRAS, you should make 1200 μL of the 6 μM KRAS-GTP.
11. E1-ClipTip matrix settings: Fill 20 μL; dispense 20 μL, manual mixing; and purge.
12. The plate can be read for a time frame of your choosing. However, 8 h was selected for mutants that are slow to hydrolyze GTP.
13. Use the standard equation $y = mx + b$, where $y$ is the signal, $m$ is the slope, and $b$ is the intercept to convert fluorescence to mM GTP hydrolyzed. Excel can be used to perform these calculations.
14. For reactions which do not reach completion during the experimental time, GTPase rates can be calculated by constraining the final plateau value to 3 μM.

## Acknowledgments

The authors wish to acknowledge Alison Champagne, Peter Frank, Randy Gapud, José Sánchez Hernández, Jennifer Mehalko, Simon Messing, Shelley Perkins, Nitya Ramakrishnan, Mukul Sherekar,

Kelly Snead, Troy Taylor, and Vanessa Wall for cloning, protein expression, protein purification, and electrospray ionization mass spectroscopy. This project was funded in part with federal funds from the National Cancer Institute, National Institutes of Health Contract 75N91019D00024. The content of this publication does not necessarily reflect the views or policies of the Department of Health and Human Services, and the mention of trade names, commercial products, or organizations does not imply endorsement by the US government.

## References

1. McGrath JP, Capon DJ, Goeddel DV et al (1984) Comparative biochemical properties of normal and activated human ras p21 protein. Nature 310:644–649
2. Der CJ, Finkel T, Cooper GM (1986) Biological and biochemical properties of human rasH genes mutated at codon 61. Cell 44:167–176
3. John J, Frech M, Wittinghofer A (1988) Biochemical properties of Ha-ras encoded p21 mutants and mechanism of the autophosphorylation reaction. J Biol Chem 263:11792–11799
4. Hiratsuka T (1985) A chromophoric and fluorescent analog of GTP, 2',3'-O-(2,4,6-trinitrocyclohexadienylidene)-GTP, as a spectroscopic probe for the GTP inhibitory site of liver glutamate dehydrogenase. J Biol Chem 260: 4784–4790
5. John J, Sohmen R, Feuerstein J et al (1990) Kinetics of interaction of nucleotides with nucleotide-free H-ras p21. Biochemistry 29: 6058–6065
6. Moore KJ, Webb MR, Eccleston JF (1993) Mechanism of GTP hydrolysis by p21N-ras catalyzed by GAP: studies with a fluorescent GTP analogue. Biochemistry 32:7451–7459
7. Ahmadian MR, Hoffmann U, Goody RS et al (1997) Individual rate constants for the interaction of Ras proteins with GTPase-activating proteins determined by fluorescence spectroscopy. Biochemistry 36:4535–4541
8. Mazhab-Jafari MT, Marshall CB, Smith M et al (2010) Real-time NMR study of three small GTPases reveals that fluorescent 2'(3')-O-(N-methylanthraniloyl)-tagged nucleotides alter hydrolysis and exchange kinetics. J Biol Chem 285:5132–5136
9. Carter SG, Karl DW (1982) Inorganic phosphate assay with malachite green: an improvement and evaluation. J Biochem Biophys Methods 7:7–13
10. Webb MR, Hunter JL (1992) Interaction of GTPase-activating protein with p21ras, measured using a continuous assay for inorganic phosphate release. Biochem J 287(Pt 2): 555–559
11. Webb MR (1992) A continuous spectrophotometric assay for inorganic phosphate and for measuring phosphate release kinetics in biological systems. Proc Natl Acad Sci U S A 89:4884–4887
12. Brune M, Hunter JL, Corrie JE et al (1994) Direct, real-time measurement of rapid inorganic phosphate release using a novel fluorescent probe and its application to actomyosin subfragment 1 ATPase. Biochemistry 33: 8262–8271
13. Phosphate Sensor for sensitive detection of GTPase activity (2011) ThermoFisher Scientific
14. Mondal S, Hsiao K, Goueli SA (2015) A homogenous bioluminescent system for measuring GTPase, GTPase activating protein, and guanine nucleotide exchange factor activities. Assay Drug Dev Technol 13:444–455
15. Kopra K, Rozwandowicz-Jansen A, Syrjanpaa M et al (2015) GTP-specific fab fragment-based GTPase activity assay. Anal Chem 87: 3527–3534
16. Kopra K, Seppala T, Rabara D et al (2018) Label-free time-gated luminescent detection method for the nucleotides with varying phosphate content. Sensors (Basel) 18:3989
17. Hoepflinger MC, Geretschlaeger A, Sommer A et al (2015) Molecular analysis and localization of CaARA7 a conventional RAB5 GTPase from characean algae. Traffic 16:534–554
18. Waybright T, Stephen AG (2024) Nucleotide exchange on RAS proteins using hydrolysable and non-hydrolysable nucleotides. In: Stephen AG, Esposito D (eds) KRAS: Methods and Protocols, Methods in Molecular Biology

19. Wall VE, Garvey LA, Mehalko JL et al (2014) Combinatorial assembly of clone libraries using site-specific recombination. Methods Mol Biol 1116:193–208
20. Taylor T, Denson JP, Esposito D (2017) Optimizing expression and solubility of proteins in E. coli using modified media and induction parameters. Methods Mol Biol 1586:65–82
21. Kopra K, Vuorinen E, Abreu-Blanco M et al (2020) Homogeneous dual-parametric-coupled assay for simultaneous nucleotide exchange and KRAS/RAF-RBD interaction monitoring. Anal Chem 92:4971–4979

# Chapter 8

# Affinity Measurement of Non-covalent Interactions of the Covalent KRAS G12C GDP Inhibitor MRTX849 to RAS Isoforms Using Surface Plasmon Resonance

**Patrick Alexander and Andrew G. Stephen**

## Abstract

Surface plasmon resonance (SPR) is an optical effect at an electron-rich surface that enables affinity measurements of biomolecules in real time. It is label free and versatile, not limited to proteins, nucleic acids, and small molecules. SPR is a widely accepted method to measure not only affinity of molecular interactions but also association and dissociation rates of such interactions. In this chapter, we describe a general method to measure the affinity of a small molecule drug, MRTX849, to GDP bound HRAS, KRAS, and NRAS.

**Key words** Surface plasmon resonance, KRAS, MRTX849, Adagrasib, Affinity measurement

## 1 Introduction

SPR is a method that allows one to determine how strongly a ligand and analyte interact. Ligands are defined as the binding partner attached to the sensor surface. Analytes are in the mobile phase flowing over the surface. The technique is based on refractive index changes at the surface of a sensor chip that occurs as mass accumulates when ligand and analyte interact. Ligands are often proteins, but ligands can be nucleic acids, lipids, liposomes, polysaccharides, cells, or viruses. Measurement of dissociation constants ($K_D$) with ranges from picomolar to millimolar, small sample consumption, and numerous ligand analyte combinations are several assets that have made SPR a common, if not preferred, method to characterize molecular interactions.

RAS is a family of small GTPases that function as molecular switches regulating signal transduction pathways for cell growth, differentiation, and survival. The three human isoforms of the RAS family, HRAS, KRAS, and NRAS, are active and signal in the guanosine triphosphate (GTP)-bound state. They are inactive in

Andrew G. Stephen and Dominic Esposito (eds.), *KRAS: Methods and Protocols*, Methods in Molecular Biology, vol. 2797,
https://doi.org/10.1007/978-1-0716-3822-4_8,

the guanosine diphosphate (GDP)-bound state. Misregulated signaling of RAS isoforms can lead to cancer: The three RAS genes are the most common oncogenes in human cancer responsible for 20% of malignancies [1]. Mutations in KRAS are responsible for 90% of pancreatic cancers, 40% of colorectal cancer, and 32% of lung cancer [1]. The most common KRAS mutation is glycine residue 12, followed by glycine residue 13 and less commonly glutamine residue 61 [2]. Mutations at all of these positions significantly decrease the sensitivity to GTPase-activating proteins (GAPs) that stimulate the GTPase activity, resulting in increased levels of active KRAS [3]. High frequency of mutations made KRAS an attractive cancer target that for decades has eluded effective treatment earning it the reputation "undruggable," until recently [4]. G12C mutations comprise 13% of non-small cell lung cancers and 3% of colorectal cancers [5]. Several small molecule electrophiles covalently modify KRAS GDP G12C, effectively sequestering the protein in the off state unable to active downstream signaling [6]. Sotorasib was the first approved G12C covalent inhibitor therapy followed by adagrasib, also known as MRTX849 [7].

Here we describe a method to measure the affinity of MRTX849 to HRAS, KRAS, and NRAS GDP. The proteins are wild type, containing a glycine at residue 12; therefore, electrophilic covalent modification is impossible. The affinity measured is the reversible non-covalent sum of the favorable binding interactions between the ligands and MRTX849. Although the focus of this chapter is to describe how to measure the interaction between MRTX849 and wild-type RAS proteins by SPR, it is applicable to evaluating small molecules that are putative KRAS binders.

## 2 Materials

Prepare all solutions with ultra pure18.2 MΩ-cm reverse osmosis purified water and at least reagent grade solutions. Always filter buffers through 0.22 μm cellulose acetate membranes to remove particulates before introducing into the instrument, excluding dimethyl sulfoxide solutions which are incompatible with cellulose acetate membranes. Although this method was designed and run on the Cytiva S200 platform, it can easily be adapted to the Cytiva 8 K and other SPR platforms.

### 2.1 Sensor Chip Preconditioning

1. HEPES-buffered saline (HBS): 20 mM HEPES, 150 mM NaCl, pH 7.4. Make 500 mL of HBS by diluting 20 mL 0.5 M HEPES/KOH pH 7.4 and 15 mL 5 M NaCl to 500 mL with water, check the pH, and correct if necessary. Vacuum filter HBS through a 0.2 μm cellulose acetate filter (*see* **Note 1**).

2. 50 mM sodium hydroxide: Add 10 mL of 1 M sodium hydroxide into a final volume of 200 mL of water.
3. Cytiva CM5 series S sensor chip.

### 2.2 Covalent Immobilization of Neutravidin via Primary Amines to Sensor Chip Surface Using EDC/NHS

1. 1-ethyl-3-(3-dimethylaminopropyl) carbodiimide hydrochloride, (EDC): Dissolve 750 mg EDC in 10 mL water.
2. N-hydroxysuccinimide (NHS): Dissolve 115 mg in 10 mL of water (*see* **Note 2**).
3. Ethanolamine HCl 1 M, pH 8.5 (*see* **Note 3**).
4. Neutravidin: 200 μg/mL in 10 mM sodium acetate, pH 4.5: 50 mM sodium acetate, pH 4.5 can be made by diluting 17.4 M glacial acetic acid to 50 mM with water and adjusting the pH to 4.5 with 5 M NaOH. Vacuum filter solution through a 0.2 μm cellulose acetate membrane and store at 4 °C. Lyophilized neutravidin powder can be solubilized in 50 mM sodium acetate, pH 4.5, and then diluted to 10 mM sodium acetate, pH 4.5 to a concentration of 200 μg/mL with water using A280 1.66 = 1 mg/mL. The lyophilized powder will not readily dissolve in 10 mM sodium acetate pH 4.5 (*see* **Note 4**).

### 2.3 Ligand Capture

1. Running buffer: 20 mM HEPES, 150 mM NaCl, 5 mM $MgCl_2$, 5 μM GDP, 0.05% Tween 20, 1 mM TCEP, and 5% DMSO, pH 7.4: For 0.5 L of running buffer, mix 20 mL 0.5 M HEPES pH 7.4, 15 mL 5 M NaCl, 2.5 mL 1 M $MgCl_2$, 2.5 mL 10% v/v Tween 20, 1 mL 0.5 M TCEP pH 7, and 25 μL 100 mM GDP. Bring up to a final volume of 475 mL with water in a graduated cylinder. This is equivalent to 1.05× No DMSO running buffer used in the analyte preparation (*see* Subheading 2.4, **item 2**). Vacuum filter this solution through a 0.2 μm cellulose acetate membrane into a 500 mL Kimax or Duran glass bottle. Remove 9.5 mL of 1.05× No DMSO running buffer from bottle, and save for analyte preparation (*see* Subheading 3.4, **step 2**). Add 24.5 mL of 100% DMSO using a glass pipet to the filtrate in the bottle. Aspirate buffer several times to rinse residual DMSO adhering to pipet into the bottle. Mix thoroughly (*see* **Note 5**).
2. Ligands: Avi-HRAS 1-169 GDP, Avi-KRAS 1-169 GDP, and Avi-NRAS 1-169 GDP (*see* **Note 6**).

### 2.4 Analyte Preparation

1. MRTX849: 20 mM stock in 100% dimethyl sulfoxide (DMSO): Dissolve 10 mg of MRTX849 in 827.6 μL 100% DMSO.
2. 1.05× No DMSO running buffer: 21 mM HEPES, 157.5 mM NaCl, 5.25 mM $MgCl_2$, 5.25 μM GDP, 0.0525% Tween 20, and 1.05 mM TCEP, pH 7.4. *See* Subheading 2.3, **item 1**.

3. DMSO solvent correction curve (5.8% to 4.5% DMSO). The solvent correction curve should bracket the concentration of the HBS 5% DMSO running buffer. Make a four-point curve from 5.8% to 4.5% DMSO: Sample #1, 5.8% DMSO: Mix 70 μL DMSO and 1130 μL of the 1.05× No DMSO buffer. Sample #4, 4.5% DMSO: Mix 54 μL DMSO 1146 μL of the 1.05× No DMSO buffer. Sample #2: Mix 500 μL of Sample #1 and 200 μL of Sample #4. Sample #3: Mix 300 μL of Sample #1 and 400 μL of Sample #4.

## 3 Methods

### 3.1 Sensor Chip Preconditioning

1. Bring all solutions and sensor chip to room temperature (RT).
2. Insert the CM5 chip into the Biacore S200, and remove buffer line A from the buffer bottle. Wipe the buffer line dry with a Kim wipe, and insert into HBS-containing bottle.
3. Prime the instrument twice with HBS buffer.
4. Run a method with a single injection of 10 mM of NaOH with a 60 s contact and 60 s dissociation time at a flow of 30 μL/mL (*see* **Note 7**).

### 3.2 Covalent Immobilization of Neutravidin via Primary Amines to Sensor Chip Surface Using EDC/NHS

1. Open the surface preparation folder in the Biacore S200 control software, and click immobilization icon.
2. Use the default "Amine method" with settings of 420 s for the contact time and a flow rate of 10 μL/min to immobilize neutravidin to all four flow cells (*see* **Note 8**).
3. Fill the appropriate tubes with solutions, and populate the rack according to the template generated by the software.
4. Save the method and initiate the program (*see* **Note 9**).

### 3.3 Ligand Capture

1. Change buffer line from the bottle containing the HBS to one containing the running buffer.
2. Prime the instrument twice using the running buffer. After the prime step has been completed, the instrument can run in standby mode.
3. Dilute ligands Avi-KRAS, Avi-HRAS, and Avi-NRAS to ~20 μg/mL into individual tubes using running buffer.
4. Capture approximately 1100 RU of each Avi-RAS ligand on flow cells two through four. Using a flow rate of 50 μL/min, allow baseline to stabilize for about 20 s. Inject Avi-HRAS with a contact time of 2 s, and record the level of capture. Repeat injections adjusting contact times and/or flow rates to capture approximately 1100 RU (*see* **Note 10**). Allow the buffer to flow for an additional 10 min to establish a postinjection

baseline, and measure the final amount of ligand captured (*see* **Note 11**).

5. Repeat the capture procedure on flow cell three and four with Avi-KRAS and Avi-NRAS, respectively.
6. Let system run on standby while preparing the analyte solutions.

### *3.4 Analyte Preparation*

1. Dilute analyte MRTX849 1:50 from a 20 mM 100% DMSO stock into 100% DMSO by mixing 1 μL of the 20 mM stock with 49 μL of 100% DMSO. This is 20x the highest concentration of analyte in the titration (*see* **Note 12**).
2. Make 270 μL of 20 μM MRTX849 from 20x analyte to match running buffer by adding 256.5 μL of 1.05x. No DMSO running buffer to 13.5 μL 400 μM MRTX849 in 100% DMSO (*see* **Note 13**).
3. Centrifuge sample 18,000× *g* for 5 min at RT in benchtop micro centrifuge (*see* **Note 14**).
4. Make six threefold serial dilutions by mixing 90 μL of 20 μM MRTX849 into 180 μL of running buffer (*see* **Note 15**).

### *3.5 Preparation of Biacore S200 Program*

1. In the Kinetics/Affinity folder of the S200 Control software, use the LMW kinetics/affinity multicycle wizard to prepare a program to run the analytes.
2. Set up a 3-cycle program consisting of 15 high-performance startup injections of running buffer at 30 μL/min flow with 60 s contact (*see* **Note 16**) and dissociation times, a 4-point solvent correction curve from 5.8% DMSO to 4.5% DMSO, and high-performance MRTX849 injections in HBS DMSO from 0 to 20 μM with 60 s contact and 120 s dissociation. Include 50% DMSO wash after injection and a carry-over control in the sample.
3. The solvent correction curve should bracket the concentration of DMSO in the running buffer. We routinely use a four-point curve from 5.8% to 4.5% DMSO.
4. Fill the appropriate tubes with solutions, populate the rack according to the template generated by the software, and start the program.

### *3.6 Steady-State Analysis of $K_D$*

1. Once the program is complete, the results can be viewed in the S200 Evaluation Software. The first step is to apply the solvent correction that corrects for any differences in refractive index caused by differences in the analyte samples and running buffer DMSO concentrations.
2. Examine the injections, and confirm that the signals reach equilibrium before dissociation begins (Fig. 1).

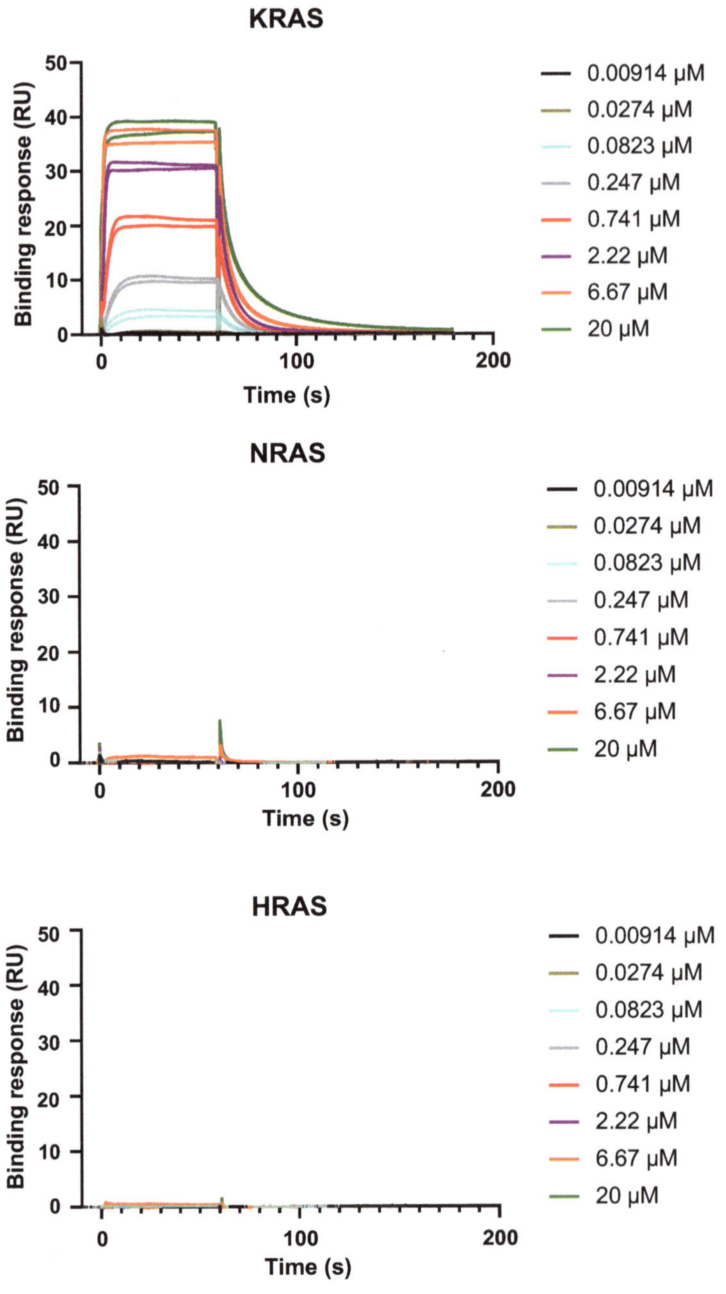


**Fig. 1** SPR sensorgrams of MRTX849 binding to KRAS, HRAS, and NRAS. Representative sensorgrams of a single experiment with duplicate titrations. All data was solvent corrected and double referenced by subtracting buffer and analyte binding to the neutravidin only reference channel from the ligand channels. Saturable binding was observed to KRAS. Little or no binding of MRTX849 was observed to NRAS or HRAS. This differential binding behavior is explained due to isoform differences in the MRTX849 binding site (*see* **Note 18**)

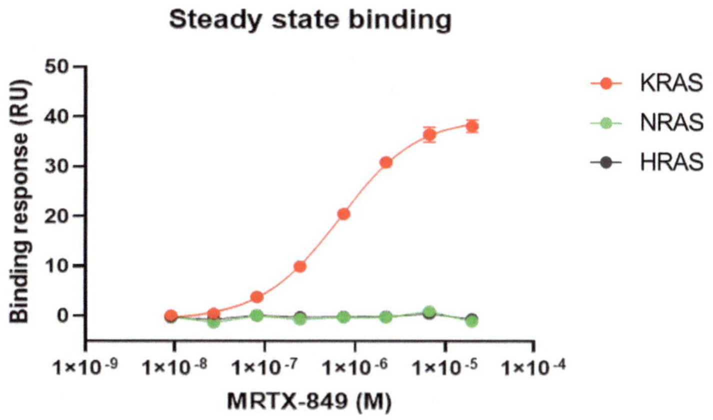


**Fig. 2** Steady-state binding isotherm of MRTX849 binding to KRAS, NRAS, and HRAS. Three experiments with duplicate titrations fit to a 1:1 steady-state model in the S200 evaluation software yield an average $K_D = 6.42 \times 10^{-7}$ M $\pm$ $3.4 \times 10^{-8}$ M

3. Fit the data using steady-state affinity model (*see* **Note 17**). Steady-state binding isotherms are shown in Fig. 2.

## 4 Notes

1. HBS can be refrigerated and used for multiple conditioning and immobilization steps until it is consumed. If the solution is not clear or turns yellow, discard and make fresh.
2. EDC and NHS solutions can be stored as 100 μL aliquots at −20 °C for at least 3 months.
3. Ethanolamine is provided in an amine coupling kit at 1 M pH 8.5 available from Cytiva.
4. Low ionic strength aids interaction of the positively charged neutravidin with the negatively charged dextran surface, analogous to ion exchange chromatography. Aliquots of neutravidin may be frozen, stored at −20 ° C and used for several months.
5. The amount of running buffer will vary depending on the experiment. The control software estimates a minimum amount needed. In general, if one makes 1.5 times the volume estimate from the software, there should be plenty of buffer for priming lines, capturing ligands, and standby operation to allow sensor surface equilibration. All DMSO solutions should be made fresh for individual experiments and not reused. One hundred percent DMSO stocks should be purchased in small volumes and sealed under argon after opening to minimize water absorption from the atmosphere. The use of polystyrene tubes with DMSO should be avoided. Polypropylene tubes are

recommended. Dissolve GDP in 50 mM Tris-HCl pH 8 to minimize acid hydrolysis of the β phosphate group and quantitate photometrically at 254 nm $\varepsilon = 13{,}700\ M^{-1}\ cm^{-1}$. Store aliquots at −80 °C. Never filter DMSO solutions through cellulose acetate membranes. One may have to remove more than 9.5 mL of 1.05× No DMSO buffer before addition of DMSO to accommodate analyte preparation and solvent correction curves. Ensure removal of enough 1.05× No DMSO running buffer by calculating volume needed in analyte preparation (*see* Subheading 3.4, **step 2**) and solvent correction curve (*see* Subheading 3.5, **step 2**) prior to addition of DMSO.

6. Protein expression constructs of the form His6-MBP-tev-protein of interest were created using the outlined protocols [8]. All HRAS proteins were expressed using the Dynamite media protocol with overnight induction at 16 °C [9]. NRAS proteins were expressed in *Vibrio natriegens* (manuscript in preparation). Proteins were purified using the protocols described for KRAS [10].
7. This step deprotonates the carboxymethyl dextran surface and gives it a negative charge to interact with the positively charged neutravidin at pH 4.5. This electrostatic interaction facilitates the covalent immobilization to the surface, outlined in Subheading 3.2.
8. These settings and protein concentration ensure a large surface density of neutravidin is created. If a less dense surface is required, one can decrease neutravidin concentration and or contact time.
9. Avoid buffers with primary amines. The primary amine in Tris (hydroxymethyl)aminomethane buffer will compete with neutravidin lysines for the negatively charged carboxymethyl dextran of the CM5 chip in the immobilization reaction. When covalently coupling protein via EDC/NHS chemistry to carboxymethyl dextran chips, the pH of the protein should be at least one pH unit below its isoelectric point, so it is positively charged, but above pH 3.5, the pKa of the carboxymethyl group on the dextran. This pH range is accessible for many but the most acidic proteins. Immobilized neutravidin chips can be sealed in their foil pouch and refrigerated for a couple of days before use but are best used the same day. Typically, ~18,000 response units (RU) of neutravidin are immobilized. If less neutravidin is desired, the concentration or the contact time can be decreased. We make our neutravidin chips, but streptavidin chips may be purchased.
10. The amount of ligand to be captured depends on the molecular ratio between the analyte and ligand based on the following relationship $R_{max} = R_{ligand}\ MW_{analyte}/MW_{ligand} \times S$,

where $R_{ligand}$ is the amount of ligand on the surface, $S$ is the stoichiometry of the ligand analyte interaction, and MW is the molecular weight in g/mol. $R_{max}$ is theoretical. It assumes 100% ligand activity, with all molecules capable of binding analyte, and infinite analyte concentration. Nevertheless, because the SPR signal is proportional to mass accumulation at the surface, the expected analyte response is proportional the amount of ligand captured. The smaller the *MW* ratio of analyte to ligand, the more ligand one must capture for a given $R_{max}$ value. Another consideration is that at higher ligand densities, the mass transport of the analyte can impact the kinetics and hence the affinity measurements. A good $R_{max}$ value is between 20 and 50 RU for steady-state affinity measurements. The S200 is capable of accurately measuring a few response units (RU), but consideration must be given to affinity of the interaction and ligand activity. Weak affinity and low activity necessitate denser ligand capture levels to allow detection of analyte binding.

The Avi HRAS, KRAS, and NRAS ligands are biotinylated at the unique lysine on the 15 amino acid Avi tag on the N-terminus of the protein. Neutravidin binds biotin with an affinity of $K_D = 10^{-15}$ M and is nearly irreversible. We assume little or no loss of ligand once captured by neutravidin. Complete biotinylation of the Avi tag ligand is not required. Non-biotin-labeled protein will wash over the surface uncaptured by neutravidin. The MW of ligands is approximately 21,250 Daltons (Da) and the analyte 604 Da. Using the Rmax equation above and assuming a stoichiometry of one, 1000 RU of ligand corresponds to an Rmax of about 30. Proteins were diluted to ~20 μg/mL in running buffer using $\varepsilon = 25{,}185$ $M^{-1}$ $cm^{-1}$ for Avi KRAS, 26,675 for both Avi HRAS and Avi NRAS. Twenty μg/mL is a good starting point to capture Avi-tagged proteins, but dilution, flow rates, and contact times are determined empirically. Assuming equivalent concentrations, biotinylation levels, flow rate, and contact time, a 21,250 MW protein and a 75,000 MW protein will yield different signal because of their MW differences. More mass will accumulate per molecule with the larger protein yielding more signal.

11. If ligand is precious and in short supply, use slower flow rates (10 μL/min) to conserve ligand.
12. The amount of 20x analyte will vary with experimental parameters of flow rate and contact time. Store unused 20x stock at −20 °C for reuse. Reuse depends on the stability of the compound in 100% DMSO.
13. The amount of the highest concentration of analyte will vary with each experiment. Always add the larger volume of 1.05× No DMSO running buffer to the smaller volume 100% DMSO

20× analyte, and mix by rapidly aspirating and dispensing solution repeatedly with the pipet tip to minimize locally high concentrations of compound. The aspiration dispensation cycles may prevent compound precipitation or aggregation. Precipitation is concentration dependent. One may check solubility of the highest concentration in the titration as a first step to determine compound solubility in running buffer before proceeding with an experiment. If the compound is insoluble in running buffer, the entire procedure can be run with higher DMSO concentrations in the running buffer (maximum of 10% DMSO); however, ligand stability in 10% DMSO may be an issue.

14. An insoluble pellet may be visible after centrifugation. If the pellet is compact and dense, the soluble fraction may be used for dilution, but the concentration is unknown. It is often worth continuing with such a sample instead of starting anew, but any $K_D$ determination would have to be considered an estimate due the uncertainty of the analyte concentration.
15. Dilutions and amounts will vary with individual experiments. Always include a zero-analyte concentration initial injection as the buffer blank for double referencing. An ideal titration would encompass three orders of magnitude in concentration, tenfold above and below the $K_D$. Depending on the affinity, tenfold concentrations above the $K_D$ may not be attainable due to compound solubility. If no information is available about the affinity of the ligand analyte interaction, a titration using fourfold dilutions with a wide concentration series can be a "searching" titration. Once the range has been narrowed, ten twofold dilutions may be done to get more data around the inflection point of the curve for increased $K_D$ accuracy.
16. Startup injections help equilibrate the flow system. They minimize refractive index differences between samples and running buffer that result in better referenced data.
17. Alternatively to fitting data to a steady-state model where the Rmax value is varied, data can be fit to a steady-state constant Rmax model. For this fitting to be accurate, a well-behaved positive control that is known to have a 1:1 stoichiometry with the ligands being evaluated is required. Based on the control molecular weight and measured signal at saturation, the analyte data can be fitted that constrains Rmax at a 1:1 stoichiometry. A poor fit to a such a constant Rmax model is an indication the assumption of 1:1 stoichiometry for analyte and ligand is not valid.
18. The X-ray crystal structure of the binding pocket of MRTX849 covalently bound to KRAS G12C reveals key side-chain interactions between the compound and KRAS [11]. Because the

ligands do not contain the G12C mutation, no covalent bond can form, and we cannot assume the hydrogen bond from the amine nitrogen of lysine 16 and the acrylamide carbonyl of MRTX849 seen in the crystal structure contributes to the binding we measure by SPR. However, it is interesting to note that a polar interaction seen in the crystal structure between the Nε2 nitrogen of His95 and N-1 of the pyrimidine ring of MRTX849 is only possible in KRAS. HRAS has glutamine, which is likely too small, and NRAS has leucine at position 95 which cannot form hydrogen bonds with its aliphatic side chain. Other side chains involved in compound interactions are identical in the three ligands. One cannot assume MRTX849 does not bind HRAS and NRAS at higher concentrations than tested here. Weaker binding would be predicted by the crystal structure and lack of histidine at position 95. This can easily be examined using this same method at higher analyte concentrations.

## Acknowledgments

The authors wish to acknowledge J-P Denson, Matt Drew, Peter Frank, Brianna Higgins, Min Hong, Jennifer Mehalko, Simon Messing, Ashley Mitchell, Shelley Perkins, Nitya Ramakrishnan, Rosamilia Reyes, Matt Smith, Kelly Snead, Troy Taylor, Vanessa Wall, and Stephanie Widmeyer for cloning, protein expression, protein purification, and electrospray ionization mass spectroscopy and Timothy Waybright and Dara Sontan for nucleotide analysis.

This project was funded in part with federal funds from the National Cancer Institute, National Institutes of Health Contract 75N91019D00024. The content of this publication does not necessarily reflect the views or policies of the Department of Health and Human Services, and the mention of trade names, commercial products, or organizations does not imply endorsement by the US government.

## References

1. Prior IA, Hood FE, Hartley JL (2020) The frequency of ras mutations in cancer. Cancer Res 80:2969–2974
2. Stephen AG, Esposito D, Bagni RK et al (2014) Dragging ras back in the ring. Cancer Cell 25:272–281
3. Trahey M, McCormick F (1987) A cytoplasmic protein stimulates normal N-ras p21 GTPase, but does not affect oncogenic mutants. Science 238:542–545
4. Moore AR, Rosenberg SC, McCormick F et al (2020) RAS-targeted therapies: is the undruggable drugged? Nat Rev Drug Discov 19:533–552
5. Hong DS, Fakih MG, Strickler JH et al (2020) KRAS(G12C) inhibition with sotorasib in advanced solid tumors. N Engl J Med 383:1207–1217
6. Ostrem JML, Shokat KM (2022) Targeting KRAS G12C with covalent inhibitors. Annu Rev Cancer Biol 6:49–64

7. Kwan AK, Piazza GA, Keeton AB et al (2022) The path to the clinic: a comprehensive review on direct KRAS(G12C) inhibitors. J Exp Clin Cancer Res 41:27
8. Wall VE, Garvey LA, Mehalko JL et al (2014) Combinatorial assembly of clone libraries using site-specific recombination. Methods Mol Biol 1116:193–208
9. Taylor T, Denson JP, Esposito D (2017) Optimizing expression and solubility of proteins in *E. coli* using modified media and induction parameters. Methods Mol Biol 1586:65–82
10. Kopra K, Vuorinen E, Abreu-Blanco M et al (2020) Homogeneous dual-parametric-coupled assay for simultaneous nucleotide exchange and KRAS/RAF-RBD interaction monitoring. Anal Chem 92:4971–4979
11. Fell JB, Fischer JP, Baer BR et al (2020) Identification of the clinical development candidate MRTX849, a covalent KRAS(G12C) inhibitor for the treatment of cancer. J Med Chem 63: 6679–6693

# Chapter 9

# 1D and 2D NMR for KRAS:Ligand Binding

**Gabriel Cornilescu**


## Abstract

Fragment-based screening by ligand-observed 1D NMR and binding interface mapping by protein-observed 2D NMR are popular methods used in drug discovery. These methods allow researchers to detect compound binding over a wide range of affinities and offer a simultaneous assessment of solubility, purity, and chemical formula accuracy of the target compounds and the $^{15}N$-labeled protein when examined by 1D and 2D NMR, respectively. These methods can be applied for screening fragment binding to the active (GMPPNP-bound) and inactive (GDP-bound) states of oncogenic KRAS mutants.


**Key words** NMR screening, FBS, FBDD, WaterLOGSY, STD, HSQC titration, CSP, KRAS

## 1 Introduction

A major goal of any RAS-related drug discovery program is to develop selective inhibitors targeting the oncogenic mutants of KRAS. Routine 1D and 2D NMR spectroscopy fragment-binding screening (FBS) methods can be used in conjunction with virtual screening, crystallography, SPR, and other techniques, to attempt to identify promising fragments that interact with oncogenic KRAS.

WaterLOGSY [1], STD [2], and $T_2$ [3] NMR experiments are well-established 1D-NMR ligand-observed experiments that typically allow the detection of small fragments that bind weakly to a protein, i.e., with binding constant in the mM to μM range. Millimolar quantities of the ligand and 10–100 μM of protein are typically required, and the total experimental time is approximately 1 h per NMR sample. The relative intensities of the STD NMR signals can additionally provide information about the binding pose, as protons exposed to the solvent will have less intense peaks when compared to protons buried in the ligand pocket [4]. Similarly, the solvent accessibility of ligands bound to proteins can be estimated from the relative peak intensities in WaterLOGSY

Andrew G. Stephen and Dominic Esposito (eds.), *KRAS: Methods and Protocols*, Methods in Molecular Biology, vol. 2797,
https://doi.org/10.1007/978-1-0716-3822-4_9,

[5]. While usually a separate ligand-only WaterLOGSY spectrum is required for comparison with the WaterLOGSY (difference) spectrum in the presence of the protein, occasionally comparing the WaterLOGSY with the STD reference spectrum allows observation of a differential attenuation among the WaterLOGSY signals.

For the most promising binders, including those which result in successful crystal structure determination, one can then perform 2D-FHSQC titrations to map the chemical shift perturbations (CSP) on the KRAS crystal structure. This allows mapping the ligand-binding interface on the protein surface or confirming the binding pocket(s) identified by crystallography.

## 2 Materials

All buffers need to be prepared with ultrapure deionized water and analytical-grade reagents for any NMR experiments.

1. Norell® Standard Series™ 5 mm NMR tubes (Sigma-Aldrich), 7 inch tubes are appropriate for Bruker's Sample Case automation system.
2. $D_2O$ (deuterium oxide 99.9 atom % D).
3. DMSO-*d6* (dimethyl sulfoxide-*d6* "100%," 99.96 atom % D).
4. 1× sodium phosphate (NaPi) buffer, pH 7.4.
5. KRAS NMR buffer: 20 mM HEPES pH 7.4, 150 mM NaCl, 2 mM $MgCl_2$, 1 mM TCEP, and 7% $D_2O$.
6. Purified KRAS proteins include residues 1–169 of KRAS G13D and were purified with $^{15}N$ incorporation [6], $^{13}C/^{15}N$ incorporation [7], or highly deuterated $^{13}C$-methyl/$^{15}N$ incorporation [8, 9].

## 3 Methods

### 3.1 1D-NMR Ligand Screening of Fragment Binding to KRAS G13D

1. Prepare 50 mM stocks of each fragment compound by dissolving in DMSO-d6. If a fragment is not fully soluble at first, gentle heating to 40 °C followed by vortexing may help.
2. Prepare a stock of sodium phosphate (NaPi) buffer (*see* **Note 1**) with a pH of 7.0–7.4 containing 7% $D_2O$ (*see* **Note 5**); you will need 425 μL of buffer per fragment screened.
3. Dilute purified KRAS protein in 500 μL of NaPi+$D_2O$ buffer to a final concentration of 30 μM. Avoid impurities when exchanging buffers or concentrating the protein (*see* **Note 4**).

4. Transfer 491 μL of the NaPi + $D_2O$ + KRAS solution to a 1.5 mL microfuge tube, add 9 μL of the 50 mM fragment stock, and mix it thoroughly by pipetting in and out. This mixture will be a 30:1 molar ratio of ligand:protein (*see* **Note 2**).
5. Slowly pipet the mixture into the 5 mm NMR tube to let the first droplet wet the tube wall. If a column of liquid forms at the top of the NMR tube, cap the tube and flick it (or swing it down) to push the liquid column to the bottom of the tube. Alternatively, use a long Pasteur glass pipette with a pipet gun or a rubber teat for the transfer.
6. Using a Bruker 500 MHz Avance Neo spectrometer with a 24-tube SampleCase and a Prodigy cryoprobe, insert all NMR tubes in sequence, and, using IconNMR, queue up three experiments for the first sample: SCREEN_WLOGSY, SCREEN_STD, and SCREEN_T2.
7. Give a descriptive name to the screening set and select $H_2O$ + $D_2O$ as the solvent. In the parameter column, for a ~ 0.9 mM ligand concentration, double the default number of scans, i.e., use $NS = 256$ for SCREEN_WLOGSY and $NS = 128$ for SCREEN_STD and SCREEN_T2. Optionally, use the low priority "night-time mode" to schedule all samples.
8. In IconNMR, use "Average" to copy all three experiments for all other sample locations, and adjust the title for each sample. Increase by two the experiment number of each SCREEN_T2 experiment to leave two consecutive empty experiment numbers after each pseudo2D experiment (STD and T2) to allow saving locations for the default automatic processing. Submit all experiments for all samples in the automatic queue.
9. On a processing computer with topspin installed, use multiple display (.md command) to stack and compare various 1D spectra (*see* **Note 6**). Figure 1 shows examples of WaterLOGSY spectra of four ligands.
10. Compare the peak intensities in the STD and WaterLOGSY spectra to assess the presence of binding (*see* **Note 7**) and inspect the less unambiguous $T_2$ experiment, as exemplified in Fig. 2.
11. Overlay the pair of STD spectra (reference and STD), and scale to normalize the peak intensities, as exemplified in Fig. 3. If evident differential attenuation exists among the aromatic peaks, the binding is likely specific; otherwise, the binding is nonspecific (*see* **Note 3**). Nonspecific binders often show strong WaterLOGSY and STD peak intensities but no differential attenuation.

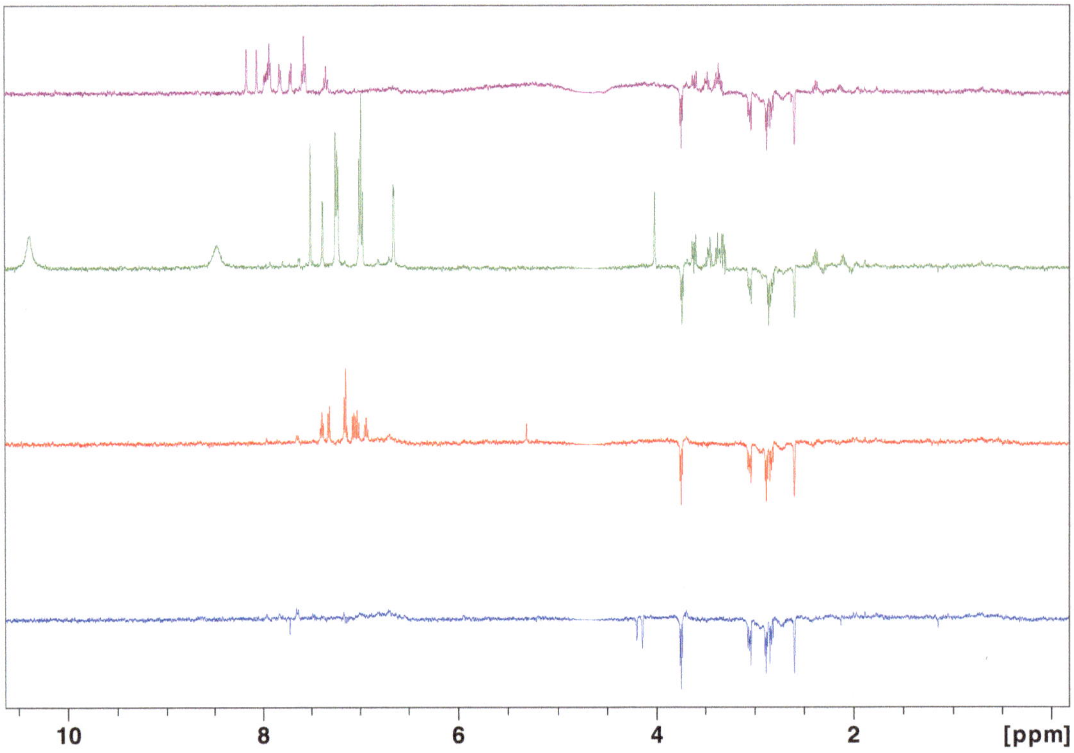


**Fig. 1** Overlay of WaterLOGSY spectra of four ligands in the presence of KRAS G13D-GMPPNP. The spectra are phased with negative intensities of the solvent peak so that binding ligands exhibit positive peaks

### *3.2 2D-NMR Titrations*

1. For 2D-NMR titrations, mix 0.2 mM of uniformly $^{15}$N-labeled KRAS (*see* **Note 8**) in KRAS NMR buffer to fragments diluted from 50 or 100 mM stocks. Suggested dilutions should range from 0:1 ligand:protein up to 80:1 in 4–5 steps.
2. Transfer the mixture of protein and ligand to 3 mm NMR tubes.
3. Acquire the NMR spectra on a Bruker 700 MHz Avance Neo spectrometer with a 5 mm high-resolution cryoprobe at 25 °C. Tune and shim the spectrometer on the sample without ligand, set the transmitter on the water signal, calibrate the 90° 1 h pulse, and leave all parameters unchanged during the titration.
4. After each fragment addition, perform a quick 1D automatic re-shimming.
5. Acquire 10 min fast $^{1}$H-$^{15}$N HSQC experiments (Bruker standard parameter set FHSQCF3GPPH) for each titration addition using nonuniform sampling (NUS) to record only 64 out of the usual 128 increments in the indirect dimension.

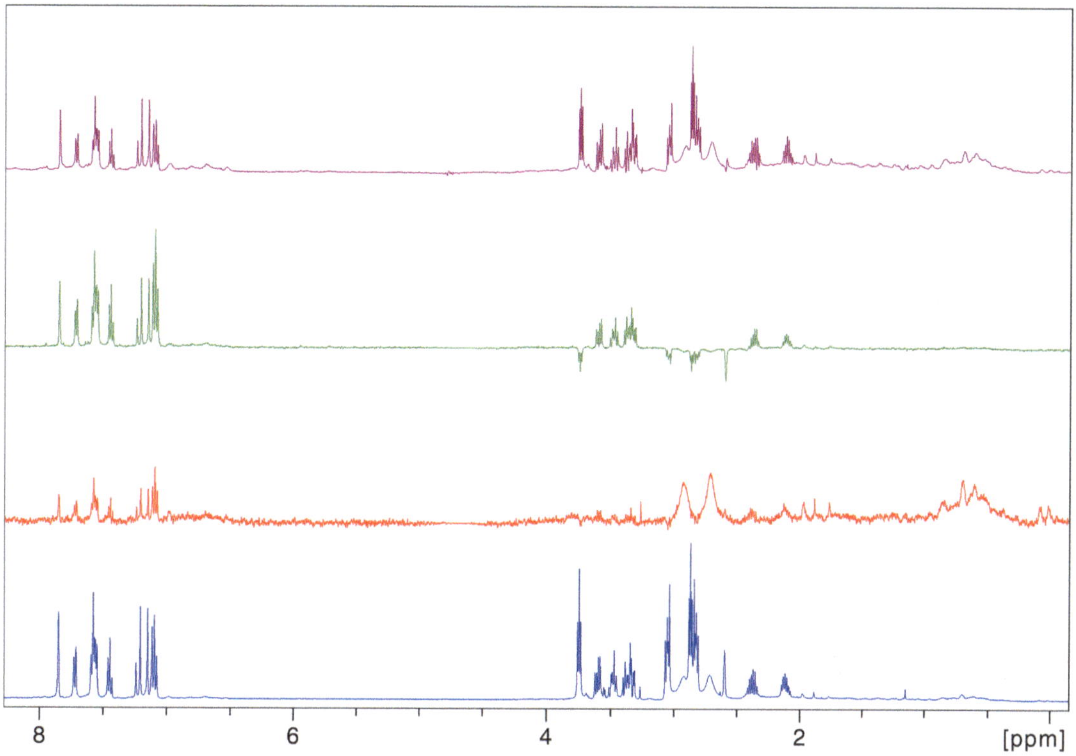


**Fig. 2** Typical spectra of a fragment binding to KRAS G13D-GMPPNP. The reference STD spectrum (equivalent to a $^{1}$H spectrum) of ~900 μM fragment in the presence of ~24 μM KRASG13D in blue, the STD spectrum in red, the WaterLOGSY spectrum in green, and the $T_2$ difference spectrum in magenta

6. Spectra can be processed automatically using a batch NMRPipe [10] script adapted for NUS (available upon request from gabriel.cornilescu@nih.gov) and analyzed in Poky [11]. The KRAS amide assignments (BMRB ID: 51642) were initially determined using standard 3D NMR experiments on a 0.8 mM $^{13}$C-methyl/$^{15}$N-labeled KRAS sample.
7. Assignments can be mapped automatically in Poky by copy/paste stepwise from the apo spectrum to each titration spectrum in the order of increasing ligand concentration. The Poky commands used are *pa/oc/op/pc* (peak all/ornament copy/ornament paste/peak center).
8. The peak assignments in regions of overlap can be inspected (using the *lt* peak list command to check amides in sequence) and adjusted manually, if needed.
9. Use the *np* (nonlinear plot) Poky command to calculate the CSPs, plot them along the protein sequence, and map them on the 3D structure in PyMOL [12]. Figure 5 illustrates an example of the output.

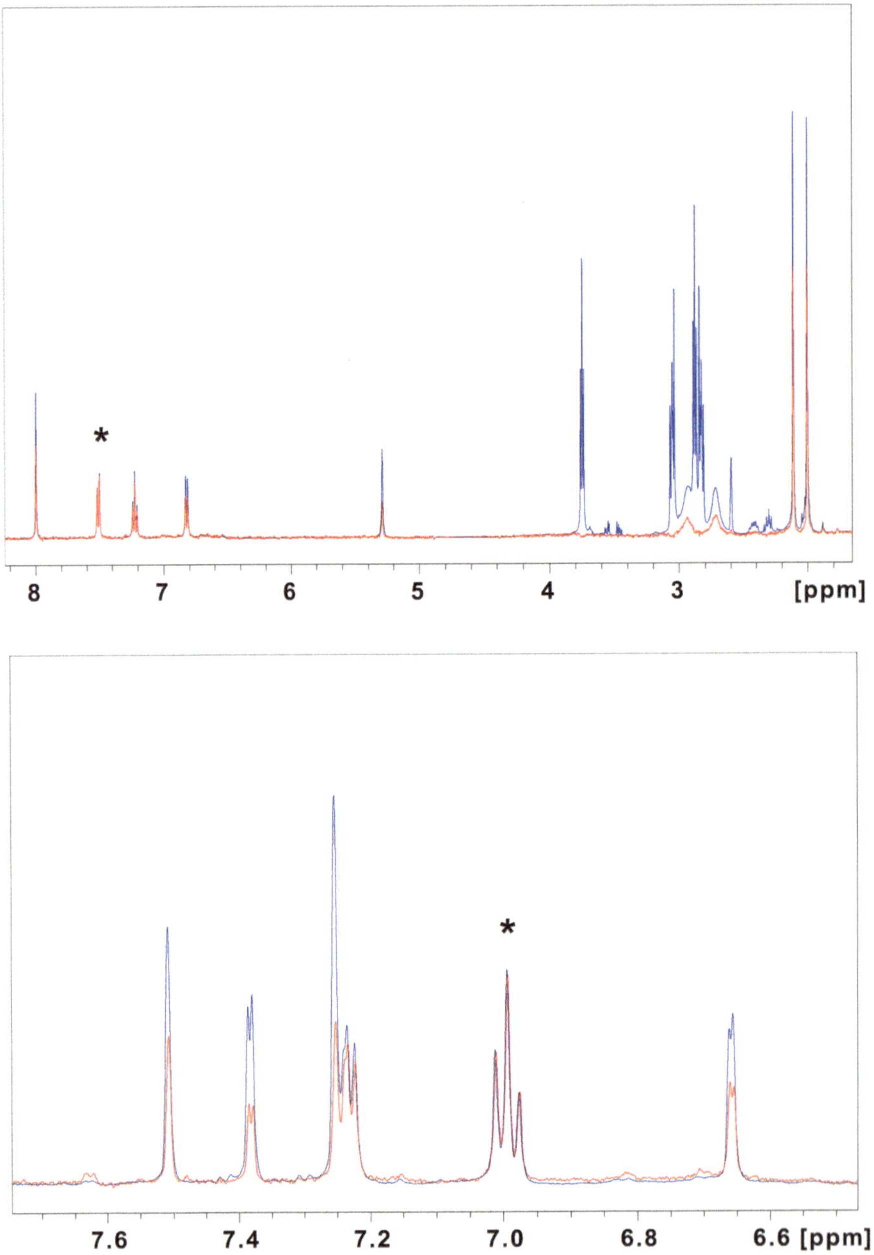


**Fig. 3** Overlay of STD spectra of two ligands in the presence of KRAS-G13D-GMPPNP showing differential signal attenuation of the protons buried in a pocket. Asterisks indicate the peak used to scale the STD spectrum (in red) to the STD reference spectrum (in blue) to normalize their intensities

## 4 Notes

1. Use a nonprotonated buffer such as phosphate (or alternatively a perdeuterated buffer) to avoid strong buffer proton signals that would mask the fragment signals and may overload the spectrometer's receiver. Use the minimal amount of salt

required for protein stability (salt decreases NMR sensitivity). Using more than 10% DMSO in the final protein solution may perturb the protein stability and even the protein structure and thus affect ligand binding. Therefore, a good fragment solubility is vital to afford a high concentration (50 mM) of the fragment stock.

2. The strength of the STD NMR signal increases with the amount of saturated protein protons, i.e., with the protein concentration. The WaterLOGSY signals for weak binders similarly increase with the protein concentration, but a higher protein concentration also increases the likelihood of nonspecific binding. A ligand molar excess of 20- to 100-fold is generally acceptable, with WaterLOGSY requiring 20- to 50-fold and the STD requiring 50- to 100-fold. For KRAS (19.4 kDa), we use a ligand:protein molar ratio of ~30:1 as a reasonable compromise for sensitivity (sufficient ligand available to get saturated during binding events but not in a large excess that would result in a large contribution from the unbound ligand).

3. In rare cases, we observe peaks from aromatic protons with vanishingly small intensity (Fig. 1, blue spectrum), while peaks from aliphatic protons show negative intensities (when spectra are phased to have negative solvent signals). These cases, stemming from a cancelation effect specific to a particular fragment: protein size and molar ratios, can be mistakenly interpreted as nonbinding (i.e., false negatives). Since aromatic peaks in WaterLOGSY should also be negative for nonbinding fragments (similar to the ligand-only WaterLOGSY spectra or the reference STD spectrum), vanishingly small intensities still indicate binding [5]. We occasionally acquire separate experiments on a ligand-only sample to clarify these cases. Alternatively, repeating the WaterLOGSY spectrum on a sample with a substantially different fragment:protein molar ratio (e.g., 10:1 instead of 30:1) also removes the ambiguity since the coincidental cancelation will no longer occur.

4. When using concentrators to exchange buffers or concentrate the protein, make sure to properly wash the membrane by centrifuging 2–3 times the maximum concentrator volume with distilled water to remove any glycerol present. Otherwise, strong glycerol signals will appear at 4.3–3.6 ppm (Fig. 4).

5. 7% $D_2O$ is sufficient for most modern spectrometers; using more than that lowers the sensitivity of the NMR experiments.

6. For spectral phasing consistency of the WaterLOGSY signals (protons from bound fragments show signals with opposite phase from those of nonbinding molecules, e.g., the solvent signal), save a properly phased spectrum named

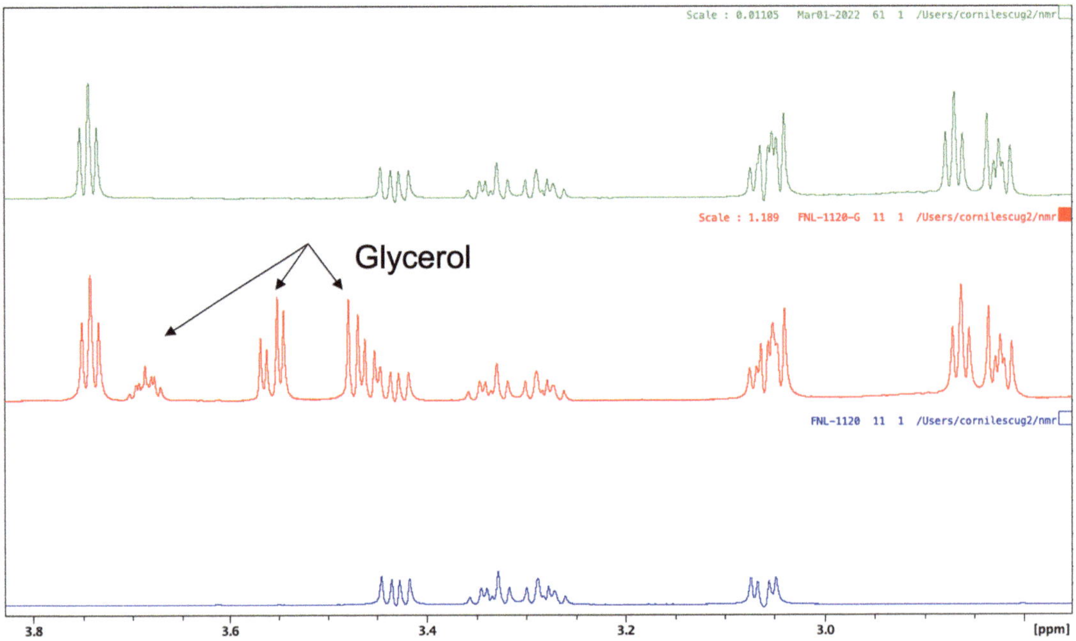


**Fig. 4** Glycerol impurity signals. A fragment-only reference spectrum is shown in blue, and the same fragment in the presence of KRAS G13D-GMPPNP and KRAS G13D-GDP is shown in red and green, respectively

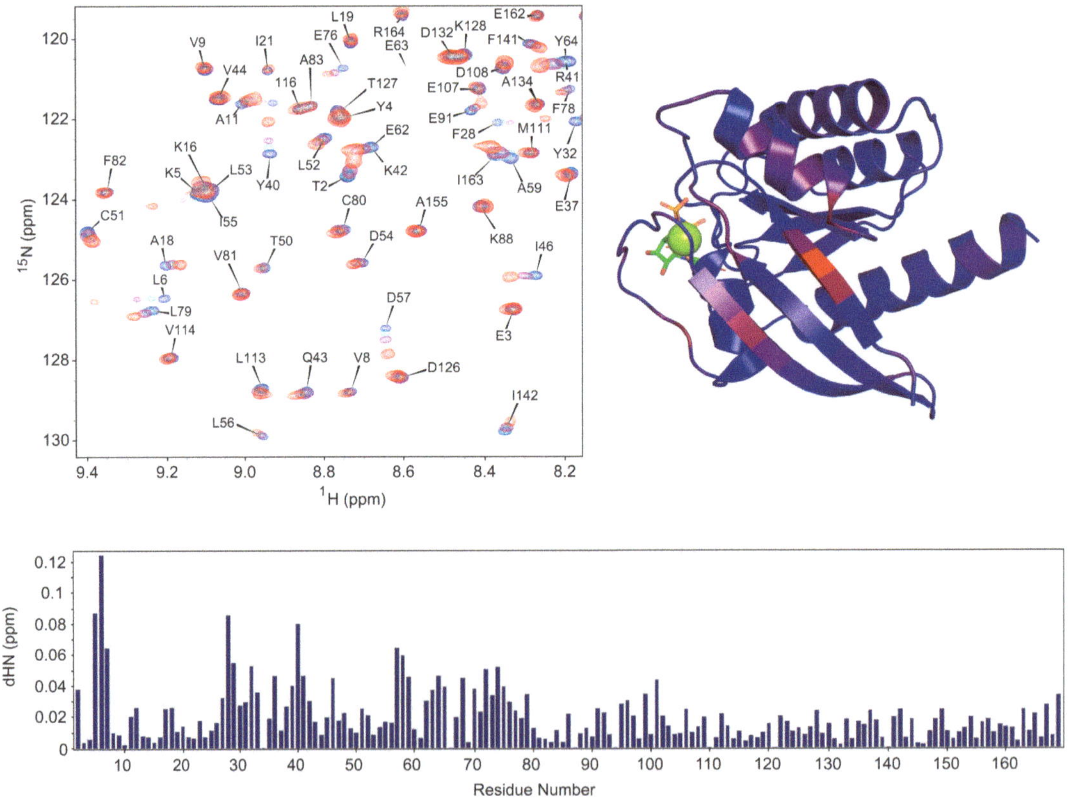


**Fig. 5** Data from 2D-NMR studies of KRAS. Top left: a selected region of the 2D-HSQC titration of a fragment on KRAS G13D/T35S-GMPPNP. Top right: CSP plotted on the 3D structure of KRAS G13D-GMPPNP (PDB ID: 8EBZ). Bottom: amide chemical shift perturbation (CSP) plot against the sequence of KRAS G13D/T35S-GMPPNP sequence

WLOGSY_PREP in the same disk location as the screening data. The software will phase all spectra using the phase correction parameters (phc0 and phc1) in this dataset (e.g., if the solvent signal is phased with a negative intensity, the protons of a binding fragment will show positive WaterLOGSY peaks).

7. Broad WaterLOGSY signals from exchangeable ligand protons (e.g., NH, OH) should be ignored since the exchange with water always results in signals with opposite phase relative to the solvent peak, regardless of the presence of the protein (false positive).

8. For fragments expected to bind close to the KRAS switch regions, we used a KRAS construct containing the additional T35S mutation, known to shift the State 1/State 2 conformational equilibrium towards State 1 affecting the otherwise broadened-out signals of the highly mobile switch regions in KRAS. This mutation renders their amides detectable by NMR, providing more probes into the switch regions.

## Acknowledgments

This project was funded in part with federal funds from the National Cancer Institute, National Institutes of Health under contract 75N91019D00024. The content of this publication does not necessarily reflect the views or policies of the Department of Health and Human Services, and the mention of trade names, commercial products, or organizations does not imply endorsement by the US government.

### References

1. Dalvit C, Fogliatto G, Stewart A, Veronesi M, Stockman B (2001) WaterLOGSY as a method for primary NMR screening: practical aspects and range of applicability. J Biomol NMR 21(4):349–359
2. Mayer M, Mayer B (1999) Characterization of ligand binding by saturation transfer difference NMR spectroscopy. Angew Chem Int Ed 38: 1784–1788
3. Hajduk PJ, Olejniczak ET, Fesik SW (1997) One-dimensional relaxation- and diffusion-edited NMR methods for screening compounds that bind to macromolecules. J Am Chem Soc 119(50):12257–12261
4. Cala O, Krimm I (2015) Ligand-orientation based fragment selection in STD NMR screening. J Med Chem 58(21):8739–8742
5. Raingeval C, Cala O, Brion B, Le Borgne M, Hubbard RE, Krimm I (2019) 1D NMR WaterLOGSY as an efficient method for fragment-based lead discovery. J Enzyme Inhib Med Chem 34(1):1218–1225
6. Agamasu C, Ghirlando R, Taylor T, Messing S, Tran TH, Bindu L et al (2019) KRAS prenylation is required for bivalent binding with calmodulin in a nucleotide-independent manner. Biophys J 116(6):1049–1063
7. Travers T, López CA, Van QN, Neale C, Tonelli M, Stephen AG et al (2018) Molecular recognition of RAS/RAF complex at the membrane: role of RAF cysteine-rich domain. Sci Rep 8(1):8461
8. Chao FA, Dharmaiah S, Taylor T, Messing S, Gillette W, Esposito D et al (2022) Insights

into the cross talk between effector and allosteric lobes of KRAS from methyl conformational dynamics. J Am Chem Soc 144(9):4196–4205

9. Li J, Byrd RA (2022) A simple protocol for the production of highly deuterated proteins for biophysical studies. J Biol Chem 298(8): 102253

10. Delaglio F, Grzesiek S, Vuister GW, Zhu G, Pfeifer J, Bax A (1995) NMRPipe: a multidimensional spectral processing system based on UNIX pipes. J Biomol NMR 6(3):277–293

11. Lee W, Rahimi M, Lee Y, Chiu A (2021) POKY: a software suite for multidimensional NMR and 3D structure calculation of biomolecules. Bioinformatics 37(18): 3041–3042

12. DeLano WL (2002) The PyMOL molecular graphics system, Version 2.0. Schrödinger, LLC

# Chapter 10

# Luminescence-Based Techniques for KRAS Thermal Stability Monitoring

## Kari Kopra

## Abstract

Various biochemical methods have been introduced to detect and characterize KRAS activity and interactions, from which the vast majority is based on luminescence detection in its varying forms. Among these methods, thermal stability assays, using luminophore-conjugated proteins or external environment sensing dyes, are widely used. In this chapter, we describe methods enabling KRAS stability monitoring in vitro, with an emphasis on ligand-induced stability. This chapter focuses mainly on luminescence-based techniques utilizing external dye molecules and fluorescence detection.

**Key words** External dye, Fluorescence, GloMelt™, KRAS, Label-free, Protein-probe, SYPRO Orange™, Thermal shift assay (TSA)

## 1 Introduction

Luminescence and especially photoluminescence is widely used signaling technique in bioanalytical assays. Photoluminescence can be further divided into different classes, and the most frequently used methods are based on fluorescence or time-resolved luminescence (TRF). In this chapter, we mainly concentrate on homogeneous luminescence-based assays for KRAS thermal stability monitoring. Methods based on nonluminescent techniques or stability assays using chemical denaturants are not discussed in detail.

KRAS is a small guanosine triphosphate (GTP)-binding protein, which is a known oncogene and regulated depending on its nucleotide bound form, GTP-loaded active and guanosine diphosphate (GDP)-bound inactive form [1, 2]. KRAS is not a highly stable protein, and for both for its activity and stability, KRAS needs magnesium ($Mg^{2+}$), which helps KRAS to preserve its nucleotide loading and activity/stability [3, 4]. This sets certain requirements

Andrew G. Stephen and Dominic Esposito (eds.), *KRAS: Methods and Protocols*, Methods in Molecular Biology, vol. 2797, https://doi.org/10.1007/978-1-0716-3822-4_10,

on assay conditions to enable reliable and reproducible KRAS studies and comparison between assays and methodologies.

The most common KRAS assays are performed utilizing enzyme activity of KRAS, mainly nucleotide exchange and GTPase activity [5–7]. These assays can provide information about KRAS interaction with its ligands (protein-ligand interaction, PLI), like inhibitor compounds. However, sometimes it is more relevant to monitor direct PLI and not KRAS activity, and for that, surface plasmon resonance (SPR) is one option [7]. However, SPR needs a specialized equipment and KRAS attachment to the surface, which makes the method out of reach for some researchers. For larger audience, homogeneous wash-free assays performed in a multi-well plate are most suitable. Thermal shift assays (TSAs), also known as differential scanning fluorimetry (DSF), fulfill these criteria [4, 8]. As KRAS contains no tryptophan moieties enabling the full use of intrinsic fluorescence, external environmental sensitive dyes are most often used [4, 9]. External TSA dye are often fluorescent molecules sensing the structural change in the studied protein, and whose fluorescence properties change upon binding to structurally unordered target, here KRAS. TSA methods are easy to use and automate, without any major instrumental demands or the need for protein labeling, in contrast to differential scanning calorimetry (DSC), circular dichroism (CD), or nanoDSF [10–12]. While DSC is the recognized gold standard method for measuring protein stability, the requirement for specialized equipment means it is often not available for many researchers. CD suffers from low sensitivity and limited throughput which make this technique unsuitable for analysis of multiple PLIs. Thus, in this section, we concentrate on the TSA method, which, in comparison to all other methods for KRAS stability monitoring, enables the use of standard multi-well plates, plate reader, and thermal cycler or additionally a real-time PCR system (qPCR). Several methods, commercial or noncommercial, utilizing external dyes to study KRAS (thermal) stability, are introduced in the following section. In addition, considerations to obtain reliable and reproducible data enabling KRAS mutant and ligand analysis are discussed.

## 2 Materials

Prepare all solutions using ultrapure water (prepared by purifying deionized water, to attain a sensitivity of 18 MΩ-cm at 25 °C) and analytical grade reagents. Prepare and store all KRAS proteins at −80 °C and other reagents at +4 °C (unless indicated otherwise). Diligently follow all waste disposal regulations when disposing of waste materials.

### 2.1 Commercial TSA Methods for KRAS Stability Monitoring

#### 2.1.1 SYPRO Orange™

1. SYPRO assay buffer: 20 mM HEPES (pH 7.5), 100 mM NaCl, and 5% glycerol. Weigh 0.48 g HEPES (238.3 g/mol), 0.6 g NaCl (58.44 g/mol), and 5.0 g glycerol (92.1 g/mol), and dilute in 100 mL of water (*see* **Note 1**). Adjust pH to 7.5 using NaOH.
2. KRAS protein (*see* **Notes 1** and **2**).
3. SYPRO Orange™ dye: 5000X stock (*see* **Note 3**).
4. KRAS inhibitors, activators, or interactor molecules (*see* **Note 4**).
5. Multi-well plate (*see* **Note 5**).

#### 2.1.2 GloMelt™

1. GloMelt™ assay buffer: 20 mM HEPES (pH 7.5), 100 mM NaCl, and 5% glycerol. Weigh 0.48 g HEPES (238.3 g/mol), 0.6 g NaCl (58.44 g/mol), and 5.0 g glycerol (92.1 g/mol), and dilute in 100 mL of water (*see* **Notes 1** and **6**). Adjust pH to 7.5 using NaOH.
2. KRAS protein (*see* **Note 2**).
3. GloMelt™ dye: 200X stock (*see* **Note** 7).
4. Ras inhibitors, activators, or interactor molecules (*see* **Note 4**).
5. Multi-well plate (*see* **Note 5**).

#### 2.1.3 Protein Thermal Shift™ Dye Kit

1. Protein Thermal Shift™ buffer and the selected assay buffer (*see* **Note 8**).
2. KRAS protein (*see* **Note 2**).
3. Protein Thermal Shift™ dye: 1000X stock (*see* **Note 9**).
4. KRAS inhibitors, activators or interactor molecules (*see* **Note 4**).
5. Multi-well plate (*see* **Note 5**).

### 2.2 Alternative TSA Dyes for KRAS Stability Monitoring

#### 2.2.1 1-Anilinonaphthalene-8-Sulfonate (ANS)

1. ANS assay buffer: 20 mM HEPES (pH 7.5), 20 mM NaCl, and 0.001% Triton X-100. Weigh 0.48 g HEPES (238.3 g/mol), 0.12 g NaCl (58.44 g/mol), and 1 mg Triton X-100, and dilute in 100 mL of water (*see* **Notes 1** and **10**). Adjust pH to 7.5 using NaOH.
2. KRAS protein (*see* **Note 2**).
3. ANS dye: 5 mM stock in DMSO (*see* **Note 11**).
4. KRAS inhibitors, activators, or interactor molecules (*see* **Note 4**).
5. Multi-well plate (*see* **Note 5**).

#### 2.2.2 Nile Red

1. Nile red assay buffer: 20 mM HEPES (pH 7.5), 20 mM NaCl, and 0.001% Triton X-100. Weigh 0.48 g HEPES (238.3 g/mol), 0.12 g NaCl (58.44 g/mol), and 1 mg Triton X-100,

and dilute in 100 mL of water (*see* **Notes 1** and **10**). Adjust pH to 7.5 using NaOH.

2. KRAS protein (*see* **Notes 2** and **10**).
3. Nile red dye: 1 mM stock in DMSO (*see* **Note 12**).
4. KRAS inhibitors, activators, or interactor molecules (*see* **Note 4**).
5. Multi-well plate (*see* **Note 5**).

### 2.3 Noncommercial TSA Methods for KRAS Stability Monitoring

#### 2.3.1 Protein-Probe

1. Protein-probe assay buffer: 10 mM HEPES (pH 7.5), 20 mM NaCl, and 0.001% Triton X-100. Weigh 0.24 g HEPES (238.3 g/mol), 0.12 g NaCl (58.44 g/mol), and 1 mg Triton X-100, and dilute in 100 mL of water (*see* **Note 13**). Adjust pH to 7.5 using NaOH.
2. Protein-probe detection buffer: 7.7 mM $Na_2HPO_4$, 6.1 mM citric acid (pH 4), 0.01% Triton X-100. Weigh 0.11 g $Na_2HPO_4$ (141.96 g/mol), 0.12 g citric acid (192.12 g/mol), and 10 mg Triton X-100, and dilute in 100 mL of water (*see* **Note 14**).
3. Peptide-probe assay buffer: 10 mM HEPES (pH 7.5), 10 mM NaCl, 0.01% Triton X-100. Weigh 0.24 g HEPES (238.3 g/mol), 0.06 g NaCl (58.44 g/mol), and 10 mg Triton X-100, and dilute in 100 mL of water (*see* **Note 15**). Adjust pH to 7.5 using NaOH.
4. FRET-probe assay buffer: 10 mM HEPES (pH 7.5) and 0.001% Triton X-100. Weigh 0.24 g HEPES (238.3 g/mol) and 1 mg Triton X-100, and dilute in 100 mL of water (*see* **Note 16**). Adjust pH to 7.5 using NaOH.
5. KRAS protein (*see* **Note 17**).
6. $Eu^{3+}$-probe and modulators or FRET-probe (*see* **Note 18**).
7. KRAS inhibitors, activators, or interactor molecules (*see* **Note 4**).
8. Multi-well plate (*see* **Note 5**).

#### 2.3.2 C-Probe Dye

1. C-probe assay buffer and the selected assay buffer (*see* **Notes 8** and **19**).
2. KRAS protein (*see* **Note 19**).
3. C-probe dye (*see* **Note 20**).
4. KRAS inhibitors, activators, or interactor molecules (*see* **Note 4**).
5. Multi-well plate (*see* **Note 5**).

## 3 Methods

Carry out all procedures at room temperature (RT) unless otherwise specified. KRAS and other possible proteins should be kept on ice at all times before addition to plate. Luminophores should be protected from light at all times, and all reactions should be sealed during thermal cycling unless otherwise indicated. All concentrations given are calculated to final 20 μL volume unless otherwise specified (*see* **Note 21**). All concentrations given are indicative depending on the KRAS construct, mutant, and batch used.

### 3.1 Commercial TSA Methods for KRAS Stability Monitoring

#### 3.1.1 SYPRO Orange™

1. Calculate the volume needed, and add 10% extra volume of all KRAS proteins and stabilizing molecule prepared (*see* **Note 22**). Prepare all assay components in the selected assay buffer, by starting from the most stable component. KRAS should be kept on ice at all times before addition to the plate or reaction in tube (*see* **Note 23**). KRAS stability reaction preincubation can be performed either in plate or tube in 18 μL, and the detection is added in 2 μL in plate (*see* **Note 24**). SYPRO Orange™ is added from 10X stock calculated from the selected label concentration (*see* **Note 24**). A simplified presentation of the KRAS thermal stability assay using external fluorescent dye is depicted in Fig. 1.
2. Add 2 μL of KRAS inhibitor, activator, or interactor molecules first to the plate (*see* **Note 25**).
3. Add 16 μL of KRAS protein (5 μM) to the plate and incubate for 15 min (*see* **Note 26**).
4. Add 2 μL of 10X SYPRO Orange™ to the PCR-plate, perform programmed heating cycle, and monitor the signal (*see* **Notes 3** and **27**).

#### 3.1.2 GloMelt™

1. Calculate the volume needed, and add 10% extra volume of all KRAS proteins and stabilizing molecules prepared (*see* **Note 22**). Prepare all assay components in the selected assay buffer, by starting from the inhibitor to keep KRAS on ice as long as possible (*see* **Note 23**). KRAS stability reaction preincubation can be performed either in plate or tube in 18 μL, and the detection is added in 2 μL as a final step in plate (*see* **Note 28**). Glomelt™ is added from 10X stock from the selected label concentration (*see* **Note 28**). Expected results from KRAS thermal stability assay using SYPRO Orange™ and GloMelt™ is depicted in Fig. 2, to highlight detergent sensitivity of these dyes.
2. Add 2 μL of KRAS inhibitor, activator, or interactor molecules first to the plate (*see* **Note 25**).

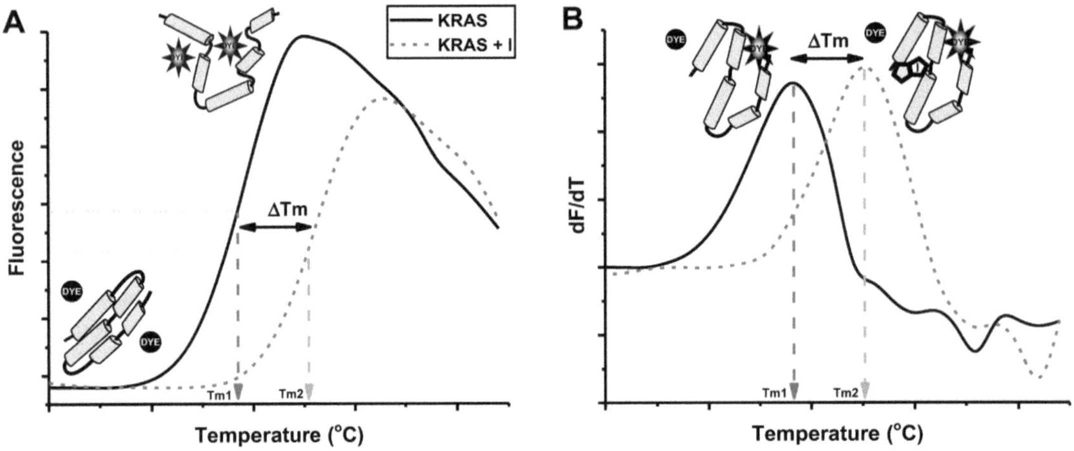


**Fig. 1** KRAS thermal stability assay using external fluorescent dye. In TSA, studied protein (here KRAS) is heated together with an external dye molecule (here SYPRO Orange™) in the absence or presence of interacting molecule (I). (**a**) Increase in temperature causes KRAS structure unfolding, enabling sensor dye binding and increase in fluorescence. Protein-ligand interaction (PLI) can be monitored as a shift in melting temperature (ΔTm), corresponding temperature in which half of the KRAS population is unfolded. Data observed from the measurement is fluorescence versus temperature. (**b**) Data can be additionally blotted as a first derivative (slope) of the fluorescence curve (dF/dT) versus temperature. This presentation can be used to visualize the data, showing the Tm as a peak value and ΔTm as a distance between the peaks. Presentation does not change the Tm value monitored, but only simplifies the visual presentation

3. Add 16 μL of KRAS protein (5 μM) to the plate and incubate for 15 min (*see* **Note 26**). SYPRO Orange™ and GloMelt™ results showing the importance of KRAS construct selection in are depicted in Fig. 3.
4. Add 2 μL of 10X GloMelt™ to the plate, perform heating cycle, and monitor the signal (*see* **Notes 7**, **27**, and **29**).

*3.1.3 Protein Thermal Shift™ Dye Kit*

1. Calculate the volume needed with 10% extra volume for all assay components (*see* **Note 22**). Prepare all assay components in the selected/provided assay buffer, keeping KRAS on ice as long as possible (*see* **Note 23**). KRAS stability reaction can be performed either in plate or tube in 12.5 μL, and the assay buffer and detection dye are added in 5 μL and 2.5 μL as a final step in plate, respectively (*see* **Note 30**). Protein Thermal Shift™ Dye is added from 8X stock to achieve the selected label concentration at the final volume (*see* **Note 30**).
2. Add 2.5 μL of KRAS inhibitor, activator, or interactor molecules first to the plate (*see* **Note 25**).
3. Add 10 μL of KRAS protein (10 μM) to the plate and incubate for 15 min (*see* **Note 26**).

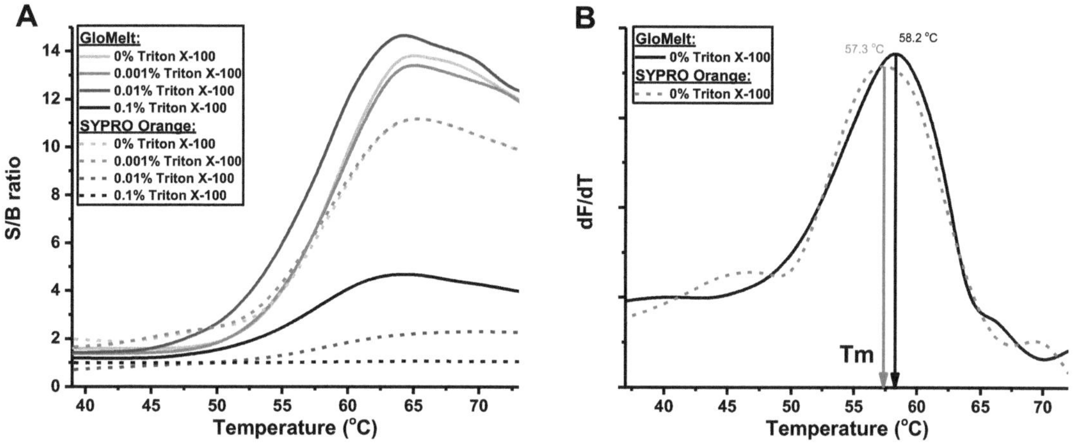


**Fig. 2** Detergent concentration effect on SYPRO Orange™ and GloMelt™. (**a**) Detergent effect varies between assay methods, which need to be considered when selecting the assay. Different Triton X-100 tolerability is here visualized using GloMelt™ and SYPRO Orange™. Using 10 μM of KRAS(WT), GloMelt™ can be used in the presence of 0.1% Triton X-100, as SYPRO Orange™ cannot be used even in the presence of 0.01% Triton X-100 concentration. In both cases, the background signal has a tendency to increase with higher detergent concentrations, but in the case of GloMelt™, it does not dramatically change the assay function. (**b**) Alternative data presentation to visualize the melting temperature (Tm) and thermal shift (ΔTm). TSA data can be additionally blotted as a first derivative (slope) of the fluorescence curve (dF/dT) versus temperature. Using this presentation, which some of the qPCR equipment can calculate directly, small changes in Tm are easier to visualize. This makes comparison of data simpler in comparison to actual melting curves. As seen with GloMelt™ and SYPRO Orange™, TSA dyes often gives similar Tm values, when assayed in the same conditions and concentrations

4. Add 5 μL of Protein Thermal Shift™ Buffer and 2.5 μL of Protein Thermal Shift™ Dye to the plate, perform heating cycle, and monitor the signal (*see* **Notes 9**, **27**, and **31**).

### *3.2 Alternative TSA Dyes for KRAS Stability Monitoring*

#### *3.2.1 1-Anilinonaphthalene-8-sulfonate (ANS)*

1. Calculate the volume needed with 10% extra volume for all assay components, and prepare the components (*see* **Notes 22** and **23**). KRAS stability reaction can be performed either in plate or tube in 18 μL, and the detection is added in 2 μL as a final step in plate (*see* **Note 32**). ANS dye is added from the prepared stock to achieve selected final concentration (*see* **Note 32**).
2. Add 2 μL of KRAS inhibitor, activator, or interactor molecules first to the plate (*see* **Note 25**).
3. Add 16 μL of KRAS protein (5 μM) to the plate and incubate for 15 min (*see* **Notes 26** and **32**).
4. Add 2 μL of 10 μM ANS to the plate, perform heating cycle, and monitor the signal (*see* **Notes 11**, **27**, and **32**).

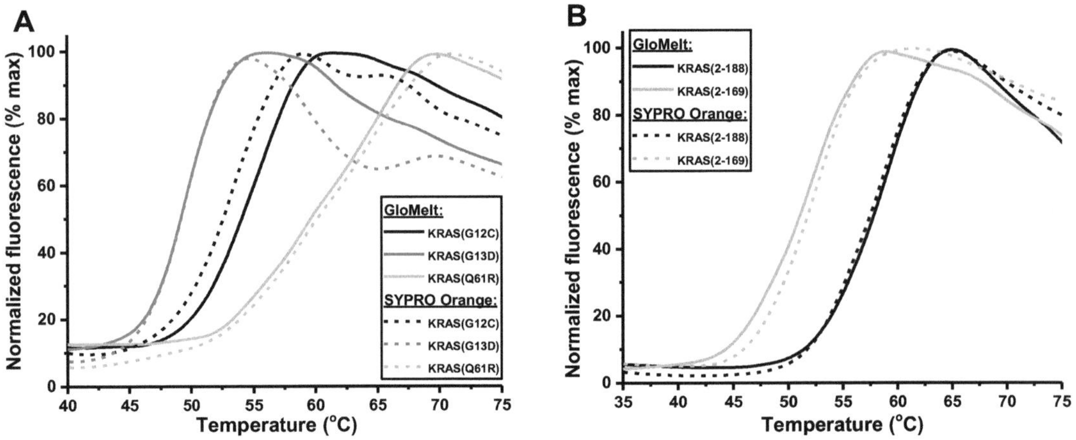


**Fig. 3** KRAS construct and mutant comparison using SYPRO Orange™ and GloMelt™. Assay conditions, as the used KRAS protein, buffer, and heating rate, changes the observed Tm value and detectability. (**a**) KRAS (G13D), KRAS(G12C), and KRAS(Q61R) all have different Tm value (49.5 °C, 54.0 °C, and 61.7 °C) when assayed in a same buffer (20 mM HEPES (pH 7.5), 100 mM NaCl, 5% glycerol). Even no $MgCl_2$ is added in the buffer, the final $Mg^{2+}$ concentration is still 20–27 μM originated from the KRAS storage buffer. This $Mg^{2+}$ concentration is already sufficient to stabilize the used 5 μM of KRAS. However, changes in KRAS concentration at these conditions will also change the observed Tm, as $Mg^{2+}$ concentration in the final reaction will change. Thus, to equalize the results, one must consider the dilution factor and KRAS stock $MgCl_2$ concentration or run the assay in buffer supplemented with $MgCl_2$ (0.5–2 mM). (**b**) KRAS construct length also has on effect on the observed thermal stability. Typically, KRAS assays are performed either with full-length KRAS (2-188) or truncated version (2-169) having high variable tail region removed. Shorter KRAS constructs are less stable than the full-length KRAS, as shown with 10 μM of KRAS proteins. In addition, the detectability of these constructs might be different depending on the external TSA dye used. In most of the cases, full-length KRAS is preferred

*3.2.2 Nile Red*

1. Calculate the volume needed with 10% extra volume for all assay components, and prepare the components (*see* **Notes 22** and **23**). KRAS stability reaction can be performed either in plate or tube in 18 μL, and the detection is added in 2 μL as a final step in plate (*see* **Note 33**). Nile red is added from the prepared stock to achieve selected final concentration (*see* **Note 33**). Measured fluorescent spectra for Nile red and two other TSA dyes is depicted in Fig. 4, to highlight the importance of the emission and excitation filter selection.
2. Add 2 μL of KRAS inhibitor, activator, or interactor molecules first to the plate (*see* **Note 25**).
3. Add 16 μL of KRAS protein (10 μM) to the plate and incubate for 15 min (*see* **Notes 26** and **33**).
4. Add 2 μL of 10 μM Nile red to the plate, perform heating cycle, and monitor the signal (*see* **Notes 12**, **27**, and **33**).

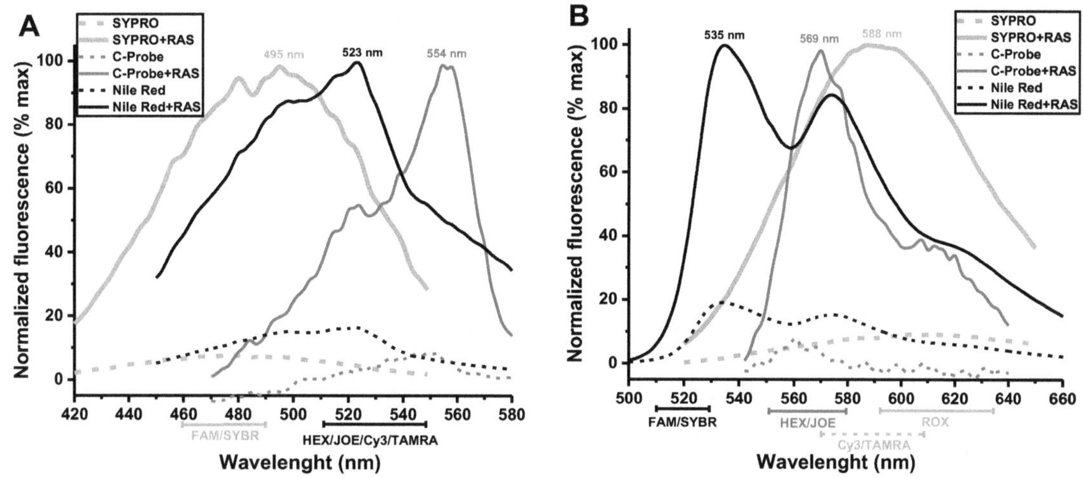


**Fig. 4** Excitation and emission spectra for selected TSA dyes. In the case of the most TSA dyes, excitation and emission maximums are well separated, simplifying the measurement even with suboptimal qPCR filter setting. In aqueous solutions, most of the dyes possess low fluorescence signal (dashed lines) alone in comparison when bound to unfolded KRAS (heated for 3 min at 60 °C). Spectral measurements are performed with 5X SYPRO Orange™ (10 μM of KRAS), 1 μM Nile red (10 μM of KRAS), and 0.17X C-probe (200 nM of KRAS). (**a**) For most of the external dyes, excitation spectra is broad, and excitation maximum is not dependent on dye interaction with its target. As an exception, C-probe excitation spectra is more sharp (dark gray) and GloMelt™ excitation maximum red shifts approximately 25 nm (not shown). (**b**) Emission maximum is more often affected when dye interacts with unfolded KRAS. The change in maximum can be either blue-shifted like in the case of SYPRO Orange™ (~20 nm) or red-shifted like with C-probe (~10 nm). These small spectral changes can affect the assay outcome in the case of suboptimal measurement filter sets in qPCR. Moreover, fluorescence sensitivity differences are significant when qPCR equipments are compared, which has dramatic effect on TSA functionality and outcome

### *3.3 Noncommercial TSA Methods for KRAS Stability Monitoring*

#### *3.3.1 Protein-Probe*

1. Calculate the volume needed with 10% extra volume for all assay components, and prepare the components (*see* **Notes 22**, **23**, and **34**). KRAS stability reaction can be performed either in plate or tube in 8 μL, and the detection is added in 12 μL (peptide-probe and FRET-probe) or 65 μL (protein-probe) as a final step in plate (*see* **Notes 34** and **35**). Freshly prepared detection solution is added to achieve selected final concentration (*see* **Notes 35** and **36**).
2. Add 2 μL of KRAS inhibitor, activator, or interactor molecules first to the plate (*see* **Note 25**).
3. Add 6 μL of KRAS protein (25–500 nM) to the plate and incubate for 15 min (*see* **Notes 26** and **37**).
4. Add 12 μL (peptide-probe and FRET-probe) or 65 μL (protein-probe) detection solution to the plate, perform heating cycle, and monitor the signal every step (*see* **Notes 18**, **27**, **36**, and **38**).

*3.3.2 C-Probe*

1. Calculate the volume needed with 10% extra volume for all assay components, and prepare the components (*see* **Notes 22** and **23**). KRAS stability reaction can be performed either in plate or tube in 25 μL, and the detection is added in 5 μL as a final step in plate (*see* **Note 39**). C-probe is added from the prepared stock (1X) to achieve selected final concentration (*see* **Note 39**). Expected TSA results from a KRAS PLI measurement, using C-probe and FRET-probe, are depicted in Fig. 5.
2. Add 5 μL of KRAS inhibitor, activator, or interactor molecules first to the plate (*see* **Note 25**).
3. Add 20 μL of KRAS protein (200 nM) to the plate and incubate for 15 min (*see* **Notes 26** and **40**).
4. Add 5 μL C-probe (1X stock) to the plate, perform heating cycle, and monitor the signal (*see* **Notes 20**, **27**, and **41**).

## 4 Notes

1. SYPRO assay buffer is selected to enable error free functionality in TSA. SYPRO Orange™ and similar types of dyes are sensitive to high detergent and protein preservative concentrations, which need to be kept in mind during assay optimization (*see* Fig. 2). The buffer can be supplemented with low detergent concentration (e.g., 0.001% Triton X-100) to lower the signal obtained at room temperature (RT). Additionally, assay buffer without glycerol is applicable, but the KRAS stability and Tm values are lower.
2. Non-modified KRAS is needed in high stock concentration. KRAS can be either full-length (FL) or truncated, but the form will affect the thermal stability (*see* Fig. 3). In addition, different KRAS mutants have varying stability. High-concentration stock is required to enable thermal stability assay with the external labels, without major KRAS storage buffer carryover. Comparison between KRAS constructs having different stock concentration is challenging, especially due to $MgCl_2$ carryover.
3. There are a high number of commercial assays utilizing external probes. SYPRO Orange™ is widely used as it enables the use of qPCR equipment (Ex/Em 485/590 nm) (*see* Fig. 4). SYPRO Orange™ usually comes in 5000X concentrate (stored at RT) in DMSO (e.g., S5692, Merck, SYPRO Orange™ protein gel stain or S6650, ThermoFisher, SYPRO™ protein gel stains), but not all products are equal. Multiple structural SYPRO Orange™ variants having slightly different binding and signaling properties exist.

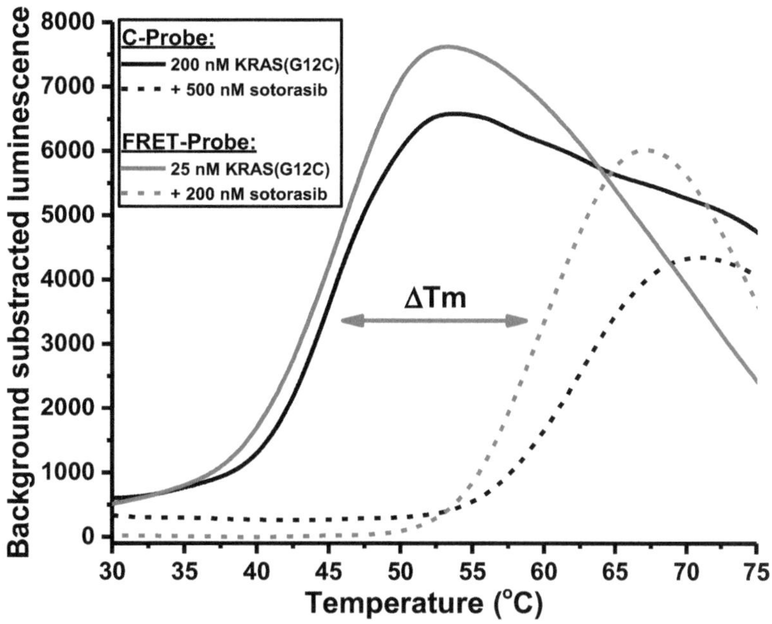


**Fig. 5** TSA method for protein-ligand interaction (PLI) studies. TSA is often used to study interactions, and in the case of KRAS, especially PLI assays for screening purposes are the most common. For these purposes, the selection of TSA method depends on the wanted outcome. TSA usually can give information about binding, but not about the affinity of the studied compound below micromolar (μM) concentration. With the most common methods, KRAS is used at 5–20 μM concentrations, limiting its usefulness with inhibitors having better affinity than the used KRAS concentration. In the case of KRAS, high-affinity molecules are often acting by forming covalent bond with KRAS mutant, like KRAS(G12C). To study these interactions, higher-affinity TSA dyes are needed, which on the other hand also lowers the material consumption enabling to run larger sample panels. C-probe (black) and FRET-probe (dark gray) can enable PLI assays with KRAS at nanomolar (nM) concentrations, enabling the use of lower inhibitor concentrations. These methods can better rank the potential high-affinity binders as shown using PLI assay with sotorasib, monitored using the C-probe (ΔTm 16.9) and FRET-probe (ΔTm 14.7)

4. Small molecules affecting KRAS activity should be prepared in 100% DMSO to ensure the solubility and functionality. In the assay, DMSO in all wells should be adjusted to the same concentration, usually not more than 1–2%, to enable direct comparison between treated and non-treated KRAS. Too high a DMSO concentration will have a negative effect on assay performance; possibly leading to KRAS lowered stability or even denaturation.
5. The assay should be performed in black or white assay plates suitable for qPCR measurements or similar temperature-cycling device. Three eighty-four-well PCR plates

(e.g., Hard-Shell® 384-well PCR plate, BioRad) are preferred with larger sample panels, but assay can be equally performed in a 96-well plate (e.g., Hard-Shell® 96-well PCR plate, BioRad). There are no major differences between the PCR-plate options from different manufacturers, but some plate surfaces might lower the assay sensitivity. In addition, white plates increase the signal observed in comparison to black ones. Clear plates are incompatible for fluorescence measurements.

6. GloMelt™ (Biotium) assay buffer selection is less critical as with SYPRO Orange™, as it is not as sensitive to high detergent or protein preservative concentrations (*see* Fig. 2). This enables the use of different types of buffers, which simplifies assay optimization.
7. GloMelt™ is applicable for measurements performed with qPCR equipment (Ex/Em 468/507 nm). GloMelt™ Dye come in 200X stock (stored at −20 °C) and is used in 1X final concentration. GloMelt™ Dye is possible to be used with ROX reference dye (Ex/Em 575/600 nm) to improve reproducibility and replicate consistency.
8. The Protein Thermal Shift™ kit comes with a buffer suitable for the assay. In addition, you can add your selected buffer with additional components important for the assay, for example, $MgCl_2$.
9. The Protein Thermal Shift™ assay can be used with qPCR equipment suitable for Ex/Em 580/620 nm measurements. Dye comes in 1000X stock and is used in 1X final concentration.
10. Similar to SYPRO Orange™, ANS is sensitive for high detergent and protein preservative concentrations. However, low detergent concentration simplifies pipetting, lowers nonspecific binding, and improves the assay quality.
11. ANS (1,8-ANS, 1-anilinonaphthalene-8-sulfonic acid, for example, Thermo Fisher, A47) cannot be used with most common qPCR equipment due to its low wavelength excitation (Ex/Em 360/470 nm). Blue range can be selected from some equipment, for example, Rotor-Gene Q, Qiagen (Ex 365 ± 20 nm). 4,4'-Dianilino-1,1′-binaphthyl-5,5-′-disulfonic acid (Bis-ANS, Thermo Fisher, B153) can be used similarly with improved functionality. Solid ANS can be purchased, and stock solution can be prepared in Milli Q-water or DMSO (e.g., 5 mM stock, weight 15.8 mg ANS and solubilize in 10 mL of Milli-Q water).
12. Nile red (e.g., Thermo Fisher, N1142) can be used with qPCR equipment (Ex/Em 540/640 nm) (*see* Fig. 4). Solid Nile red can be purchased, and stock solution can be prepared in

DMSO (e.g., 1 mM stock, weight 3.18 mg Nile Red and solubilize in 10 mL of DMSO).

13. The assay is sensitive to high salt concentrations, but low detergent concentration are preferred to simplify pipetting, lower nonspecific binding, and improve the assay quality. Buffer can be supplemented with $MgCl_2$ or other relevant components, but strong buffer solutions with high pH should be avoided.
14. The protein-probe detection is functional only at low pH, and thus the assay buffer cannot contain, for example, strong buffer with high pH. The assay is always performed in two steps, and thus assay is not highly suitable for melting curves. Detection buffer is always prepared fresh, as Triton X-100 containing low pH buffer cannot be stored. Also detection components, $Eu^{3+}$-probe and HIDC, are freshly mixed prior the assay.
15. The peptide-probe prefers low salt concentration buffer with added detergent. Detergents other than Triton X-100 can be used, for example, Brij 30, and buffer can be supplemented with relevant buffer components as $MgCl_2$.
16. Similar to protein-probe and peptide-probe buffers, the FRET-probe prefers buffer having relatively low salt concentration and low detergent concentration. All assay components are prepared in the same buffer.
17. FL KRAS construct (2–188) is preferred to be used to obtain full low nanomolar sensitivity of the protein-probe and FRET-probe assays. Peptide-probe has no KRAS length preference, but the sensitivity with KRAS is at mid to high nanomolar level.
18. All three systems come with a different detection component setting. Protein-probe is not optimal for thermal curve measurement, but for single temperature assays. In the assay, detection components ($Eu^{3+}$-probe, stored at −20 °C, and HIDC iodide) are added in special low-pH modulation buffer added after KRAS heating. Peptide-probe ($Eu^{3+}$-probe and Q14) and FRET-probe (dual $Eu^{3+}$-probe-Cy5, stored at −20 °C) can be used in a normal real-time TSA. Assays are measured using time-resolved $Eu^{3+}$-setting (ex/em 340/620 nm, protein-probe and peptide-probe) or TR-FRET setting (ex/em 340/665 nm, FRET-probe).
19. Similar to Protein Thermal Shift™ kit, C-probe beta kit (QRET Technologies) comes with selected assay buffer. In the case of the C-probe beta kit, two preferred assay buffer +/− $MgCl_2$ are provided at a 5X concentration.
20. C-probe (stored at −20 °C) is compatible with qPCR equipments with filters for Cy3 (ex/em 535/575 nm). The C-probe

beta kit (QRET Technologies) is not openly available, but has been provided for several laboratories for testing.

21. The total volume of 20 μL is the lowest recommended for real-time TSA. In most cases, assay volume can be freely selected between 20 and 100 μL without a major effect on results. However, volumes below 20 μL are prone to error especially at the highest temperatures due to evaporation. Error can be minimized by using appropriate plate, sealing, and higher volume.
22. The volumes needed for each assay component are calculated with 10% extra solution to enable pipetting. The studied ligand is added first in 2 μL volume prior KRAS addition in 16 μL. Example: The need for 1.25 × KRAS solution for 100 individual assay points is 1760 μL, assuming the need for 10% extra solution for pipetting. If KRAS concentration in well (20 μL) is 10 μM, the concentration in 16 μL of added 1.25 × KRAS working solution is 12.5 μM. Assuming that the KRAS stock concentration is 10 mg/ml (~430 μM), 51.2 μL of KRAS stock should be mixed with 1709 μL of the selected assay buffer, to enable the assay with 10 μM of KRAS in final volume. All other assay components should be calculated and prepared similarly.
23. KRAS and possible other studied proteins are stored at −80 °C in aliquots appropriate for a single assay. KRAS aliquots can withstand approximately three freeze-thaw cycles without losing its activity and structural properties. KRAS can be significantly more sensitive to a number of freeze-thaw cycles when stored in nonoptimal buffer conditions, without nucleotide loading or with nucleotide analogs. In addition, KRAS mutational status might affect the stability and optimal storage conditions. During the assay component preparation, KRAS should be stored on ice at all times. Depending on assay buffer, KRAS should be ideally used within 60 min of its preparation in assay buffer. However, in some cases, 30 min or longer pre-incubation (KRAS + compound) is needed, and thermal cycling can take up to 60 min. In these cases, KRAS stability must be studied before the main assay. Especially in buffers with detergent, for example, Triton X-100 or buffer without $MgCl_2$, KRAS stability is significantly reduced already after 60 min. In thermal stability assays, KRAS is often used in approximately 5–10 μM concentration, which increases its stability. In all cases, incubations at RT for longer than 60 min should be avoided, or the effect needs to be carefully studied beforehand.
24. SYPRO Orange™ is a widely used dye due to its low price and compatibility with most of the qPCR equipment (*see* Fig. 4). However, not all qPCR equipment have similar sensitivity,

which might affect the selected dye concentration. SYPRO Orange™ method are based on fluorescence signal increase upon dye binding to the hydrophobic regions exposed to solution after protein unfolding. Thus, depending on the buffer, additional components, and the equipment used for signal readout, the optimal SYPRO Orange™ concentration might vary. Often, 5X SYPRO Orange™ (1–5X) in final volume is the most suitable for KRAS.

25. External probes/dyes for TSA are environment sensitive, and thus changes in buffer composition between and within the assays should be minimized. KRAS and its binders should all be added in the same buffer, and nonspecific binding should be minimized with appropriate plates suitable for high temperatures (*see* **Note 5**). In addition, some buffer components, for example, detergents, have different effects on different TSA dyes, and thus simplified HEPES-based assay buffers are preferred. Small molecule inhibitors are often stored in DMSO, and in titrations and other assays, the DMSO concentration must be equalized (*see* **Note 4**).
26. The sensitivity for TSA dyes is relatively low, and high KRAS stock concentration is needed to avoid artifacts, especially when different KRAS concentrations are compared (*see* **Note 2**). KRAS storage buffers often contains components, which may change the thermal denaturation curve, for example, GDP or $MgCl_2$. Thus, sufficiently high dilution to assay buffer or buffer exchange can ensure comparable results if KRAS proteins are from different sources and/or different buffer composition are compared. A preincubation step prior to detection solution addition is also often needed to produce optimal results. With non-covalent KRAS binders, 15 min incubation time is sufficient. However, 30 min or longer preincubation time might be needed for covalent binders, and longer incubations should be performed at +4 °C.
27. KRAS TSAs can be performed with one- or two-step protocol and with or without automated measurement and heating (qPCR). In all cases the plate should be sealed with a compatible optical film and spun (e.g., 1000 rpm for 1 min) after final component addition. In a one-step protocol, the TSA dye is added before the heating cycle, and all cycles are performed using the same reaction. This enables the use of qPCR equipment with compatible TSA dye and continuous measurement during the heating cycle. The use of qPCR enables constant heating with the selected speed, for example, gradient from 20 to 100 °C with speed 2 °C/min and signal monitoring every 0.2 min. With a separate plate reader and thermal cycler, one cannot use a gradient, but the heating is performed in stepwise manner, for example, from 20 to 100 °C with 2 °C

steps each for 1.5 min before signal measurement. Plate readers are significantly more sensitive compared to qPCR, but temperature change during long measurements (over 48 wells) might affect the result. Selected heating speed or the method always affects the melting temperature monitored, and thus assays performed with or without qPCR cannot be directly compared. In a two-step protocol, the TSA dye is added after heating cycle and signal is monitored in end-point fashion. This is not commonly used for thermal curves due to the high material consumption, but enables single temperature assays. KRAS concentration applicable for the assay is dye specific and also dependent on concentration of the TSA dye, buffer composition, and measurement device. In the case of KRAS, 3–15 μM protein concentration and 1–5X SYPRO Orange™ concentration at the final volume are sufficient (*see* **Note 3**). Depending on thermocycler used, measurement should be performed using excitation at 485 and emission at 590 nm or by using FAM excitation filter and ROX emission.

28. Glomelt™ is less sensitive to buffer compounds like detergents than, for example, SYPRO Orange™. GloMelt™ dye comes in 200X stock, and optimal final concentration for KRAS is 1X in final 20 μL volume (*see* Fig. 2). The assay is preferred to be performed with a passive ROX reference dye to improve the assay quality. ROX reference dye (stock 40 μM) can be purchased with the GloMelt™ Thermal Shift Protein Stability Kit (Biotium). Needed ROX concentration is qPCR equipment dependent and must be confirmed from the kit manual or experimentally with the used qPCR device.

29. With the Glomelt™, thermal ramping rates similar to SYPRO Orange™ (0.01–0.05 °C/s) are commonly used. Depending on thermocycler, the measurement should be performed at the blue channel or green channel (Ex/Em 470/510 nm), or the reporter should be set to SYBR® Green or FAM™.

30. The Protein Thermal Shift™ dye comes in a 1000X stock, from which 8X solution is prepared to obtain 1X final concentration optimal for KRAS.

31. The Protein Thermal Shift™ kit provides a buffer enabling correct functionality for the dye, but also the selected assay buffer of interest can be used as a supplement. A thermal ramping rate similar to Glomelt™ and SYPRO Orange™ is optimal also for the Protein Thermal Shift™ dye. Depending on the thermocycler used, measurements should be performed using excitation at 580 nm and emission at 620 nm.

32. ANS dye has slightly improved sensitivity in comparison to SYPRO Orange™ when used in 1 μM of final concentration. For that, prepare 10 μM of ANS solution in an assay buffer to

obtain 1 μM of final concentration in plate. Optimal KRAS concentration also in the case of ANS is 5 μM. ANS is not compatible for most common qPCR equipment, which makes it less used in comparison to, for example, SYPRO Orange™.

33. Nile red dye can be used in 1 μM of solution in final volume with 10 μM of KRAS. Nile red is compatible for most common qPCR equipment (Ex/Em 540/640 nm) (*see* Fig. 4).
34. There are three versions of the protein-probe family, from which two are well suitable for TSA thermal curves [4, 13–16]. The original protein-probe can be used only with the two-step protocol, making it less suitable for thermal ramping [4, 13, 14]. Peptide-probe and FRET-probe can be used for real-time TSA at neutral pH [15, 16].
35. The final volume in the peptide-probe and FRET-probe is 20 μL, and concentrations are calculated in that volume. Detection and reaction buffer can be the same, so also addition volume can be freely selected. In protein-probe, the reaction volume is 8 μL in which all the reaction components are calculated, making detection solution (65 μL) separate part from reaction.
36. The protein-probe detection solution contains two sensing element, $Eu^{3+}$-probe and HIDC, and it is added in a special modulation solution with a low pH (*see* **Note 14**) [4, 13, 14]. For the detection, it is essential that the buffer is fresh and Triton X-100, HIDC, and $Eu^{3+}$-probe carefully mixed with low pH citrate/phosphate buffer 5–30 min before use. The peptide-probe also contains two detection components, $Eu^{3+}$-probe and Q14 peptide modulator, which are mixed before using in the used assay buffer [15]. The FRET-probe is a single added component added directly in assay buffer [16].
37. FL KRAS protein concentration in a protein-probe assay is optimally 50 nM (in 8 μL), but it can be performed with even 10 nM or additionally using 1 μM of KRAS concentrations. Low KRAS concentrations are optimal for high-affinity covalent inhibitors as high-protein concentration is more robust for low affinity and difficult binders and comparative studies with HRAS and NRAS [4, 13]. Optimal KRAS concentration for the peptide-probe is 500 nM (in 20 μL), and especially in the case of truncated KRAS forms, even lower concentration can be used [15]. For the FRET-probe, 25 nM of FL KRAS (in 20 μL) is optimal for TSA, and it also can be used with chemical KRAS denaturation using 1–2 M of Urea [16]. The FRET-probe is highly suitable for nanomolar affinity covalent binders (*see* Fig. 5).
38. The protein-probe family assays are not suitable for qPCR equipment due to their TRF-signal detection, which in the

case of protein-probe and peptide-probe is directly through the $Eu^{3+}$-channel (ex/em 340/620 nm). TR-FRET signal (ex/em 340/665 nm) is used for FRET-probe detection. In the case of peptide-probe and FRET-probe, thermal ramping can be performed using conventional PCR-equipment for heating and standard TRF compatible plate reader for detection. For this reason the heating is performed in a stepwise manner (e.g., 1.5 min every 2 °C), which affects the Tm observed in comparison to assays performed using qPCR and constant heating ramp.

39. C-Probe Thermal Stability Assay Kit is a beta test phase TSA method (QRET Technologies) and highly suitable for KRAS. The assay kit is provided with appropriate assay buffers with or without $MgCl_2$. Hard-Shell® 96-well PCR plate white (HSP9655, BioRad) is recommended by the manufacturer, but the kit is functional also in 384-well plate format. A black plate is not recommended with low-dye concentration due to low signal level. The assay is performed in 30 μL of final volume using 0.17X final C-probe concentration optimal for FL KRAS. C-probe kit is compatible for most common qPCR equipment with Cy3/TAMRA channel (Ex/Em 535/575 nm) (*see* Fig. 4).
40. C-probe can be used in 0.17X final concentration by adding 5 μL of the prepared 1X dye solution (stock 100X). The assay is significantly more sensitive in comparison to SYPRO Orange™, and optimal concentration for FL KRAS is 200 nM.
41. The C-probe can be used with the most common qPCR equipment with Cy3/TAMRA option. Thermal ramping can be performed using thermal ramping rates similar to SYPRO Orange™ (20–80 °C). KRAS TSA with high thermal shift-induced molecules (e.g., covalent inhibitors) are preferred to be assayed in $Mg^{2+}$ free buffer (provided by the manufacturer) (*see* Fig. 5).

## Acknowledgments

This work was supported by the Academy of Finland grants 323433, 329012, and 353324.

### References

1. Wennerberg K, Rossman KL, Der CJ (2005) The Ras superfamily at a glance. J Cell Sci 118: 843–846. https://doi.org/10.1242/jcs.01660
2. Huang L, Guo Z, Wang F et al (2021) KRAS mutation: from undruggable to druggable in cancer. Signal Transduct Target Ther 6(1):386. https://doi.org/10.1038/s41392-021-00780-4
3. Hunter JC, Manandhar A, Carrasco MA et al (2015) Biochemical and structural analysis of common cancer-associated KRAS mutations.

Mol Cancer Res 13(9):1325–1335. https://doi.org/10.1158/1541-7786.MCR-15-0203
4. Kopra K, Valtonen S, Mahran R et al (2022) Thermal shift assay for small GTPase stability screening: evaluation and suitability. Int J Mol Sci 23(13):7095. https://doi.org/10.3390/ijms23137095
5. Eberth A, Ahmadian MR (2009) In vitro GEF and GAP assays. Curr Protoc Cell Biol 14:Unit 14.9. https://doi.org/10.1002/0471143030.cb1409s43
6. Rubio I, Prior I (eds) (2021) Ras activity and signaling. Humana New York, New York
7. Esposito D, Stephen AG, Turbyville TJ et al (2019) New weapons to penetrate the armor: novel reagents and assays developed at the NCI RAS initiative to enable discovery of RAS therapeutics. Semin Cancer Biol 54:174–182. https://doi.org/10.1016/j.semcancer.2018.02.006
8. Huynh K, Partch CL (2015) Analysis of protein stability and ligand interactions by thermal shift assay. Curr Protoc Protein Sci 79:28.9.1–28.9.14. https://doi.org/10.1002/0471140864.ps2809s79
9. Zhang R, Monsma F (2010) Fluorescence-based thermal shift assays. Curr Opin Drug Discov Devel 13(4):389–402
10. Lang BE, Cole KD (2017) Differential scanning calorimetry and fluorimetry measurements of monoclonal antibodies and reference proteins: effect of scanning rate and dye selection. Biotechnol Prog 33(3):677–686. https://doi.org/10.1002/btpr.2464
11. Greenfield NJ (2006) Using circular dichroism collected as a function of temperature to determine the thermodynamics of protein unfolding and binding interactions. Nat Protoc 1(6):2527–2535. https://doi.org/10.1038/nprot.2006.204
12. Langer A, Bartoschik T, Cehlar O et al (2022) A new spectral shift-based method to characterize molecular interactions. Assay Drug Dev Technol 20(2):83–94. https://doi.org/10.1089/adt.2021.133
13. Vuorinen E, Valtonen S, Eskonen V et al (2020) Sensitive label-free thermal stability assay for protein denaturation and protein-ligand interaction studies. Anal Chem 92(5):3512–3516. https://doi.org/10.1021/acs.analchem.9b05712
14. Valtonen S, Vuorinen E, Kariniemi T et al (2020) Nanomolar protein-protein interaction monitoring with a label-free protein-probe technique. Anal Chem 92(24):15781–15788. https://doi.org/10.1021/acs.analchem.0c02823
15. Malakoutikhah M, Mahran R, Gooran N et al (2023) Nanomolar protein thermal profiling with modified cyanine dyes. Anal Chem 95(50):18344–18351. https://doi.org/10.1021/acs.analchem.3c02844
16. Mahran R, Vello N, Komulainen A, Malakoutikhah M et al (2023) Isothermal chemical denaturation assay for monitoring protein stability and inhibitors interactions. Sci Rep 13(1):20066. https://doi.org/10.1038/s41598-023-46720-w

# Chapter 11

# MALDI-TOF Mass Spectrometry-Based Assay for Measuring Covalent Target Engagement of KRAS G12C Inhibitors

**Marcin Dyba, John-Paul Denson, and Anna E. Maciag**

## Abstract

MALDI-TOF mass spectrometry enables high-throughput screening of covalent fragment libraries and SAR compound progressions of selective KRAS G12C inhibitors. Using the MALDI-TOF platform instead of the more traditional ESI-MS TOF/orbitrap instrumentation can radically shorten sample acquisition time, allowing up to 384 samples to be screened in 30 min. The typical throughput for a covalent library screen is 1152 samples per 8 h, including processing, calculation, and reporting steps. The throughput can be doubled without any significant assay modification.

**Key words** MALDI-TOF MS, High-throughput screening, Covalent screening, Fragment-based screening, RAS covalent inhibitors, KRAS G12C, Sotorasib, Adagrasib

## 1 Introduction

Cysteine mutations at codon 12 of KRAS (G12C) represent the most frequent KRAS alteration in non-small cell lung cancer (NSCLC) and are also commonly observed in colorectal cancers. Covalent inhibitors have been developed to specifically react with the thiol group in the cysteine residue of the mutant KRAS G12C protein [1]. These compounds dock to the Switch II/Helix3 pocket in KRAS and covalently react with C12 in the GDP-bound state of KRAS G12C, trapping it in an inactive conformation [2, 3].

Covalent KRAS G12C inhibitors sotorasib (AMG510) and adagrasib (MRTX849) entered clinical trials in 2018 and 2019, respectively. Sotorasib was approved for use in KRAS G12C mutant lung adenocarcinoma in May 2021 and adagrasib followed in December 2022. These compounds demonstrate clinical benefit in lung cancer, with response rates of 30–40% [4]. Like earlier preclinical compounds, both sotorasib and adagrasib target the inactive (GDP)-bound protein.

Andrew G. Stephen and Dominic Esposito (eds.), *KRAS: Methods and Protocols*, Methods in Molecular Biology, vol. 2797, https://doi.org/10.1007/978-1-0716-3822-4_11,

Most patients with KRAS G12C mutant NSCLC experience clinical benefits from selective KRAS G12C-GDP inhibition, whereas patients with colorectal cancer bearing the same mutation rarely respond. The cause of this limited efficacy in colorectal cancers has been investigated. Unlike NSCLC, KRAS G12C colorectal cancers have high basal receptor tyrosine kinase (RTK) activation and are responsive to growth factor stimulation. In colorectal cancer lines, KRAS G12C inhibition induces higher phospho-ERK rebound than in NSCLC cells. This in turns induces the synthesis of new KRAS G12C that is in the GTP-bound conformation and is insensitive to KRAS G12C inactive state inhibitors [5]. Developing KRAS G12C inhibitors capable of binding and inhibiting both the active GTP-bound form of KRAS G12C and the inactive GDP-bound protein could have therapeutic advantage over KRAS G12C GDP-only inhibitors.

A variety of assays can be used to determine covalent engagement of small molecules to KRAS Cys 12. The most prevalent assays include NMR analysis, crystallography, and mass spectrometry [6–11]. Mass spectrometry-based assays have been proven to provide reliable and accurate compound engagement data with reasonable throughput and sample preparation requirements. Both top-down and bottom-up proteomics methods are used for in vivo binding screens, while simpler intact protein mass measurements can be used with in vitro covalent compound progression assays [3, 12, 13]. Small covalent fragment library screens and covalent compound structure-activity relationship (SAR) progression assays typically utilize liquid chromatography-mass spectrometry (LC-MS) approaches using time-of-flight (TOF) or orbitrap instruments. Although those methods can be optimized for high throughput, the usual instrument time is measured in minutes rather than seconds per sample. In contrast, the matrix-assisted laser desorption ionization time-of-flight mass spectrometry (MALDI-TOF)-based assay described below requires on average less than 5 s of instrument time per sample. Here, using MALDI-TOF MS, we compare commercial KRAS G12C inhibitors known to hit the inactive (GDP) form of KRAS, and the novel, dual GTP/GDP KRAS G12C-targeting compound developed in a collaboration with TheRas, Inc.

## 2 Materials

Prepare all solutions using Type 1 ultrapure water (prepared by purifying water, to attain a resistivity of 18.2 MΩ × cm at 25 °C) and high-purity or ultrapure grade reagents (mass spectrometry, HPLC, analytical, sequencing, or for molecular biology grade reagents). Prepare and store all reagents at room temperature, except the buffers which are stored in refrigerator at 4–6 °C. We do not add sodium azide to reagents. All protein stocks and solutions used are free from reducing agents—3,3′,3′′-phosphanetriyltripropanoic acid (TCEP), 2-sulfanylethan-1-ol (BME), and (2*S*,3*S*)-1,4-bis(sulfanyl)butane-2,3-diol (DTT).

### 2.1 MALDI-TOF Mass Spectrometry

1. Bruker rapifleX® Tissuetyper® MALDI-TOF/TOF.
2. Bruker MPT 384 ground steel BC target plate with Bruker MTP target frame III holder (*see* **Note 1**).
3. Bruker flexAnalysis (4.0 Build 14) software for data processing and exporting.
4. A flexAnalysis macro that extracts the peak information (m/z, peak area, and peak height) from the spectra into an Excel sheet (*see* **Note 2**) [14, 15].
5. Beckman Coulter Biomek® FX$^P$ Laboratory Automation Workstation with P30/384 head.
6. Alpillo® Plate Cushion (*see* **Note 3**).
7. 384 Well round or V-bottom polypropylene plates.
8. Target pretreatment solution—saturated (2*E*)-3-(4-hydroxy-3,5-dimethoxyphenyl)prop-2-enoic acid (sinapinic acid) in acetonitrile (ACN): Dissolve 1.1 g sinapinic acid in 100 mL ACN (*see* **Note 4**).
9. MALDI matrix solution—saturated sinapinic acid in 1:1 ACN: water, 0.15% trifluoroacetic acid (TFA): Dissolve 8.5 g sinapinic acid in 250 mL 1:1 ACN:water containing 0.15% TFA (*see* **Note 5**).

### 2.2 Covalent Protein Modification Assay

1. Echo® 555 acoustic liquid handler.
2. Echo® Qualified 384-well cyclic olefin copolymer (COC) source microplate.
3. 384 low-volume assay plate.
4. 1 M 2-[4-(2-Hydroxyethyl)piperazin-1-yl]ethane-1-sulfonic acid (HEPES), pH 7.3.
5. 5 M sodium chloride (NaCl).
6. 2 M magnesium chloride ($MgCl_2$).
7. Dimethyl sulfoxide (DMSO).
8. HEPES assay buffer: 20 mM HEPES, 150 mM NaCl, 2 mM $MgCl_2$, pH 7.3 (*see* **Note 6**).

## 3 Methods

All procedures are carried at room temperature. Buffers stored in refrigerators are aliquoted and allowed to warm to room temperature before assay.

### *3.1 MALDI Target Pretreatment*

1. Wash the MALDI target plate for the assay (use MPT or disposable MALDI target plates, *see* **Note 1**).

   Washing disposable MALDI target plates: Scrub the target plate using lint-free paper tissue wetted with methanol. Rinse the target with methanol and dry by wiping with lint-free paper tissue.

   Washing MPT MALDI target plates: Scrub the surface of the target plate with a paper towel under a stream of tap water to remove previous deposits. Rinse the target plate with Type 1 water, and then place in a 1 L beaker filled with 2-propanol. Cover the beaker tightly with aluminum foil and sonicate for 20 min. Rinse the target plate using methanol, and then place it in a second beaker filled with Type 1 water. Next, sonicate for 20 min, rinse with Type 1 water, and dry by lint-free paper tissues.
2. Dispense 30–40 mL of target plate pretreatment solution to flat-bottom, full-size reservoir placed on Biomek® FXP workstation (*see* **Note 7**).
3. Wet the target plate with clean, lint-free paper tissue wetted with Type 1 water. Dry the target plate with another clean, lint-free paper tissue (*see* **Note 8**).
4. Mount disposable target plate with holder or bare MPT 384 ground steel BC target plate on Alpillo® plate cushion.
5. Dispense 0.65 μL target plate pretreatment solution on a freshly washed MALDI target plate using P30/384 head.
6. Air dry target plates.

### *3.2 Covalent Protein Modification Assay*

1. Prepare the ECHO source plate by dispensing 10 μL of 1 mM DMSO stock solutions of tested compounds into the appropriate wells (*see* **Note 9**).
2. Dispense 12 μL of 1 μM protein stock in HEPES assay buffer into the appropriate wells in a low-volume assay plate. Centrifuge for 2 min at 930× *g* (*see* **Note 10**).
3. Prepare 384-well plates filled with 15 μL of MALDI matrix for 2 early timepoints (5 and 15 min). Seal plates with adhesive sealing. For the remaining 30- and 60-min timepoints, prepare matrix plates just before sampling the reaction mixture (*see* **Note 11**).

4. Dispense 30 nL of tested compounds from the ECHO source plate to the assay plate using the ECHO liquid handler to achieve a final concentration of 2.5 μM of compound. Seal the plate and centrifuge at 930× *g* for 1 min.
5. Pipette 2 μL of reaction mixtures into 15 μL of MALDI matrix solutions dispensed in 384-well polypropylene plates at 5-, 15-, 30-, and 60-min timepoints.
6. Mix the contents of the wells by pipetting up and down four times. Seal the plate and centrifuge at 930× *g* for 1 min.
7. Dispense 1.5 μL of the resulting solution onto a pretreated MALDI target plate.
8. Dry the MALDI target plate in a desiccator connected to a vacuum (*see* **Note 12**). Scan target plates.

### 3.3 MALDI Measurements and Processing

1. Setup the MALDI-TOF instrument for linear mode and a mass range from −700 to +1700 Da as the protein mass. Adjust all other parameters accordingly to achieve good sensitivity and resolution for the assay protein (*see* **Note 13**).
2. Setup automated data collection using an AutoXecute method (*see* **Note 14**).
3. Collect the mass chromatograms.
4. Process the mass chromatograms by smoothing spectrograms with the Savitzky-Golay algorithm using a width of 15 m/z and four cycles. Subtract the baseline using the median algorithm with a flatness of 1 and a median level of 0.01. Use a centroid algorithm for peak detection according to the following parameters: signal-to-noise threshold: 8, relative intensity threshold: 4%, peak width: 15 m/z, height: 75%, and the maximum number of peaks: 8. Example of processed mass chromatogram is given in Fig. 1.
5. Inspect processed mass chromatograms and extract data to Excel using custom macro (*see* **Note 2**).

### 3.4 Data Calculation

Extracted data is processed in an Excel spreadsheet by finding peaks for the unmodified protein and proteins modified by up to five compound molecules. The peak height is used instead of the peak area for quantification. Data is expressed in percentages as total protein modification, defined as the sum of all observed protein species modified by the compound, versus the sum of all modified proteins plus the unmodified protein peak (Tables 1, 2 and 3). If multiple compound additions are observed, modification is also reported as the percentage of individual protein species modified by single and multiple compound additions (up to five compound additions are taken into consideration) (*see* **Note 15**).

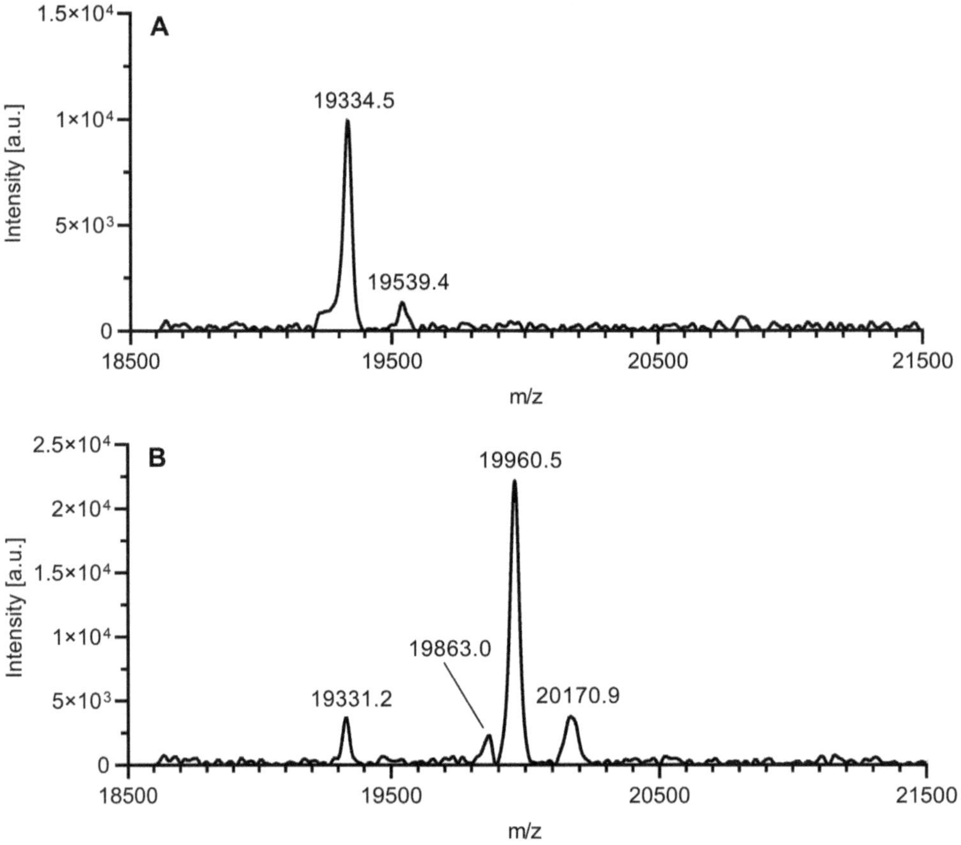


**Fig. 1** Mass chromatograms for 5-min reaction timepoints of GDC6036 at 2.5 μM and 1 μM KRAS4b(1-169) G12/C118S GppNHp (**a**) and GDP (**b**). Peaks around 19,333 m/z correspond to unmodified protein. Small peaks on the right side of the predominant peaks are artifacts related to matrix interaction with protein. The panel **a** shows no modification observed for the GppNHp-loaded protein. The panel **b** shows GDC6036 modified GDP-loaded protein at 19,960.5 m/z. A small peak at 19,863.0 m/z corresponds to protein modified by an impurity/decomposition product related to GDC6036

## 4 Notes

1. For large-scale screens, use Plain HTS MALDI plates. 1.0 mm disposable target plates in conjunction with Bruker MTP Sheet Metal Adapter 1.0 mm holders can be used to avoid time-consuming target plates' recovery procedures between assays. From our experience, it is a viable solution for screens utilizing proteins at 5 μM or higher concentrations. For protein concentration in 1–5 μM range, standard target plates are preferred due to generally higher quality of mass spectrograms obtained.
2. A macro is used to extract m/z, peak area, and peak height from the processed mass chromatograms to Excel for future processing. Alternatively, R or Python can be used to extract

**Table 1**
**MALDI-TOF reported relative modification of KRAS4b(1-169) G12C/C118S GppNHp by a panel of selected compounds**

| KRAS4b(1-169) G12C/C118S GppNHp; 1 μM protein, 2.5:1 compound to protein ratio; $n = 3$ | | | | | | | | |
|---|---|---|---|---|---|---|---|---|
| | 5 min | | 15 min | | 30 min | | 1 h | |
| Compound | % | ±SD | % | ±SD | % | ±SD | % | ±SD |
| AMG510 | 0.0 | | 0.0 | | 0.0 | | 0.0 | |
| MRTX849 | 0.0 | | 0.0 | | 0.0 | | 0.0 | |
| GDC6036 | 0.0 | | 0.0 | | 6.2 | 0.9 | 11.9 | 2.5 |
| Compound A | 100.0 | | 100.0 | | 100.0 | | 100.0 | |

Protein concentration was 1 μM and compound concentration was 2.5 μM. All samples were run in triplicates

**Table 2**
**MALDI-TOF reported relative modification of KRAS4b(1-169) G12C/C118S GDP by a panel of selected compounds**

| KRAS4b(1-169) G12C/C118S GDP; 1 μM protein, 2.5:1 compound to protein ratio; $n = 3$ | | | | | | | | |
|---|---|---|---|---|---|---|---|---|
| | 5 min | | 15 min | | 30 min | | 1 h | |
| Compound | % | ±SD | % | ±SD | % | ±SD | % | ±SD |
| AMG510 | 86.1 | 0.4 | 87.8 | 1.1 | 88.6 | 1.0 | 90.6 | 0.8 |
| MRTX849 | 83.1 | 0.8 | 80.9 | 0.8 | 85.7 | 1.3 | 87.3 | 0.7 |
| GDC6036 | 86.4 | 0.8 | 86.1 | 0.9 | 88.7 | 0.7 | 88.3 | 1.5 |
| Compound A | 93.2 | 0.8 | 96.9 | 2.1 | 100.0 | | 100.0 | |

Protein concentration was 1 μM and compound concentration was 2.5 μM. All samples were run in triplicates

**Table 3**
**MALDI-TOF reported relative modification of KRAS4b(1-169) G12C/C118S GTP by a panel of selected compounds**

| KRAS4b(1-169) G12C/C118S GTP; 1 μM protein, 2.5:1 compound to protein ratio; $n = 3$ | | | | | | | | |
|---|---|---|---|---|---|---|---|---|
| | 5 min | | 15 min | | 30 min | | 1 h | |
| Compound | % | ±SD | % | ±SD | % | ±SD | % | ±SD |
| AMG510 | 0.0 | | 0.0 | | 0.0 | | 9.6 | 1.4 |
| MRTX849 | 0.0 | | 4.4 | 0.8 | 6.2 | 0.6 | 10.9 | 0.8 |
| GDC6036 | 0.0 | | 6.0 | 1.7 | 7.5 | 1.4 | 9.9 | 0.7 |
| Compound A | 75.4 | 3.6 | 91.3 | 2.1 | 100.0 | | 100.0 | |

Protein concentration was 1 μM and compound concentration was 2.5 μM. All samples were run in triplicates. Observed small percentage of labeling from the KRAS G12C GDP inhibitors AMG510, MRTX849, and GDC6036 comes from GTP hydrolysis in this assay that we estimated as 10% per hour

the data directly from xml files of the processed mass chromatogram directories. The macro code is given as follows:

```
Attribute VB_Creatable = True
Attribute VB_Exposed = False
Attribute VB_GlobalNameSpace = False
Attribute VB_Name = "MassListToExcel"
'#Uses "D:\Methods\flexAnalysisMacroModules\FATools.obm"
Option Explicit
Option Base 1
Private aAnaSpec() As TAnaSpec
Private nSpectra%
Private Const sDelimiter$ = vbTab
Property Get MenuItem$
MenuItem = "File/Export/Mass List to Excel"
End Property
Property Get EnableMenuItem As Boolean
EnableMenuItem = Tools.GetSelectedAnaSpec(aAnaSpec) > 0
End Property
Sub Main
Execute
End Sub
Sub Execute
On Error GoTo Failed
MassListToExcel
Exit Sub
Failed:
MsgBox Err.Description
End Sub
Private Sub MassListToExcel
' Transfer the mass list to Excel
nSpectra = Tools.GetSelectedAnaSpec(aAnaSpec)
If nSpectra = 0 Then
Err.Raise 1,,"No spectrum selected"
End If
' Open Excel application
Dim Excel As Object
Set Excel = OpenExcel
Dim i%
For i = 1 To nSpectra
Excel.Sheets(1).Activate
Dim sName$
sName = Replace(aAnaSpec(i).oAnalysis.Name, "\", "_")
sName = Replace(sName, " ", "_")
sName = Left(sName, 31)
Dim Paster$
```

```
Paster = "A" + CStr(22*i-21) + ":" + "A" + CStr(22*i-21)
CopyMassListToClipboard aAnaSpec(i).oAnalysis, aAnaSpec(i).
oSpectrum
Excel.Range(Paster).Select
Excel.Sheets(1).Paste
Next i
Clipboard ""
End Sub
Private Function OpenExcel As Object
Dim ExcelApp As Object
On Error Resume Next
Set ExcelApp = GetObject(,"Excel.Application")
If ExcelApp = Null Then
Set ExcelApp = CreateObject("Excel.Application")
End If
ExcelApp.Visible = True
ExcelApp.Workbooks.Add
Set OpenExcel = ExcelApp
End Function
Private Sub CopyMassListToClipboard(oAnalysis As Analysis,
oSpectrum As Spectrum)
Dim sName$
sName =Replace(GetSpectrumName(oAnalysis, oSpectrum), "\1
1SRef Raw", "")
Dim sCsv$
sCsv =sName + vbTab
sCsv = sCsv + "m/z" +vbTab + "Intensity" +vbTab + "Area"+
vbCrLf
On Error Resume Next
Dim oPeakList As MassList
Set oPeakList = oSpectrum.MassList
Dim i%
For i = 1 To oPeakList.Count
Dim oPeak As MassPeakInfo
Dim sValue$
Set oPeak = oPeakList(i)
sValue = ""
sValue = Replace(CStr(oPeak.MassToChargeValue), ",", ".")
sCsv = sCsv +sDelimiter + sValue + sDelimiter
sValue = ""
sValue = Replace(CStr(oPeak.Intensity), ",", ".")
sCsv = sCsv + sValue + sDelimiter
sValue = ""
sValue = Replace(CStr(oPeak.Area), ",", ".")
sCsv = sCsv + sValue + vbCrLf
Next
Clipboard sCsv
End Sub
```

```
Private Function GetSpectrumName(oAnalysis As Analysis, oS-
pectrum As Spectrum)
GetSpectrumName = oAnalysis.Name + " " + oSpectrum.Name
End Function
```

3. The Alpillo® plate cushion was modified by grinding part of the bottom right holder tabs to accommodate the bare Bruker MPT 384 ground steel BC target plate or disposable target plates mounted on MTP Sheet Metal Adapter 1.0 mm holder.
4. To prepare the target plate pretreatment solution, use a 100 mL capacity glass bottle. Use gentle heating (less than 50–55 °C) and sonication to obtain a clear solution. Leave the bottle closed and protected from the light at room temperature until the crystals are visible at the bottom of the bottle. Store in a cool, dark place in a tightly closed bottle. The faintly yellow solution is stable for this assay for 2 weeks. Discard the solution earlier if any orange hue discoloration is visible.
5. To prepare the MALDI matrix solution, use a 250 mL capacity glass bottle. Use gentle heating (less than 50–55 °C) and sonication to obtain a clear solution. Leave the bottle closed and protected from the light at room temperature until the crystals are visible at the bottom of the bottle (it may take up to 2 days to initiate crystallization). Using the solution prematurely, while oversaturated, may result in the tips becoming clogged on the Biomek platform. Store in cool, dark place in tightly closed bottle. The slightly yellow-orange solution is stable for the assay for 2 weeks. Discard earlier if the color changes to darker yellow-orange.
6. The HEPES assay buffer was chosen specifically for the KRAS protein family. Other buffers, such a Tris–HCl or PBS, in a wide range of pH levels, can be used if they do not interact with protein or tested compound reactivity. In general, the assay buffer is chosen based on protein properties and compounds reactivity profile. The use of detergents is acceptable, but not preferable.
7. We are using reversed lids from BioMek 384 tip boxes because the robot's original full-size reservoirs are not compatible with the 384-tip head. Any clean, standard plate-sized, polypropylene, flat-bottom reservoir without dividers can be used.
8. We found that this procedure significantly improves spotting quality on the MALDI target plate by eliminating overflow of the pretreatment solution.
9. Keep the source plate sealed. Thermal sealing is preferred, but for short-term storage, adhesive seals can be used.

10. For small scale assays use a 24-channel pipettor to dispense protein. For large-scale assays, use a BioMek robotic liquid handler. Do not exceed 930× *g* when the plate is centrifuged as the raised assay plate may shatter under its own weight. Usually, the assays are designed to have three replicates per tested compound or reaction condition. For large fragment library screenings, a single replicate may be chosen.
11. The described assay has early timepoints at 5 and 15 min and is thus restricted to a maximum of one 384-well plate due to time constraints related to the BioMek platform's pipetting speed and to a maximum of 5-6 384-well plate rows when using a manual 24-channel pipette. For assays designed with longer timepoints (e.g., 2, 8, or 24 h), the BioMek pipetting speed is a negligible factor and the assay can be extended to 2–4 plates.
12. Polypropylene/polycarbonate desiccators with polymer seals are preferred when drying the MALDI plate. A mild lab vacuum is sufficient; however, using dedicated small membrane or scroll dry pumps may significantly reduce drying time. Do not use oil vacuum pumps, as the target plate, the samples, and ultimately the MALDI mass spectrometer may become contaminated with oil.
13. In our case, the usual MALDI setup for KRAS4b(1-169) G12C/C118S in GppNHp/GDP/GTP form at 1 μM concentration is as follows: MALDI is set up for linear mode and mass range from 18.6 to 21.6 kDa. Detector gain is set to 1.24× (504 V). Sample rate is set to 5 GS/s. Real-time smoothing is set to medium (175 MHz). Laser smart beam pattern is set to "Custom" M5. Laser operational frequency is 10,000 Hz. Laser power is adjusted manually for a given sample set and re-adjusted dynamically during automated scanning by method fuzzy control mechanism.
14. The usual AutoXecute method setup used is as follows: Laser power is automatically adjusted using fuzzy control logic. The peak selection range is set to be between 19,000 and 21,500 Da. Peak evaluation uses a resolution parameter set to be higher than 400 for processed spectrograms. Centroid peak detection is used. The spectrograms are smoothed by the Savitzky-Golay algorithm using a width of 7 m/z and two cycles. The baseline is subtracted using the median algorithm with a flatness of 1 and a median level of 0.01. Fuzzy control is set to the proteins/oligonucleotides protocol with a maximum resolution of ten times above the threshold. Accumulation is guided by fuzzy control with MS/Parent mode on. Up to 40,000 satisfactory shots are collected in 10,000 shot steps. Dynamic termination is implemented to finish data collection when the peak signal-to-noise ratio reaches a value of 1000.

15. Due to the specific way that the Bruker flexAnalysis software assigns and calculates the peak area, we found that using the peak height for calculation is a more reliable method, especially with less than perfect quality of mass chromatograms. We tested both methods. While in the majority of cases the results from both are indistinguishable, the method using the peak height for the calculation proved to be more resilient and truer for mass chromatograms with lower signal-to-noise ratios. Due to Excel limitations for large-scale screening assays, we are using R for processed data extraction and calculation.

## Acknowledgments

We acknowledge TheRas, Inc. for the use of Compound A developed under a cCRADA. We thank Dominic Esposito, William Gillette, Vanessa Wall, Simon Messing, and the Protein Expression Laboratory (FNLCR) staff for their help with protein reagents. This project was funded in part with federal funds from the National Cancer Institute, National Institutes of Health under Contract 75N91019D00024. The content of this publication does not necessarily reflect the views or policies of the Department of Health and Human Services, and the mention of trade names, commercial products, or organizations does not imply endorsement by the US government.

### References

1. Ostrem JM, Peters U, Sos ML et al (2013) K-Ras(G12C) inhibitors allosterically control GTP affinity and effector interactions. Nature 503:548–551
2. Lito P, Solomon M, Li LS et al (2016) Cancer therapeutics: allele-specific inhibitors inactivate mutant KRAS G12C by a trapping mechanism. Science 351:604–608
3. Patricelli MP, Janes MR, Li LS et al (2016) Selective inhibition of oncogenic KRAS output with small molecules targeting the inactive state. Cancer Discov 6:316–329
4. Hallin J, Engstrom LD, Hargi L et al (2020) The KRASG12C inhibitor MRTX849 provides insight toward therapeutic susceptibility of KRAS-mutant cancers in mouse models and patients. Cancer Discov 10:54–71
5. Awad MM, Liu S, Rybkin II et al (2021) Acquired resistance to KRAS G12C inhibition in cancer. N Engl J Med 384:2382–2393
6. Fell JB, Fischer JP, Baer BR et al (2020) Identification of the clinical development candidate MRTX849, a covalent KRASG12C inhibitor for the treatment of cancer. J Med Chem 63: 6679–6693
7. Mathieu M, Steier V, Fassy F et al (2022) KRAS G12C fragment screening renders new binding pockets. Small GTPases 13:225–238
8. Orgován Z, Péczka N, Petri L et al (2023) Covalent fragment mapping of KRasG12C revealed novel chemotypes with in vivo potency. Eur J Med Chem 250:115212
9. Ostrem JML, Shokat KM (2016) Direct small-molecule inhibitors of KRAS: from structural insights to mechanism-based design. Nat Rev Drug Discov 15:771–785
10. Weiss A, Lorthiois E, Barys L et al (2022) Discovery, preclinical characterization, and early clinical activity of JDQ443, a structurally novel, potent, and selective covalent oral inhibitor of KRASG12C. Cancer Discov 12:1500–1517
11. Zhang T, Hatcher JM, Teng M et al (2019) Recent advances in selective and irreversible covalent ligand development and validation. Cell Chem Biol 26:1486–1500

12. Canon J, Rex K, Saiki AY et al (2019) The clinical KRAS(G12C) inhibitor AMG 510 drives anti-tumour immunity. Nature 575:217–223
13. Wijeratne A, Xiao J, Reutter C et al (2018) Chemical proteomic characterization of a covalent KRASG12C inhibitor. ACS Med Chem Lett 9:557–562
14. Bergman N, Thapper A, Styring S et al (2014) Quantitative determination of the Ru(bpy)32+ cation in photochemical reactions by matrix-assisted laser desorption/ionization time-of-flight mass spectrometry. Anal Methods 6: 8513–8518
15. Bruker Daltonics Gmb H (2009) flexanalysis 3.3 Scripting Manual, Version 3.3 (June 2009). Bruker Daltonics GmbH, pp 1–104

# Chapter 12

# KRAS4b:RAF-1 Homogenous Time-Resolved Fluorescence Resonance Energy Transfer Assay for Drug Discovery

**Erik K. Larsen, Maria Abreu-Blanco, Dana Rabara, and Andrew G. Stephen**


## Abstract

Homogenous time-resolved FRET (HTRF) assays have become one of the most popular tools for pharmaceutical drug screening efforts over the last two decades. Large Stokes shifts and long fluorescent lifetimes of lanthanide chelates lead to robust signal to noise, as well as decreased false positive rates compared to traditional assay techniques. In this chapter, we describe an HTRF protein-protein interaction (PPI) assay for the KRAS4b G-domain in the GppNHp-bound state and the RAF-1-RBD currently used for drug screens. Application of this assay contributes to the identification of lead compounds targeting the GTP-bound active state of K-RAS.


**Key words** Homogenous time-resolved FRET, RAS, RAF, Protein-protein interaction, Drug screen, Assay development, MRTX1133

## 1 Introduction

RAS, from rat sarcoma virus, is a family of proteins that originally were identified as responsible for the oncogenic properties of retroviruses in 1982 [1]. RAS is a GTPase that exists in two states: a GTP-bound "on-state" and a GDP-bound "off-state" [2, 3]. In the on-state, KRAS interacts with several downstream effectors including rapidly accelerated fibrosarcoma (RAF) kinases of the mitogen-activated protein kinase (MAPK) pathway impacting cell proliferation, cell differentiation, and cell death [4–8]. Of the three isoforms of RAS (HRAS, KRAS, and NRAS), missense mutations are most common in KRAS [9, 10]. KRAS is alternatively spliced into KRAS4a and KRAS4b, with KRAS4b as the most heavily researched splice variant [11, 12]. The abundance of mutations in deadly cancers such as colorectal, lung, and pancreatic have spurred efforts to drug KRAS; however, these endeavors were stymied for nearly four decades of research deeming RAS undruggable. Yet in 2021, Amgen's sotorasib became the first KRAS inhibitor

Andrew G. Stephen and Dominic Esposito (eds.), *KRAS: Methods and Protocols*, Methods in Molecular Biology, vol. 2797, https://doi.org/10.1007/978-1-0716-3822-4_12,

approved by the FDA, followed quickly by Mirati's adagrasib in 2022 [13–15]. Both compounds target the G12C mutation in the GDP-bound off-state, but other oncogenic mutations have drastically slower turnover rates necessitating a search for compounds targeting the GTP-bound on-state [16].

With the emergence of compounds targeting KRAS, a need for an in vitro protein-protein interaction (PPI) assay for GTP-bound KRAS:RAF interaction has become apparent. The RAS:RAF interaction has been assayed in vitro since 1993 [7], and many assays have been developed since then, including coprecipitation followed by Western blotting, and ELISA [17–19] as well as scintillation proximity assays (SPA) and fluorescence anisotropy [20, 21]. Homogenous time-resolved FRET (HTRF) assays were developed in the 1990s and have become a staple in screening efforts [22, 23]. HTRF utilizes an increased Stokes shift and long-lasting fluorescence of lanthanides to decrease the false positive hit rate caused by intrinsic fluorescence of small molecule lead compounds [24, 25]. Because FRET requires a donor/acceptor pair, signal is observed only when two tagged proteins bind, removing the need for washing steps previously encountered in assays such as ELISA, opening the possibility for automation [25].

HTRF assays for RAS/RAF PPI have been reported previously [26, 27] and are offered commercially; however, the details of these assays and how they are developed are limited. Development of the assay reported here followed the procedure outlined by PerkinElmer [28]. Briefly, donor/acceptor pairs and tag location were screened in a cross-titration experiment. RAF-1-Ras binding domain with a C-terminal 3xFLAG tag (RAF-1-RBD) and N-terminal biotinylated Avi-KRAS4b produced the best signal out of options screened. Additionally, N-terminal placement of a biotinylated Avi-tag on KRAS4b avoided protein aggregation and proteolysis seen with C-terminal tags. Next, a displacement assay was performed in the presence of increasing concentrations of untagged RAF-1-RBD. This experiment demonstrated that the FRET signal observed was specific to the tagged-KRAS:tagged-RAF-1-RBD interaction. Untagged RAF-1-RBD competed with the tagged-RAF-1-RBD and a signal decrease was observed as less tagged-KRAS:tagged-RAF-1-RBD complex and more tagged KRAS4b:untagged RAF-1-RBD complex existed (*see* Fig. 1). Concentrations of Avi-KRAS4b and RAF-1-RBD-3xFLAG as well as acceptor:donor reagent concentrations were optimized via cross-titration. Finally, a competition assay was performed with a known competitive inhibitor of RAS:RAF binding, MRTX1133 [29]. This assay demonstrated that the signal decreased with increasing concentration of inhibitor compound, its intended purpose for a drug screening effort (*see* Fig. 2). The MRTX1133 competition experiment will be used to describe the protocol for this assay.

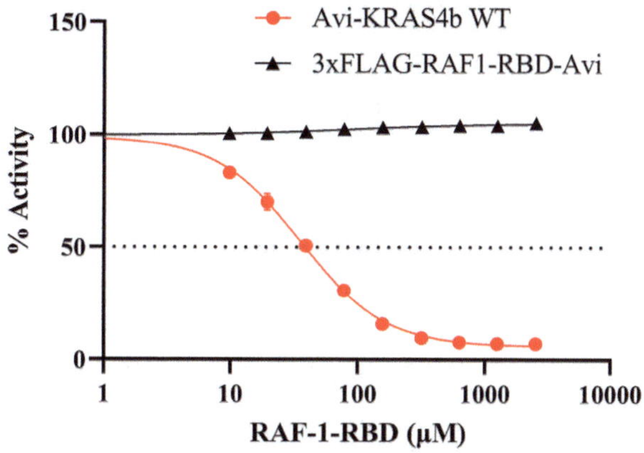


**Fig. 1** Displacement of the Avi-KRAS4b:RAF-1-RBD-3xFLAG complex with untagged RAF-1-RBD. The red circles represent % activity for Avi-KRAS4b WT, and the black triangles represent % activity for the double-tagged 3xFLAG-RAF-1-RBD-Avi control construct

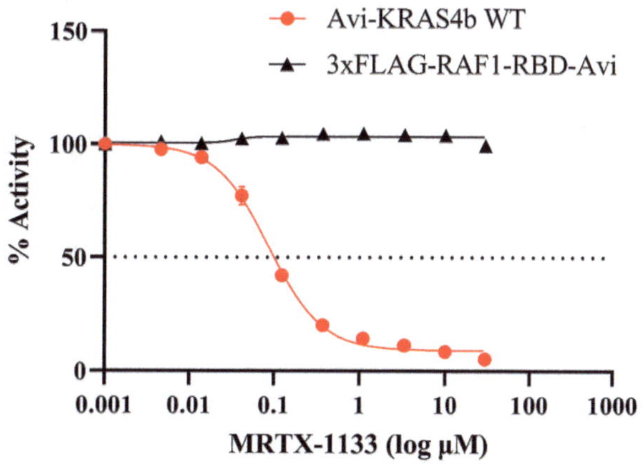


**Fig. 2** Competition of the Avi-KRAS4b:RAF-1-RBD-3xFLAG complex with MRTX1133. The red circles represent % activity for Avi-KRAS4b WT, and the black triangles represent % activity for the double-tagged 3xFLAG-RAF-1-RBD-Avi control construct

## 2 Materials

Prepare all solutions with ultrapure water with a sensitivity of 18 MΩ at room temperature. Store all reagents at room temperature unless indicated otherwise.

### 2.1 Lab Equipment

1. Envision XCite 2105 multilabel reader (*see* **Note 1**).
2. Echo® 555 Liquid Handler from Labcyte (Beckman Coulter) (*see* **Note 2**).
3. 1–30 μL E1-ClipTip electronic pipette.

4. 2–125 μL E1-ClipTip electronic pipette.
5. Thermo Scientific™ ClipTip™ filtered 384-format pipette tips.
6. Labcyte Echo Qualified 384-well low dead volume microplate.
7. Greiner Bio-one microplate, 384 well, PS, F-bottom, small volume, Hibase, med. Binding, white.
8. Plate sealer.
9. Argos Technologies pipetting reservoir: basin, polypropylene, 3 mL.
10. Heathrow Scientific™ polystyrene sterile single-wrapped reagent reservoir.
11. 1.5 mL Eppendorf tubes.
12. 50 mL Falcon tubes.
13. 0.22 μm filter units.

### 2.2 Buffer

1. Bovine serum albumin (Sigma) diluted to 10% BSA in ultra-pure water and stored in 1 mL of aliquots at −20 °C.
2. DMSO (cell signaling technology).
3. Nucleotide exchange buffer: 20 mM Tris–HCl pH 7.5, 200 mM ammonium sulfate, 100 μM zinc chloride, and 1 mM TCEP. Add 400 μL of 1 M Tris–HCl, pH 7.5, 2 mL of 2 M $(NH_4)SO_4$, and 40 μL of 500 mM TCEP to 17.56 mL of $H_2O$. Filter sterilize with 0.22 μm filter unit. This buffer can be saved up to 1 week at 4 °C. On the day of the nucleotide exchange, add 20 μL of 100 μM $ZnCl_2$.
4. Nucleotide exchange storage buffer: 20 mM HEPES pH 7.3, 150 mM sodium chloride, 2 mM magnesium chloride, and 1 mM TCEP. Add 40 mL of 1 M HEPES, pH 7.3, 4 mL of 1 M $MgCl_2$, 60 mL of 5 M NaCl, and 4 mL of 500 mM TCEP to 1.892 L of $H_2O$. Filter sterilize.
5. Assay buffer: 50 mM Tris–HCl pH 7.4, 100 mM NaCl, 5 mM $MgCl_2$, 0.1% BSA, 0.01% Tween 20, and 10% DMSO (*see* **Note 3**). Add 50 mL of 1 M Tris–HCl pH 7.5, 20 mL of 5 M NaCl, and 5 mL of 1 M $MgCl_2$ to a 1 L bottle. Add deionized water to the 1 L fill-line. Filter sterilize and store at room temperature. On the day of the assay, prepare 3 mL of assay buffer by adding 30 μL of 10% BSA by mass in water, 3 μL of 10% Tween 20 by volume in water, and 300 μL DMSO to 2.667 mL of the prepared 50 mM Tris–HCl pH 7.5, 100 mM NaCl, and 5 mM $MgCl_2$.
6. HTRF buffer: 50 mM Tris–HCl, pH 7.5 and 0.01% Tween 20. Add 50 mL of 1 M Tris–HCl pH 7.5 and 100 μL Tween 20 to a 1 L bottle. Add $H_2O$ to the 1 L fill-line.

*2.3 Protein, Nucleotide, and Compound*

1. Phosphatase, alkaline-agarose (AP) from calf intestine.
2. GppNHp-tetralithium salt (GMPPNP) (*see* **Note 4**).
3. Protein expression constructs His6-MBP-tev-Avi-KRAS4b (1-169), His6-MBP-tev-3xFLAG-RAF1(51-131)-Avi (see **Note 5**), and His8x2-tev-GS-RAF1(51-131)-3xFLAG-tev-Halo (*see* **Note 6**), were created using the protocols outlined in Wall et al. [30]. All proteins were expressed using the Dynamite media protocol with overnight induction at 16 °C as outlined in Taylor et al. [31]. Proteins were purified using the protocols described in Kopra et al. [26] for proteins of the format His6-MBP-tev-POI.
4. Avi-KRAS4b(1-169) was nucleotide exchanged in nucleotide exchange buffer at a 10x nucleotide to protein molar ratio in the presence of AP and stored long term in "storage buffer" (*see* **Note 7**) [32].
5. To 2 mg of MRTX1133 (molecular weight = 600.63 g/mol), add 333 μL DMSO to create a 10 mM stock solution.

*2.4 HTRF Reagents*

All HTRF reagents are acquired from Cisbio (PerkinElmer) (*see* **Note 8**) and stored at −20 °C.

1. Streptavidin-XLent.
2. Monoclonal antibody (Mab) anti-FLAG $M_2$-Tb cryptate.

## 3 Methods

Perform all steps at room temperature unless specified otherwise.

*3.1 Compound Dilution Series*

1. Remove Avi-KRAS4b, RAF-1-RBD-3xFLAG, and 3xFLAG-RAF-1-RBD-Avi from the −20 °C freezer, and thaw on ice.
2. Label one 384-well low dead volume microplate (LDV plate) as "compound source plate."
3. Label one LDV plate as "compound intermediate plate."
4. Dilute the 10 mM MRTX1133 stock to 3 mM in the "compound source plate" (well A1) using DMSO as the solvent (*see* **Note 9** and Fig. 3). To do this, add 3.6 μL of 10 mM MRTX to well A1 in the "compound source plate." Then, add 8.4 μL of DMSO to that same well, and pipette up and down to mix 7–10 times.
5. Centrifuge the "compound source plate" at 1000× *g* for 1 min.
6. Use the liquid handler to dispense 125 nL of the 3 mM MRTX1133 compound stock from "compound source plate" into the "compound intermediate plate" (well B1) leaving the well above it empty (A1). This will be used to create a second

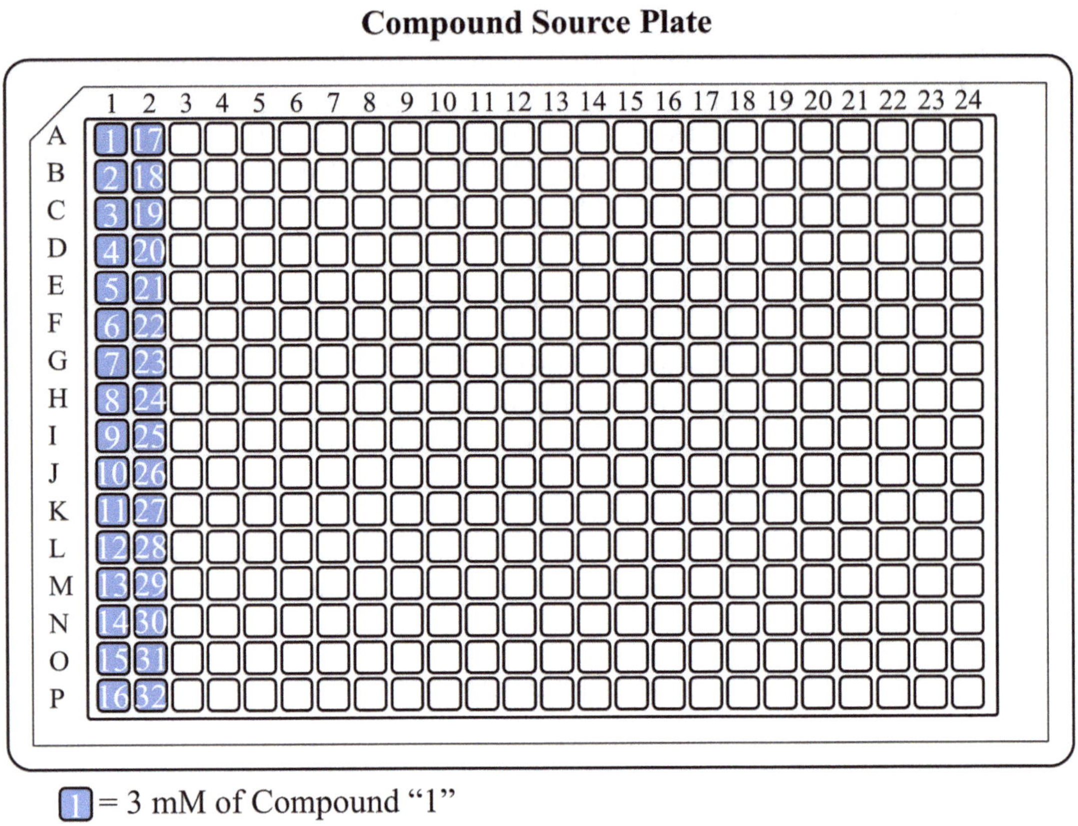


**Fig. 3** Compound source plate. Representative 384-well compound source plate layout for 32 individual compounds

dilution of MRTX1133 at a concentration of 37 μM. These two dilutions (the 3 mM and the 37 μM) will be used to create the nine dilutions used in the assay (*see* **Note 10** and Fig. 4).

7. With a few mL DMSO in the polystyrene sterile single-wrapped reagent reservoir, take the 1–30 μL electronic pipette, and dilute the 125 nL to 10 μL using DMSO as the solvent. To do this, add 10 μL of DMSO to the well containing the 125 nL of MRTX1133, and pipette up and down 7–10 times to mix (*see* **Note 11**).
8. Take the 1–30 μL electronic pipette, and transfer 10 μL of the 3 mM stock solution of MRTX1133 from the "compound source plate" to the "compound intermediate plate" in well A1 (*see* **Note 12**).
9. Fill four additional columns of wells in the "compound intermediate plate" with 10 μL DMSO to be used as solvent reservoirs by the liquid handler.
10. Centrifuge the "compound intermediate plate" at 1000× *g* for 1 min.

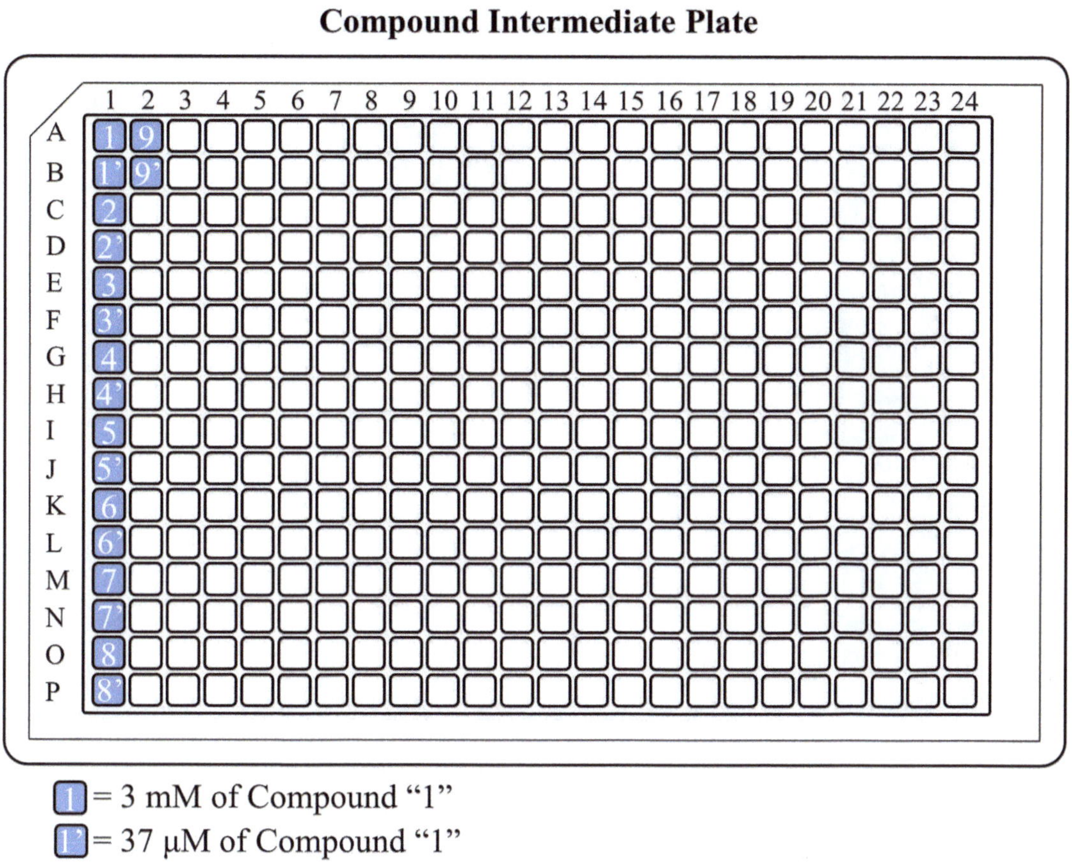


**Fig. 4** Compound intermediate plate. Representative 384-well compound intermediate plate layout for nine individual compounds

11. If desired, the "compound intermediate plate" can be sealed with the plate sealer set to 163 °C, for 3.0 s, and 45 μm, and stored over night at −20 °C at this step. If not, proceed to the next step.
12. Label one microplate with the KRAS mutant type that will be assayed (in this case "WT" for wild type), and one microplate with "control," followed by "compound destination plate" (*see* **Note 13**).
13. Use the liquid handler to dispense 9, 202.5 nL, 3× dilutions in adjacent wells in a row of the "WT compound destination plate" (A1–A9) starting at a final concentration of 30 μM in a 20 μL assay volume and ending at a final concentration of 4.57 nM in a 20 μL assay volume (*see* **Note 14**). Fill the tenth well (A10) with 202.5 nL DMSO as a negative control. Repeat this for a duplicate (B1–B10) (*see* Fig. 5).
14. Repeat **step 13** for the "control compound destination plate."

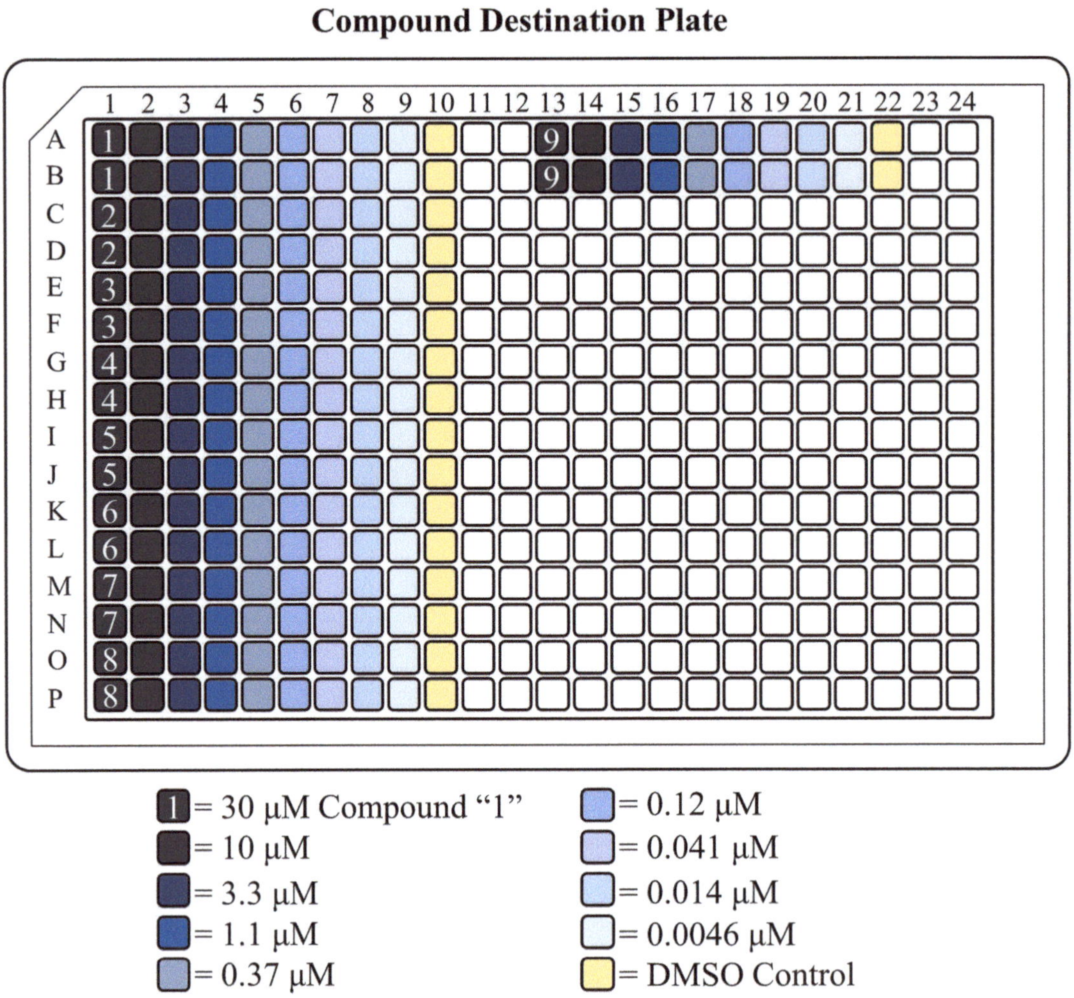


**Fig. 5** Compound destination plate. Representative 384-well compound destination plate layout for nine individual compounds assayed in duplicate

15. Set the two "compound destination plates" aside.
16. This procedure can be modified to accommodate the screening of multiple compounds (*see* **Note 15**).

### *3.2 Protein Stock Solutions*

1. Prepare at least 1 mL of the assay protein mix for each KRAS4b mutant to be assayed.
2. The "assay protein mix" contains Avi-KRAS4b and RAF-1-RBD-3xFLAG each at a concentration of 100 nM (final concentration in each well will be 50 nM) in "assay buffer" (*see* **Note 16**).
3. Prepare the "assay control mix" for the double-tagged Avi-RAF-1-RBD-3xFLAG control protein at a concentration of 100 nM (final concentration in each well will be 50 nM) in "assay buffer" (*see* **Note 17**).

4. Fill one channel of the pipetting reservoir with "assay protein mix" (*see* **Note 18**).
5. Repeat **step 4** in an adjacent channel of the pipetting reservoir for the "assay control mix" (*see* **Note 18**).
6. With the 2–125 μL electronic pipette in the 384-well plate setting, use the "matrix" method to withdraw 100 μL of "assay protein mix" from the pipetting reservoir. Two tips fit in each channel, enough for the duplicate measurement.
7. Dispense 10 μL of the "assay protein mix" over the MRTX1133 dispensed in the "WT compound destination plate." Pipette 10 μL, ten times starting from the DMSO only control wells (A10 and B10) and ending at the 30 μM of MRTX1133 wells (A1 and B1).
8. When pipetting, ensure tips are touching the side of the respective well for consistent dispensing.
9. Discard the used tips.
10. Repeat **steps 6–9** for the "assay control mix" in the "control compound destination plate."
11. Centrifuge the "compound destination plates" at 1000× *g* for 1 min.
12. Incubate at room temperature for 60 min on the microplate shaker at 700 RPM.

### 3.3 FRET Label Stock Solutions

1. During the 60 min incubation of MRTX1133 with proteins, remove aliquots of MAb anti-FLAG $M_2$-Tb cryptate and Streptavidin-XLent from the −20 °C refrigerator, and thaw on ice.
2. Once thaw, prepare the "HTRF mix" which contains MAb anti-FLAG $M_2$-Tb cryptate at a concentration of 3.4 nM and Streptavidin-XLent at a concentration of 170 nM in "HTRF buffer." Prepare by first diluting the solid in the "20,000 tests" vials of MAb anti-FLAG $M_2$-Tb cryptate and Streptavidin-XLent in 1 mL of $H_2O$. To 1 mL of "HTRF buffer," add 19.62 μL of MAb anti-FLAG $M_2$-Tb cryptate and 9.96 μL Streptavidin-XLent.
3. Fill one channel of the pipetting reservoir with "HTRF mix."
4. After the 60-min incubation of MRTX1133 with proteins, with the 2–125 μL E1-ClipTip electronic pipette in the 384-well plate setting, use the "matrix" method and withdraw 100 μL of "HTRF mix" from the pipetting reservoir (*see* **Note 19**).
5. Dispense 10 μL of the "HTRF mix" over the MRTX1133/protein mixture. Pipette 10 μL, ten times starting from the DMSO only control wells and ending at the 30 μM MRTX1133 well.

6. When pipetting, ensure the tips are touching the side of the respective well for consistent dispensing.
7. Discard the used tips.
8. Repeat **steps 5–7** for the "control compound destination plate."
9. Centrifuge the "compound destination plates" at 1000× *g* for 1 min.
10. Incubate at room temperature for 60 min on the microplate shaker at 700 RPM.

### *3.4 TR-FRET Measurement*

1. Read the plate on the multilabel reader (*see* **Note 20**) using the following parameters:

   Operation type: normal measurement.

   Excitation, emission and second emission are from the top.

   Top mirror: LANCE/DELFIA Dual Enh D400/D630.

   Bottom mirror: None.

   Excitation filter: UV2 (TRF) 320 – BW = 75 nm, Tmin = 70%.

   Emission filter: APC 665 – BW = 7.5 nm, Tmin = 75%.

   Second emission filter: Europium 615 – BW = 8.5 nm, Tmin = 80%.

   Measurement height (mm): 6.5.

   Excitation light (%): 100.

   Delay (μs): 60.

   Window time (μs): 400.

   Number of sequential windows: 1.

   Time between flashes (μs): 2000.

   Number of flashes: 100.

   Number of flashes for second detector: 100.

### *3.5 Data Fitting*

1. Export the data.
2. Take the ratio of the acceptor (665) to donor (615) values (A/D values) and multiply by 10,000 [25].
3. Divide each of the nine drug-containing $A/D$ values by the DMSO only control $A/D$ value and multiply by 100 for a % activity value.
4. Plot the % activity value vs. the compound concentration.
5. Fit the plot using the model $y = \frac{(A-D)}{\left(\frac{X}{C}\right)^B} + D$ where $A$ is the top plateau, $B$ is the hill slope, $C$ is the IC50, and $D$ is the bottom plateau [33].
6. For high-throughput data processing, access the python script at https://github.com/larse585/el_c3008_envision_dose_response.

## 4 Further Assay Validation

Increasing concentrations of MRTX1133 lead to decreasing A/D ratios which provided further validation of the assay after the displacement assay with unlabeled-RAF-1-RBD (*see* Fig. 2). The IC50 was found to be 84 nM. While IC50 values are not directly comparable between different HTRF assays [34], Mirati reported an IC50 of 112 nM with a similar HTRF PPI assay [35].

Additionally, a Z-factor was calculated from a panel of lead compounds to ensure the signal to baseline of the assay was sufficient for further screening efforts. Historically, an assay with a Z-factor above 0.5 indicates that the difference in signal intensity between the negative control and the sample is sufficient. While this cutoff is not supported statistically, the only negative aspect of using an assay with a *Z*-factor > 0.5 is that the assay may find hits for compounds that are unwanted and less active [36, 37]. The *Z*-factor is calculated by

$$Z = \frac{\left(\mu_n - \mu_p\right) - \left(3\sigma_n + 3\sigma_p\right)}{\left(\mu_n - \mu_p\right)}$$

where $\mu_n$ is the negative control mean, $\mu_p$ is the sample mean, $\sigma_n$ is the negative control standard deviation, and $\sigma_p$ is the sample standard deviation [37]. For this panel of 25 compounds, *Z*-factor = 0.68 (*see* Fig. 6).

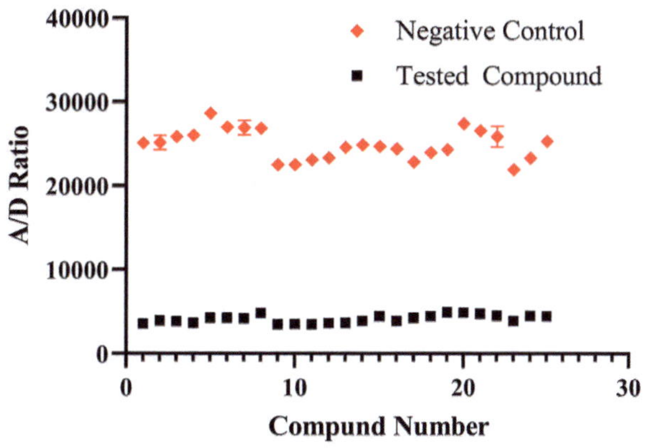


**Fig. 6** Acceptor/donor (*A/D*) ratio data for the z-factor calculation. The red diamonds represent the average *A/D* ratio of duplicate DMSO only control wells for Avi-KRAS4b WT, and the black squares represent the average *A/D* ratio of duplicate compound containing wells for Avi-KRAS4b WT

## 5 Notes

1. This assay can be read by any plate reader if it can delay observation after pulsed excitation [38, 39].
2. An Echo® 555 liquid handler is not a necessity, however, for the intermediate plate used here, accurate dispensing of volumes down to 2.5 nL is required. Increasing the number of dilutions prepared in the intermediate plate can increase this volume to 202.5 nL.
3. When optimizing a new HTRF assay, DMSO and Tween20 concentrations must be optimized for the proteins in the assay as well as the compounds to be assayed. In this case, it was found that the A/D signal started to increase beyond 10% DMSO. A high DMSO concentration limits compound solubility challenges (*see* Fig. 7).
4. The choice of nonhydrolyzable GTP mimetic can impact the signal of an HTRF assay. While GTPγS may produce tighter binding to RAF-1-RBD increasing signal [40], the intrinsic GTPase rate is faster with GTPγS than with GppNHp or GppC$H_2$p [41, 42]. Additionally, most structures of KRAS were solved using GppNHp, providing more information on potential structure-activity relationships. Furthermore, potential interactions between the nucleotide mimetic and the compounds to be screened need to be considered. Finally, cost, availability, and ease of nucleotide exchange will play a role in deciding which nucleotide mimetic will work best.

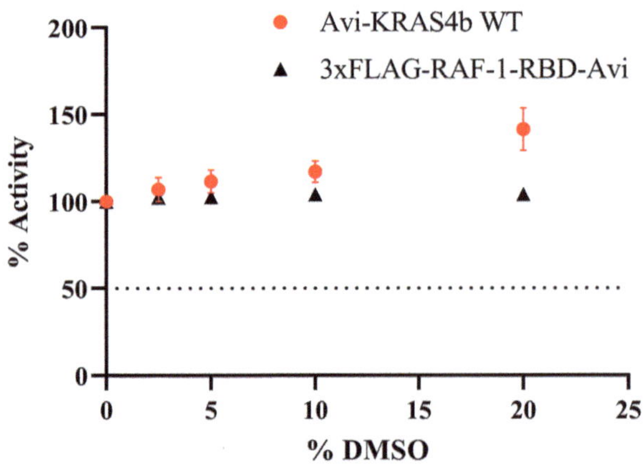


**Fig. 7** DMSO tolerance of the Avi-KRAS4b:RAF-1-RBD-3xFLAG HTRF assay. The red circles represent % activity for Avi-KRAS4b WT, and the black triangles represent % activity for the double-tagged 3xFLAG-RAF-1-RBD-Avi control construct

5. The double-labeled control is used to identify compounds whose intrinsic fluorescence interferes with the HTRF PPI assay. Generating a protein that has the same tags used in the HTRF PPI assay, but on either terminus of the same protein, creates a protein that generates a FRET signal that will not change unless the protein undergoes a conformational change. Because the lead compounds explored with this assay interact with KRAS, no such signal change is expected. If one is observed, it can be concluded that the intrinsic fluorescence of the compound is interfering with the assay.
6. We have found C-terminal tags on the G-domain of KRAS4b are prone to aggregation issues and proteolysis. In general, all tagged KRAS4b constructs are N-terminal.
7. This buffer is a standardized storage buffer for many of the KRAS proteins we work with. The volumes used in this assay of the stock protein solution are low enough where buffer exchange of the protein into assay buffer is unnecessary. Aliquots of protein should be flash frozen in liquid nitrogen prior to storage at −80 °C.
8. Tag location and type must be optimized for each assay. Cross-titrations of KRAS4b and RAF-1-RBD with different tag types (His6, His12, GST, FLAG, 3xFLAG, streptavidin, etc.) as well as location on the N-terminus, C-terminus, or covalent tags to residues within the protein will provide the concentration range available with each FRET pair [25]. Typically, a S/N of >5 is the minimum that provides a large enough window to assay IC50 values [43]. For PPI assays, the bottom of the assay occurs at half the protein concentration in the assay, so there is a balance between S/N and sensitivity of the assay [44].
9. If assaying multiple compounds, make 3 mM dilutions in adjacent wells in the same column (B1-P1). If assaying more than 16 compounds, compound 17 should start at well A2 and continue to P2 through compound 32.
10. If assaying multiple compounds, each 125 nL of 3 mM dilution from the "compound source plate" should be in every second well. For example, 125 nL of "compound source plate" well A1 is dispensed in "compound intermediate plate" well B1; 125 nL of "compound source plate" well B1 is dispensed in "compound intermediate plate" well D1. The eighth compound would be 125 nL of "compound source plate" well H1 dispensed in "compound intermediate plate" well P1. If more than eight compounds, continue to "compound intermediate plate" well B2.
11. The 1–30 μL E1-ClipTip electronic pipette is used for this step for assaying multiple compounds. With the pipette set to 96-well plate, every other well in a column can be filled with 10 uL DMSO easily in one step.

12. This is done easiest by picking up the stock solutions with the pipette tips set to 384-well plate and after picking up the solution, sliding the tip setting to 96-well plate. This lines up the tips with every second well in a 384-well plate.
13. This assay is compatible with many oncogenic mutants of Avi-KRAS4b. Duplicate the compound dilution series in a new plate for each mutant. If there are five Avi-KRAS4b mutants, duplicate the dilution series of each compound into six plates (the sixth plate is for the double-tagged Avi-RAF-1-RBD-3xFLAG control). The type of plate can also contribute to the sensitivity of the assay and minimization of edge effects.
14. The first four dilutions will come from the 3 mM MRTX1133 well of the "compound intermediate plate." A1 contains 202.5 nL of 3 mM MRTX1133, A2 contains 67.5 nL of 3 mM MRTX1133, A3 contains 22.5 nL of 3 mM MRTX1133, and A4 contains 7.5 nL of 3 mM MRTX1133. These wells (A1–A4) are then backfilled with DMSO from one of the wells of the four rows containing DMSO in the "compound intermediate plate" to 202.5 nL total. The last five dilutions will come from the 37 μM MRTX well of the "compound intermediate plate." A5 contains 202.5 nL of 37 μM MRTX1133, A6 contains 67.5 nL of 37 μM MRTX1133, A7 contains 22.5 nL of 37 μM MRTX1133, A8 contains 7.5 nL of 37 μM MRTX1133, and A9 contains 2.5 nL of 37 μM MRTX1133. These wells (A5–A9) are then backfilled with DMSO from one of the wells of the four rows containing DMSO in the "compound intermediate plate" to 202.5 nL total. Finally, fill the tenth well in that row (A10) with 202.5 nL of DMSO only as a negative control.
15. If multiple compounds are being screened, this procedure can be extended to additional wells on each plate. For example, the "compound source plate" can contain up to 384 individual compounds. These would be diluted into two "compound intermediate plates" which would then be diluted into as many adjacent "compound destination plate" rows (A1–A9, B1–B9,. . ., P1–P9) and plates as required. Each compound to be assayed should be done so in duplicate at minimum, meaning one 384-well plate can assay 16 compounds (A1–A10, B1–B10,. . ., P1–P10 and A13–A22, B13–B22,. . ., P13–P22). Typically, no more than 50 compounds are assayed at a time, but the capability for 100s of compounds is there.
16. The volumes used to prepare these solutions will change based on the concentration of the proteins after preparation as well as the number of compounds assayed. The following is an example for the Avi-KRAS4b WT protein stock solution prepared from a stock solution of 50 μM Avi-KRAS4b WT and 100 μM

RAF-1-RBD-3xFLAG: Add 997 μL of "assay buffer" to a 1.5 mL Eppendorf tube. Add 2 μL of 50 μM Avi-KRAS4b WT and 1 μL of 100 μM RAF-1-RBD-3xFLAG. Flick the bottom of the Eppendorf tube to stir.

17. The volumes used to prepare these solutions will change based on the concentration of the proteins after preparation. The following is an example for the control protein stock solution prepared from a stock solution of 100 μM 3xFLAG-RAF-1-RBD-Avi: Add 999 μL of "assay buffer" to a 1.5 mL Eppendorf tube. Add 1 μL of 100 μM 3xFLAG-RAF-1-RBD-Avi. Flick the bottom of the Eppendorf tube to stir.
18. If performing this assay with multiple compounds, fill four adjacent channels of the Argos Technologies pipetting reservoir with equal volumes of "assay protein mix" or "assay control mix." With the 2–125 μL electronic pipette set to 384 well, 2 tips fit in each channel, meaning 4 adjacent channels allows for 8 tips to fill with 100 μL each. Eight tips dispense for half of the "compound destination plate" (A–H, or I through P). **Steps 6–9** can then be repeated with clean tips for all the wells containing compound or DMSO in the "compound destination plate" for a single KRAS4b mutant (repeated 4× for one plate).
19. If performing this assay with multiple compounds, fill four adjacent channels of the Argos Technologies pipetting reservoir with equal volumes of "HTRF mix." With the 2–125 μL E1-ClipTip electronic pipette set to 384 well, two tips fit in each channel, meaning four adjacent channels allows for eight tips to fill with 100 μL each. Eight tips dispense for half of the "compound destination plate" (A–H, or I through P). **Steps 5–7** can then be repeated with clean ClipTips for all the wells containing compound or DMSO in the "compound destination plate" for a single KRAS4b mutant (repeated 4× for one plate).
20. This plate reader contains a high-stability, short-arc flash lamp. Laser excitation can improve S/N through narrower excitation pulses [45].

## Acknowledgments

We would like to acknowledge Bridge Bio for supplying the library of compounds to assay. Additionally, we would like to acknowledge Dominic Esposito, Bill Gillette, Gulcin Gulten, Jennifer Mehalko, Simon Messing, Nitya Ramakrishnan, Troy Taylor, and Vanessa Wall for cloning, protein expression, protein purification, and electrospray ionization mass spectroscopy. We would also like to acknowledge Anna Maciag for facilitating lead compound acquisition. This project was funded in part with federal funds from the National Cancer Institute, National Institutes of Health Contract 75N91019D00024. The content of this publication does not necessarily reflect the views or policies of the Department of Health and Human Services, and the mention of trade names, commercial products, or organizations does not imply endorsement by the US government.

## References

1. Cox AD, Der CJ (2010) Ras history: the saga continues. Small GTPases 1:2–27
2. Shih TY, Weeks MO, Young HA et al (1979) Identification of a sarcoma virus-coded phosphoprotein in nonproducer cells transformed by Kirsten or Harvey murine sarcoma virus. Virology 96:64–79
3. Scolnick EM, Papageorge AG, Shih TY (1979) Guanine nucleotide-binding activity as an assay for src protein of rat-derived murine sarcoma viruses. Proc Natl Acad Sci USA 76:5355–5359
4. Van Aelst L, Barr M, Marcus S et al (1993) Complex formation between RAS and RAF and other protein kinases. Proc Natl Acad Sci USA 90:6213–6217
5. Han M, Golden A, Han Y et al (1993) *C. elegans* lin-45 raf gene participates in let-60 ras-stimulated vulval differentiation. Nature 363:133–140
6. Warne PH, Viciana PR, Downward J (1993) Direct interaction of Ras and the amino-terminal region of Raf-1 in vitro. Nature 364: 352–355
7. Zhang XF, Settleman J, Kyriakis JM et al (1993) Normal and oncogenic p21ras proteins bind to the amino-terminal regulatory domain of c-Raf-1. Nature 364:308–313
8. Moodie SA, Willumsen BM, Weber MJ et al (1993) Complexes of Ras.GTP with Raf-1 and mitogen-activated protein kinase kinase. Science 260:1658–1661
9. Huang L, Guo Z, Wang F et al (2021) KRAS mutation: from undruggable to druggable in cancer. Signal Transduct Target Ther 6:386
10. Prior IA, Lewis PD, Mattos C (2012) A comprehensive survey of Ras mutations in cancer. Cancer Res 72:2457–2467
11. McGrath JP, Capon DJ, Smith DH et al (1983) Structure and organization of the human Ki-ras proto-oncogene and a related processed pseudogene. Nature 304:501–506
12. Aran V (2021) K-RAS4A: lead or supporting role in cancer biology? Front Mol Biosci 8: 729830
13. Skoulidis F, Li BT, Dy GK et al (2021) Sotorasib for lung cancers with KRAS p.G12C mutation. N Engl J Med 384:2371–2381
14. Bauml JM, Li BT, Velcheti V et al (2022) Clinical validation of Guardant360 CDx as a blood-based companion diagnostic for sotorasib. Lung Cancer 166:270–278
15. Janne PA, Riely GJ, Gadgeel SM et al (2022) Adagrasib in non-small-cell lung cancer harboring a KRAS(G12C) mutation. N Engl J Med 387:120–131
16. Hunter JC, Manandhar A, Carrasco MA et al (2015) Biochemical and structural analysis of common cancer-associated KRAS mutations. Mol Cancer Res 13:1325–1335
17. Stokoe D, McCormick F (1997) Activation of c-Raf-1 by Ras and Src through different mechanisms: activation in vivo and in vitro. EMBO J 16:2384–2396
18. Fujita-Yoshigaki J, Shirouzu M, Ito Y et al (1995) A constitutive effector region on the C-terminal side of switch I of the Ras protein. J Biol Chem 270:4661–4667
19. Ghosh S, Xie WQ, Quest AF et al (1994) The cysteine-rich region of raf-1 kinase contains zinc, translocates to liposomes, and is adjacent to a segment that binds GTP-ras. J Biol Chem 269:10000–10007
20. Skinner RH, Picardo M, Gane NM et al (1994) Direct measurement of the binding of RAS to neurofibromin using a scintillation proximity assay. Anal Biochem 223:259–265
21. Gorman C, Skinner RH, Skelly JV et al (1996) Equilibrium and kinetic measurements reveal rapidly reversible binding of Ras to Raf. J Biol Chem 271:6713–6719
22. Mathis G (1995) Probing molecular interactions with homogeneous techniques based on rare earth cryptates and fluorescence energy transfer. Clin Chem 41:1391–1397

23. Einhorn L, Krapfenbauer K (2015) HTRF: a technology tailored for biomarker determination-novel analytical detection system suitable for detection of specific autoimmune antibodies as biomarkers in nanogram level in different body fluids. EPMA J 6:23
24. Selvin PR (2002) Principles and biophysical applications of lanthanide-based probes. Annu Rev Biophys Biomol Struct 31:275–302
25. Degorce F, Card A, Soh S et al (2009) HTRF: a technology tailored for drug discovery – a review of theoretical aspects and recent applications. Curr Chem Genomics 3:22–32
26. Kopra K, Vuorinen E, Abreu-Blanco M et al (2020) Homogeneous dual-parametric-coupled assay for simultaneous nucleotide exchange and KRAS/RAF-RBD interaction monitoring. Anal Chem 92:4971–4979
27. Trinh TB, Upadhyaya P, Qian Z et al (2016) Discovery of a direct Ras inhibitor by screening a combinatorial library of cell-permeable bicyclic peptides. ACS Comb Sci 18:75–85
28. PerkinElmer (ed) (2021) Protein-protein interaction assays with HTRF
29. Wang X, Allen S, Blake JF et al (2022) Identification of MRTX1133, a noncovalent, potent, and selective KRAS(G12D) inhibitor. J Med Chem 65:3123–3133
30. Wall VE, Garvey LA, Mehalko JL et al (2014) Combinatorial assembly of clone libraries using site-specific recombination. Methods Mol Biol 1116:193–208
31. Taylor T, Denson JP, Esposito D (2017) Optimizing expression and solubility of proteins in *E. coli* using modified media and induction parameters. Methods Mol Biol 1586:65–82
32. Waybright T, Stephen AG (2024) Nucleotide exchange on RAS proteins using hydrolysable and non-hydrolysable nucleotides. In: KRAS, methods in molecular biology
33. Sebaugh JL (2011) Guidelines for accurate EC50/IC50 estimation. Pharm Stat 10:128–134
34. Burlingham BT, Widlanski TS (2003) An intuitive look at the relationship of K-i and IC50: a more general use for the Dixon plot. J Chem Educ 80:214–218
35. Christensen JG (2021) Discovery and characterization of MRTX1133, a selective non-covalent inhibitor of KRASG12D. Mirati Therapeutics, p 13
36. Bar H, Zweifach A (2020) Z' does not need to be > 0.5. SLAS Discov 25:1000–1008
37. Zhang JH, Chung TD, Oldenburg KR (1999) A simple statistical parameter for use in evaluation and validation of high throughput screening assays. J Biomol Screen 4:67–73
38. Seveus L, Vaisala M, Syrjanen S et al (1992) Time-resolved fluorescence imaging of europium chelate label in immunohistochemistry and in situ hybridization. Cytometry 13:329–338
39. Connally R, Veal D, Piper J (2004) Flash lamp-excited time-resolved fluorescence microscope suppresses autofluorescence in water concentrates to deliver an 11-fold increase in signal-to-noise ratio. J Biomed Opt 9:725–734
40. Spoerner M, Nuehs A, Herrmann C et al (2007) Slow conformational dynamics of the guanine nucleotide-binding protein Ras complexed with the GTP analogue GTPgammaS. FEBS J 274:1419–1433
41. Stumber M, Herrmann C, Wohlgemuth S et al (2002) Synthesis, characterization and application of two nucleoside triphosphate analogues, GTPgammaNH(2) and GTPgammaF. Eur J Biochem 269:3270–3278
42. Spoerner M, Nuehs A, Ganser P et al (2005) Conformational states of Ras complexed with the GTP analogue GppNHp or GppCH2p: implications for the interaction with effector proteins. Biochemistry 44:2225–2236
43. Kane SA, Fleener CA, Zhang YS et al (2000) Development of a binding assay for p53/HDM2 by using homogeneous time-resolved fluorescence. Anal Biochem 278:29–38
44. Brooks HB, Geeganage S, Kahl SD et al (2004) Basics of enzymatic assays for HTS. In: Markossian S et al (eds) Assay guidance manual. Eli Lilly & Company and the National Center for Advancing Translational Sciences, Bethesda
45. Reichstein E, Shami Y, Ramjeesingh M et al (1988) Laser-excited time-resolved solid-phase fluoroimmunoassays with the new europium chelate 4,7-bis(chlorosulfophenyl)-1,10-phenanthroline-2,9-dicarboxylic acid as label. Anal Chem 60:1069–1074

# Chapter 13

# Biophysical Characterization of RAS-SOS Complexes by Native Mass Spectrometry

Sangho Yun, Elena Scott, and Arthur Laganowsky

## Abstract

RAS is regulated by specific guanine nucleotide exchange factors, such as Son of Sevenless (SOS), that activates RAS by facilitating the exchange of inactive, GDP-bound RAS with GTP. The catalytic activity of SOS is known to be allosterically modulated by an active, GTP-bound RAS. However, it remains poorly understood how oncogenic RAS mutants interact with SOS and modulate its activity. In this chapter, we describe the application of native mass spectrometry (MS) to monitor the assembly of the catalytic domain of SOS ($SOS^{cat}$) with RAS and cancer-associated mutants. Results from this approach have led to the discovery of different molecular assemblies and distinct conformers of $SOS^{cat}$ engaging KRAS. It was also found that $KRAS^{G13D}$ exhibits high affinity for $SOS^{cat}$ and is a potent allosteric modulator of its $SOS^{cat}$ activity. $KRAS^{G13D}$-GTP can allosterically increase the nucleotide exchange rate of KRAS at the active site by more than twofold compared to the wild-type protein. Furthermore, small-molecule RAS•SOS disruptors fail to dissociate $KRAS^{G13D}$•$SOS^{cat}$ complexes, underscoring the need for more potent disruptors targeting oncogenic RAS mutants. Taken together, native MS will be instrumental in better understanding the interaction between oncogenic RAS mutants and SOS, which is of crucial importance for development of improved therapeutics.

**Key words** Native mass spectrometry, KRAS, RAS-SOS, Allostery, Drugs

## 1 Introduction

In cells, RAS possesses low intrinsic guanosine triphosphate (GTP) hydrolysis and guanidine nucleotide exchange rates [1, 2]. Therefore, regulation of RAS signaling is controlled by GTPase-activating proteins (GAPs) and guanine exchange factors (GEFs) that stimulate GTP hydrolysis and nucleotide exchange of guanosine diphosphate (GDP) for GTP, respectively [1, 3–5]. Son of Sevenless (SOS) is a GEF with catalytic domains ($SOS^{cat}$) compromising the cdc25 and RAS exchanger motifs [6]. Two RAS proteins are known to bind to $SOS^{cat}$ to facilitate the nucleotide exchange: binding of RAS-GDP at active site; RAS-GTP binding to the distal (or allosteric) site that allosterically modulates

Andrew G. Stephen and Dominic Esposito (eds.), *KRAS: Methods and Protocols*, Methods in Molecular Biology, vol. 2797, https://doi.org/10.1007/978-1-0716-3822-4_13,

$SOS^{cat}$ activity, increasing nucleotide exchange at the active site [6, 7]. The nucleotide-bound state of RAS greatly impacts the degree of the allosteric modulation [6, 8, 9]. Additionally, SOS is conformationally dynamic and binding of RAS-GTP at the allosteric site shifts the population to active conformation(s) of SOS [10].

Native or non-denaturing mass spectrometry (MS) has emerged as an indispensable biophysical technique for investigating proteins and their interactions with other molecules [11–15]. In native MS, biological samples are introduced into the mass spectrometer using nano-electrospray ionization. The instrument is carefully tuned to preserve non-covalent interactions [15]. This enables the analysis of protein-ligand interactions, subunit stoichiometry, and measurement of thermodynamics and equilibrium-binding affinities [15–17]. Recently, native MS has been used to characterize the binding of small molecules such as drugs, nucleotides, lipids, metals, and small enzyme cofactors [18–24]. Another advantage of using native MS to study protein-ligand interactions over other biophysical techniques, such as BRET/FRET-based approaches, is that natural, unmodified proteins/ligands can be used directly. Application of native MS approaches to study oncogenic proteins and the complexes they form will address key questions that would remain otherwise intractable using other biophysical techniques.

Although the interactions of RAS and SOS have been studied in the past, the interactions of oncogenic RAS mutants with SOS are not well understood. In this chapter, we describe the application of native mass spectrometry (MS) to investigate the biochemical properties of KRAS and its oncogenic mutants bound to different guanosine nucleotides. In addition, inhibitors designed to inhibit RAS, SOS, and RAS•SOS complex such as ARS-1620, Bay-293, Kobe0065, and BI-3406 can be investigated as well as monitor their disrupting efficacies.

## 2 Materials

Prepare all solutions using ultrapure, deionized water (Milli-Q) and analytical grade reagents. Prepare and store all reagents at room temperature (unless indicated otherwise). Diligently follow all waste disposal regulations when disposing waste materials. We do not add sodium azide to reagents.

### 2.1 Protein Expression and Purification

1. KRAS4B (residues 1–169, Uniprot P01116-2) plasmid for expression in *Escherichia coli* (a kind gift from K. D. Westover at the University of Texas Southwestern Medical Center Dallas, TX): modified pDNA 2.0 containing N-terminal TEV protease cleavable $His_6$ tag.

2. HRAS (residues 1–166, UniProt P01112) and NRAS (residues 1–172, UniProt P0111) codon optimized for *E. coli.*: Synthesized as gBlock gene fragments (Integrated DNA Technologies) and cloned into a modified pCDF-1b (Novagen) expression construct containing N-terminal TEV protease cleavable $His_6$ tag.
3. $SOS1^{cat}$ (residues 558–1049) plasmid (a kind gift from Prof. John Kuriyan at the University of California Berkeley): N-terminal TEV protease cleavable $His_6$ tag.
4. Lemo21-(DE3) (New England Biolabs).
5. Rosetta 2-(DE3) (Novagen).
6. Luria broth (LB).
7. β-D-1-thigalactopyranoside (IPTG) (0.5 M solution in water).
8. β-mercaptoethanol (β-ME) (14 M solution): diluted to 5 mM final in buffer of choice.
9. Microfluidizer (microfluidics M-110P).
10. 5 mL HisTrap HP column (GE Healthcare).
11. HiPrep 26/10 desalting column (GE Healthcare).
12. HiLoad 16/600 Superdex 75 pg column (GE Healthcare).
13. Centrifugal concentrator with 10 K molecular weight cutoff (MWCO).

### 2.2 Loading RAS with Nucleotides

1. Ethylene diamine tri-acetic acid (EDTA) (0.5 M solution in water): 0.19 grams of EDTA disodium salt is dissolved in 1 mL of water.
2. $MgCl_2$ (1 M solution in water): 2.03 grams of magnesium chloride hexahydrate dissolved in 1 mL of water.
3. Guanine diphosphate (disodium salt): Stock solution is made at 50 mM concentration in water. Stock solution is stored at −20 °C for long-term storage, thawed on ice before use.
4. Guanine triphosphate (disodium salt): Stock solution is made at 50 mM concentration in water. Stock solution is stored at −20 °C for long-term storage, thawed on ice before use.
5. Micro bio-spin 6 column.
6. Dilution buffer (20 mM Tris–HCl, 150 mM NaCl, 10% glycerol, pH 7.4): 17.53 grams of sodium chloride, 2.43 grams of Tris–HCl, and 126 grams of glycerol are dissolved in 800 mL of water. Buffer solution is pH adjusted with hydrochloric acid to pH 7.4. Water is added up to 1 L, and the final buffer solution is filtered (0.45 micron filter) and stored at ambient temperature.

### 2.3 Sample Preparation for Native Mass Spectrometry

1. Ammonium acetate (AA) (200 mM solution in water): 0.77 grams of ammonium acetate is dissolved in 40 mL of water. pH is adjusted to 7.4 with ammonium hydroxide (30%). Water is added up to 50 mL and the final solution is filtered and stored at 4 °C.
2. Micro bio-spin 6 column.

### 2.4 Native Mass Spectrometry

1. Thermo Q-Exactive extended mass range mass spectrometer (ThermoFisher Scientific).
2. Thermo Q-Exactive ultra-high mass range mass spectrometer (ThermoFisher Scientific).
3. Unidec 4.2.2: Refer to Marty et al. Anal. Chem. 2015, [25] and can be accessed at https://github.com/michaelmarty/UniDec/releases.
4. Borosilicate glass (Sutter Instrument, B150-86-10): Outer diameter: 1.5 mm, inner diameter: 0.86 mm, 10 cm length pulled by P-97 Flaming/Brown Micropipette Puller.
5. Platinum wire, 0.3 mm diameter, cut to ~5 cm.

### 2.5 Drug-Binding Studies

1. ARS1620 Stock solution (Chemietek): 50 mM stock solution was made with DMSO. Stock solution is stored at −20 °C for long-term storage, thawed on ice before use.
2. Kobe0065 stock solution (Tocris Bioscience): 50 mM stock solution was made with DMSO. Stock solution is stored at −20 °C for long-term storage, thawed on ice before use.
3. Bay-293 stock solution (Tocris Bioscience): 50 mM stock solution was made with DMSO. Stock solution is stored at −20 °C for long-term storage, thawed on ice before use.
4. BI-3406 stock solution (TargetMol): 50 mM stock solution was made with DMSO. Stock solution is stored at −20 °C for long-term storage, thawed on ice before use.

## 3 Methods

### 3.1 Protein Expression and Purification

1. Grow cells containing the expression plasmid in LB medium supplemented with the necessary antibiotics.
2. Induce protein expression with 500 μM IPTG once the cells reach an $OD_{600}$ of 0.6–0.8.
3. Grow cells at 18 °C overnight.
4. Harvest cells by centrifugation at 5000× *g*, and store pellets at −80 °C.

5. Thaw cell pellets on ice, and resuspend in buffer A (300 mM NaCl, 20 mM Tris–HCl pH 7.4) supplemented with 1 mM $MgCl_2$ and 5 mM β-ME.
6. Add 50 μM of nucleotide (GDP or GTP) to all buffers throughout the purification process.
7. Lyse the cell suspension with a microfluidizer operating at 20,000 psi.
8. Clarify the lysate by centrifugation at 40,000× *g* for 20 min.
9. Perform all subsequent purification steps at 4 °C.
10. Load the clarified lysate onto a 5 mL HisTrap HP column pre-equilibrated in buffer A containing 20 mM imidazole.
11. Elute the protein with buffer A containing 500 mM imidazole.
12. Pool the fractions containing protein and desalted using a HiPrep 26/10 desalting column equilibrated in buffer A.
13. Digest the desalted protein with TEV protease at 4 °C, and incubated overnight to remove the N-terminal affinity tag.
14. Pass the protein over a 5 mL HisTrap HP column equilibrated with buffer A, and collect flow-through containing tag-less protein.
15. Concentrate the protein in the flow-through using a centrifugal concentrator.
16. Perform size exclusion chromatography using a HiLoad 16/600 Superdex 75 pg column equilibrated in buffer C (20 mM Tris–HCl, pH 7.4, 150 mM sodium chloride, 5 mM β-ME, 10% w/v glycerol, 0.5 mM $MgCl_2$, ± 50 μM GDP or GTP).
17. Pool the peak fractions containing the target protein, and add glycerol to a final concentration of 20% (v/v).
18. Concentrate using a centrifugal concentrator (10 K molecular weight cutoff), and flash-freeze the final protein in liquid nitrogen. For more detailed protocols for expression and purification of the protein, refer to [26].

### 3.2 Loading RAS with Nucleotide of Choice

1. Incubate RAS or its oncogenic mutants at a concentration of 100 μM in dilution buffer supplemented with 10 mM EDTA and 2 mM GTP or GDP (30× molar excess) for 3 h at 4 °C.
2. Supplement the reaction mixture with 10 mM $MgCl_2$, and incubate for an additional 30 min at 4 °C.
3. Remove excess nucleotide, $MgCl_2$, and EDTA using a micro bio-spin 6 desalting column.
4. Invert the column to resuspend the resin inside. Centrifuge the column with the top cap and the tip still intact at 1000× *g* for 2 min.

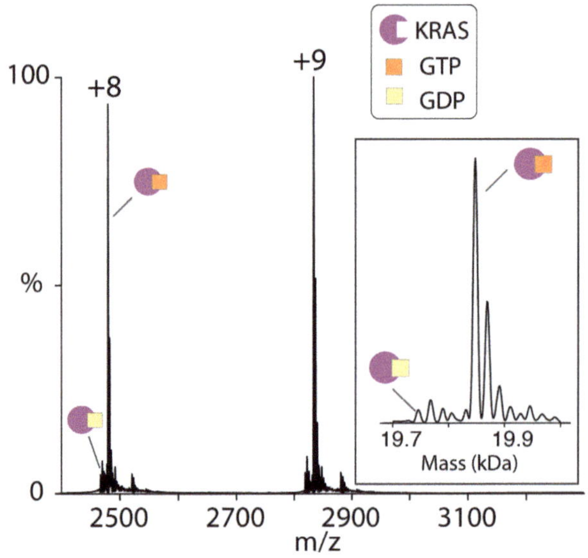


**Fig. 1** Mass spectrum of KRAS-GTP after nucleotide loading with GTP. Deconvolution of the mass spectrum of KRAS predominantly loaded with GTP

5. Snap off the tip and remove the top cap.
6. Centrifuge the column remove the resin storing solution (2 min at 1000× *g*).
7. Equilibrate with the buffer of choice by adding 450 μL of buffer and centrifuging for 1 min at 1000× *g*.
8. Discard the flow-through and repeat the procedure four times.
9. Place the column into a clean 1.6 mL tube, and add 100 μL sample solution directly onto the resin.
10. Centrifuge the column for 4 min at 1000× *g* to collect the protein of interest in the exchanged buffer.
11. Repeat this procedure until full exchange of GDP or GTP is completed as determined by native MS measurement. For example, KRAS wild type is loaded with GTP three consecutive times to reach >95% GTP-loading (*see* Fig. 1). Despite repeated loading, some oncogenic mutants of RAS are unable to reach complete or satisfactory loading of the desired nucleotide (*see* **Note 1**).

### 3.3 Sample Preparation for Nano ESI-Native Mass Spectrometry

1. Buffer exchange each individual protein to a volatile, non-adducting solvent (200 mM AA).
2. Follow instructions in Subheading 3.2, **step 3** for use of the micro bio-spin column.
3. Equilibrate the column by adding 450 μL of 200 mM AA as the exchange buffer.

4. Repeat this procedure four times.
5. Place the desalting column into a clean tube, and add 100 μL 200 mM AA directly onto the resin.
6. Centrifuge the column at 1000× *g* for 4 min to collect the eluted protein now in 200 mM AA buffer.

### 3.4 Native Mass Spectrometry Analysis

1. Load protein samples from Subheading 3.3 into pulled borosilicate glass capillaries prepared in-house [26] and electro sprayed with a voltage applied via a platinum wire inserted directly into the solution (*see* **Note 2**).
2. Introduce ionized samples into a Thermo Q-Exactive plus extended mass range orbitrap mass spectrometer (*see* **Note 3** for information on tuning parameters).
3. Calculate the mole fraction of nucleotide-bound RAS by deconvolution using Unidec [25] and calculated from abundances as follows where $i$ is the number of adducts ($A$) bound to RAS:

$$F_{[\mathrm{RAS\cdot GDP}]} = \frac{\sum_{i=0}^{3} [\mathrm{RAS\cdot GDP}]\cdot A_i}{\sum_{i=0}^{3} [\mathrm{RAS\cdot GDP}]\cdot A_i + \sum_{i=0}^{3} [\mathrm{RAS\cdot GTP}]\cdot A_i}$$

4. Calculate the specific mole fraction as follows, where $P_{\mathrm{tot}}$ represents the enzyme's total concentration:

$$[\mathrm{RAS\cdot GDP}]_{\mathrm{total}} = P_{\mathrm{tot}} F_{[\mathrm{RAS\cdot GDP}]}$$

5. Calculate the mole fraction and concentration of [RAS · GTP] in a similar fashion as [RAS · GDP].
6. Calculate the mole fraction of RAS●SOS$^{\mathrm{cat}}$ complexes based on the relative abundance obtained from deconvoluted mass spectrum where $i$ is the number of adducts ($A$) bound to ternary complexes:

$$F_{[\mathrm{RAS\cdot SOScat\cdot RAS}]} = \frac{\sum_{i=0}^{3} [\mathrm{RAS\cdot SOScat\cdot RAS}]\, A_i}{\sum_{i=0}^{3} [\mathrm{RAS\cdot SOScat\cdot RAS}]\cdot A_i + \sum_{i=0}^{3} [\mathrm{RAS\cdot SOScat\cdot RAS\_GDP}]\cdot A_i}$$

7. Use these data to analyze the details of complexes formed between RAS and SOS (*see* Fig. 2 for examples of mass spectra and *see* **Note 4** for additional discussion).
8. Optionally, confirmation of the stoichiometric measurement of the complexes can also be done via native MS through mass isolation and collision-induced dissociation experiments performed on a Waters Synapt G1 (*see* **Note 5**).

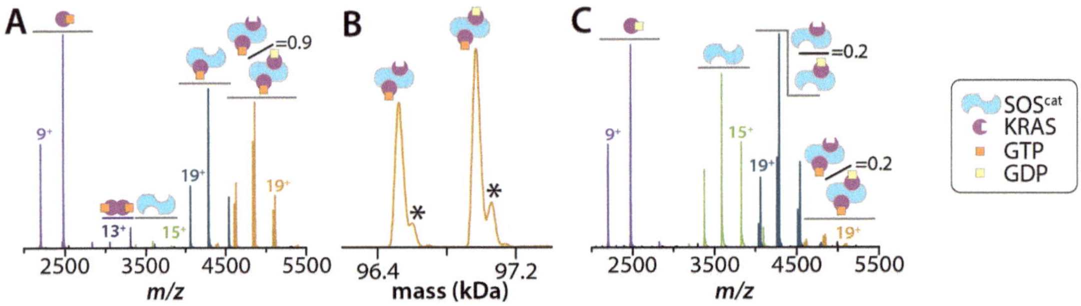


**Fig. 2** Molecular assemblies of $SOS^{cat}$ with KRAS. (**a**) Native mass spectrum of 2 μM $SOS^{cat}$ and 6 μM KRAS-GTP recorded immediately after mixing. (**b**) Deconvolution of mass spectrum in panel A and selected mass range corresponding to ternary complexes. Inorganic phosphate adducts are denoted by an asterisk. (**c**) Mass spectrum of 2 μM $SOS^{cat}$ and 6 μM KRAS-GDP recorded immediately after mixing. (Adapted from [30])

9. Competition assays can also be performed using native MS by mixing a complex of $SOS^{cat}$ with GTP-loaded $KRAS^{WT}$ harboring an N-terminal hexa-histidine affinity tag ($His_6$-$KRAS^{WT}$-GTP) in three molar excess of $SOS^{cat}$. Keeping the intact affinity tag results in additional mass aiding in the assignment of different KRAS species (*see* **Note 6**).

### 3.5 Disruption of the Interaction Between KRAS and $SOS^{cat}$

There is growing interest in the development of small molecule disruptors of RAS-SOS interactions. Drug treatment studies can assess the ability of compounds to disrupt these interactions.

1. Mix 1 μM $SOS^{cat}$ and 3 μM $KRAS^{G12C}$-GDP or $KRAS^{G12C}$-GTP and incubate for 5 min on ice.
2. Add either 10 μM ARS-1620, 2.5 μM BAY-293, 2.5 μM Kobe0065, or 2.5 μM BI-3406 to the preincubated complex.
3. Incubate on ice for 60 min.
4. Determine native mass spectra as described in Subheading 3.4.
5. Example data are shown for ARS-1620 (*see* Fig. 3a, e and **Note 7**), BAY-293 (*see* Fig. 3c, g and **Note 8**), Kobe0065 (*see* Fig. 3b, f and **Note 9**), or BI-3406 (*see* Fig. 3d, h and **Note 10**).

## 4 Notes

1. Some RAS mutants, such as G12D and G12V, may be recalcitrant to complete nucleotide exchange, regardless of the number of loading procedures (*see* Fig. 4). Although repeated loading with GTP increases overall abundance of GTP-bound species, GTP-bound state never reaches >90% for these mutants. This phenomenon seems to be consistent between isoforms of RAS only for G12D and G12V mutations. In this

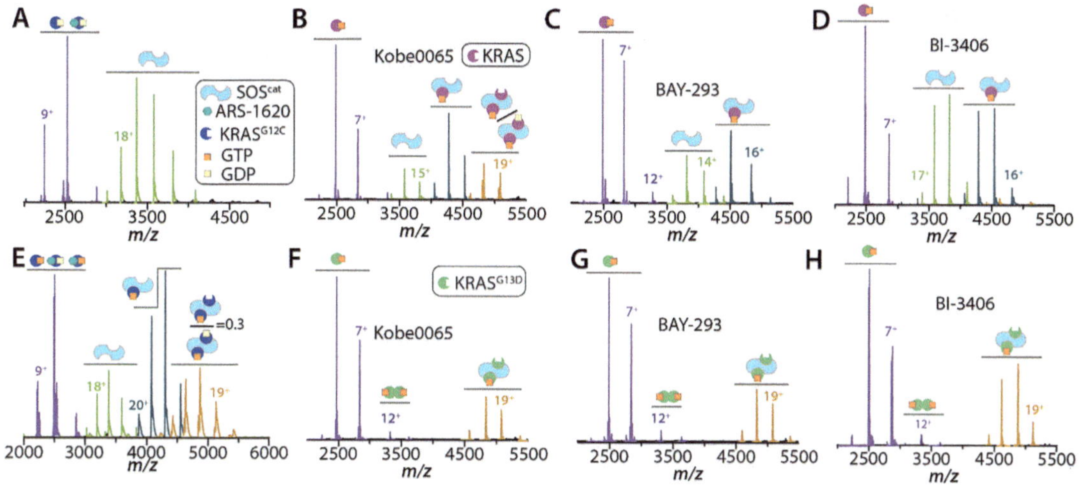


**Fig. 3** The effect of ARS-1620, BAY-293, Kobe0065, and BI-3406 on assemblies of KRAS and $SOS^{cat}$. (**a–e**) Mass spectrum recorded after adding 10 μM ARS-1620 to preincubated mixtures of 1 μM $SOS^{cat}$ with 3 μM (**a**) $KRAS^{G12C}$-GDP and (**e**) $KRAS^{G12C}$-GTP. Mass spectra are shown as described in Fig. 1. (**b–f**) Mass spectrum recorded after adding 2.5 μM Kobe0065 to preincubated mixtures of 1 μM $SOS^{cat}$ with 3 μM (**b**) KRAS-GTP and (**f**) $KRAS^{G13D}$-GTP. (**c–g**) Mass spectrum recorded after adding 2.5 μM BAY-293 to preincubated mixtures of 1 μM $SOS^{cat}$ with 3 μM (**c**) KRAS-GTP and (**g**) $KRAS^{G13D}$-GTP. (**d–h**) Mass spectrum recorded after adding 2.5 μM BI-3406 to preincubated mixtures of 1 μM $SOS^{cat}$ with 3 μM (**d**) KRAS-GTP and (**h**) $KRAS^{G13D}$-GTP. (Adapted from [30])

case, concentration of GTP-bound RAS protein is adjusted based on the molar ratio calculated from the deconvoluted mass spectrum.

2. Borosilicate nanoESI capillaries are prepared in-house using a Sutter Instrument P-97 Flaming/Brown Micropipette Puller (https://www.sutter.com/MICROPIPETTE/p-97.html). For this experiment, borosilicate glass (Sutter Instrument, B150-86-10) with outer diameter: 1.5 mm, inner diameter: 0.86 mm, and 10 cm length is used. Various size capillaries are available for different purposes, but this is a good starting point. P-97 Micropipette Puller should be optimized to control the capillaries' shape and size. Refer to *Pipette Cookbook* and product description found on the Sutter instrument website for further information regarding programmable parameters. The sample is then loaded into the capillary either by 10 μL Hamilton syringe, or a gel-loading tip. Generally, 5–10 μL sample volume should be sufficient for native MS analysis (*see* Fig. 5d). The amount of sample necessary for analysis will depend on the duration of the experiment and/or on the flow rate. More sample can be added into the capillary as needed. Sample concentration is an important parameter to be considered. Too low of a sample concentration will make it difficult to obtain a strong signal for the analyte of

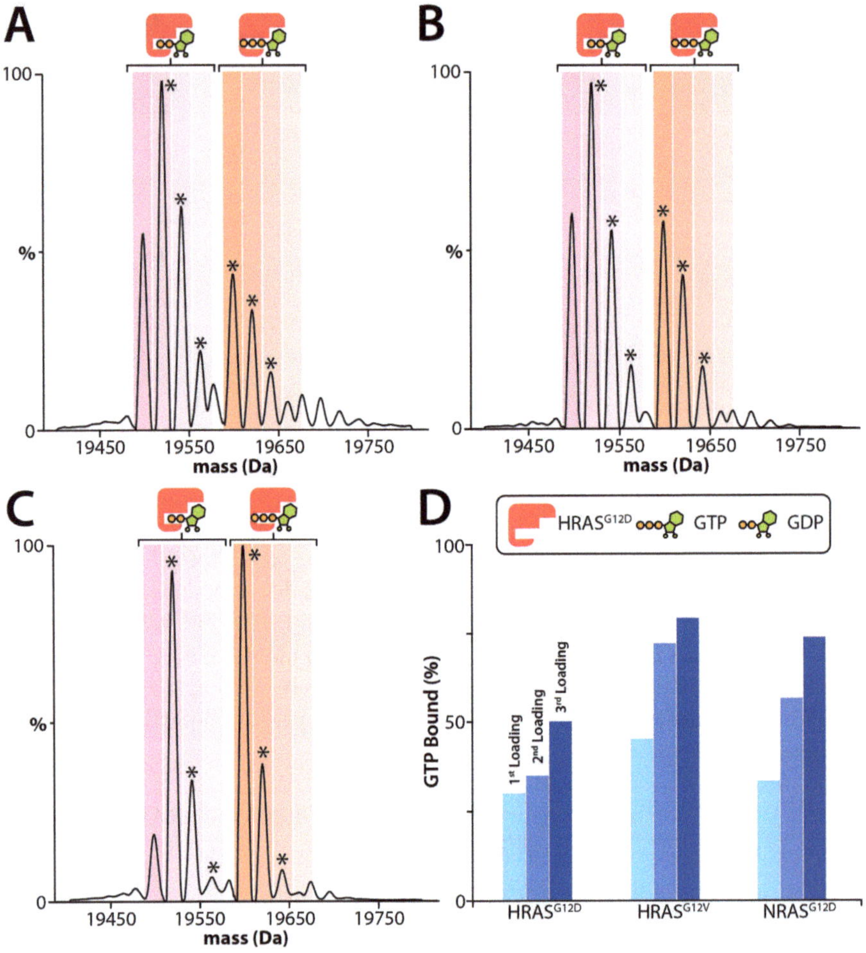


**Fig. 4** Some oncogenic RAS mutants are recalcitrant to loading with GTP. (**a**–**c**) Deconvolution of native mass spectra for each GTP loading step of HRAS$^{G12D}$. The (**a**) first, (**b**) second, and (**c**) third consecutive loading of the protein is shown. Asterisks denote sodium or magnesium-bound adducts. (**d**) Percentage of GTP bound after each loading step for HRAS$^{G12D}$, HRAS$^{G12V}$, and NRAS$^{G12D}$

interest. On the other hand, too high of a sample concentration could induce nonspecific oligomerization and/or sub-optimal ionization for the MS analysis. Generally, recommended concentration for an analyte is 1 μM, but can be optimized based on the signal-to-noise ratio.

The Thermo Nanospray Flex ion source (ES071) is used for the NanoESI platform (*see* Fig. 5). The stage can be adjusted in three dimensions via adjustable knobs (*see* Fig. 5a). Also, USB cameras (ES218) can be used for a live image of the source entry to monitor the location of the capillary tip and the size of the droplet prior to applying voltage. Platinum wire is inserted into the source head (*see* Fig. 5b) for the voltage application, and the 0.3 mm diameter platinum

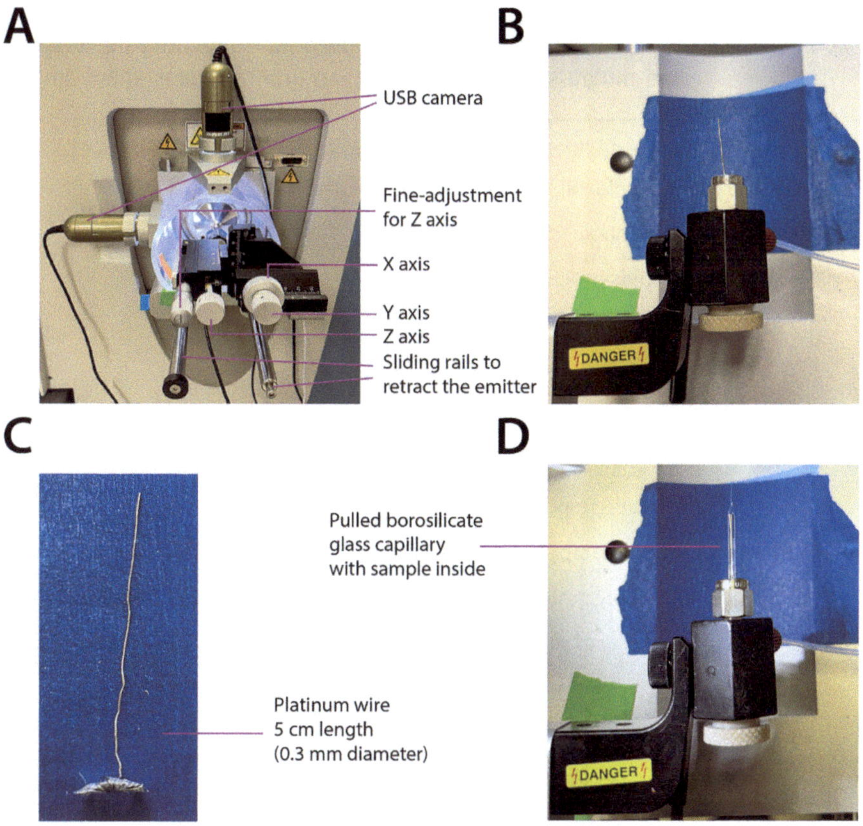


**Fig. 5** Nano-ESI source for Thermo Q-Exactive mass spectrometers. (**a**) Thermo Nanospray Flex ion source (ES071) with the USB cameras (ES218). (**b**) Source head with platinum wire inserted. (**c**) Cut platinum wire to fit into the source head. (**d**) Sample containing pulled borosilicate glass capillary inserted onto the source head with platinum wire

wire is typically used (*see* Fig. 5c). Alternatively, borosilicate glass capillaries can be gold coated to avoid using platinum as well. Lastly, sample containing capillary is inserted into the source head (*see* Fig. 5d); the capillary's length can be shortened to fit on the platinum wire using a glass cutter. Often, the capillary tip is closed during the pulling process and needs to be clipped to create the opening. The tip is gently clipped using flat-ended forceps, and ideally opening should be less than 20 μm. Larger opening with larger droplet size may require higher spraying voltage, which can disrupt native structure of the analyte. Additional useful protocols regarding native MS are also available at (https://nativems.osu.edu/protocols).

3. Optimization for instrument tuning is crucial for native MS analyses. Tuning parameter used for monitoring the assemblies of KRAS and mutants using an orbitrap mass spectrometer (*see* Table 1) provides a good starting point. Note that these parameters should be optimized based on the purpose of the

**Table 1**
**Tuned instrument parameters used for monitoring the assembly of KRAS and mutants with $SOS^{cat}$ using an orbitrap mass spectrometer**

| | |
|---|---|
| m/z range | 1000–8000 |
| Resolution | 8000 |
| In-source collision energy dissociation (V) | 100 |
| Collision energy | 10 |
| Source temperature (C) | 100 |
| Capillary voltage (kV) | 1.2–1.5 |
| Source DC offset (V) | 25 |
| Injection flatapole lens(V) | 18 |
| Inter flatapole DC (V) | 4 |
| Bent flatapole DC (V) | 5 |
| Transfer multipole DC (V) | 2 |
| Pressure (mbar) | $3 \times 10^{-10}$ |

experiment. Capillary voltage and the distance between the droplet and the source entry should be adjusted depending on the pulled borosilicate glass capillaries' opening and the consistency of the spraying.

4. Native MS can analyze various complexes formed between RAS and SOS. For example, KRAS protein loaded with the intended nucleotide mixed with $SOS^{cat}$ in 3:1 molar ratio reveals various nucleotide dependent complex formations (*see* Fig. 2). Interestingly, deconvolution of the mass spectrum reveals a KRAS●$SOS^{cat}$●KRAS-GTP ternary complex composed of ± GDP (*see* Fig. 2b). Notably, when KRAS-GDP is mixed with $SOS^{cat}$, ternary complex formation is not observed (*see* Fig. 2c), displaying the nucleotide-bound state dependence of complex formations. Most interestingly, despite KRAS being loaded >95% with GTP (*see* Fig. 1) prior to mixing with $SOS^{cat}$, ternary complex is comprised of a substantial amount of GDP bound to KRAS (*see* Fig. 2b). The slow intrinsic GTP hydrolysis of KRAS in the absence of $SOS^{cat}$ cannot account for the perplexing abundance of GDP. The appearance of GDP is most likely due to the GTPase activity of KRAS but stimulated by $SOS^{cat}$, which would result in production of GDP and inorganic phosphate ($H_2PO_4^-$) (*see* Fig. 2b).
5. The ternary complexes of KRAS and $SOS^{cat}$ can be mass isolated in the quadrupole and activated inside the trap region. The instrument parameters are set to a capillary voltage 1.8 kV, sampling cone 50 V, extraction cone 5 V, low mass resolution

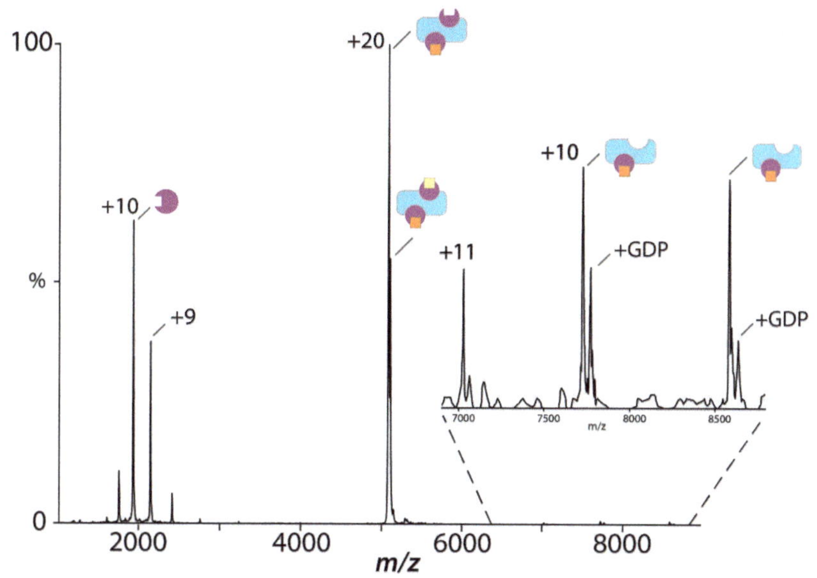


**Fig. 6** Mass spectrum of the ternary KRAS•SOS$^{cat}$ •KRAS-GTP.Mg$^{2+}$(GDP)$_1$ complex post collisional activation. The 20$^{+}$ charge state of ternary complex was isolated and subsequently activated in the collision cell. The peaks at low m/z range correspond to dissociated KRAS and those in higher m/z range represents a KRAS-stripped complex composed of SOS$^{cat}$•KRAS-GTP-Mg$^{2+}$ with a peak corresponding to bound GDP. (Adapted from [30])

3.5, trap collision energy 100 V, transfer collision energy 10 V, trap gas flow 2 mL/min, and source temperature of 90 °C. Collision-induced activated ternary complex dissociates apo KRAS (low m/z region) and a binary complex representing SOS$^{cat}$•KRAS-GTP-Mg$^{2+}$ in the high m/z region (7000–8500) (*see* Fig. 6). This dissociation pattern is consistent with a molecular stoichiometry of the complex between two KRAS and one SOS$^{cat}$.

6. Assembled complexes display similar abundances of molecular species as the tag-less protein (*see* Fig. 7a). The addition of KRAS$^{G13D}$-GTP to the mixture at a final molar ratio of 1:3:3 SOS$^{cat}$:KRAS$^{WT}$:KRAS$^{G13D}$ results in abolished binding of wild-type KRAS to SOS$^{cat}$. Observed is predominantly a ternary complex of KRAS$^{G13D}$•SOS$^{cat}$ •KRAS$^{G13D}$-GTP-Mg$^{2+}$ (*see* Fig. 7b). The experiment can also be performed in the opposite fashion, where His$_6$-KRAS$^{WT}$-GTP is added to preformed complexes of KRAS$^{G13D}$-GTP and SOS$^{cat}$, resulted in no disruption of KRAS$^{G13D}$•SOS$^{cat}$ complexes (*see* Fig. 3d).

7. Reacting ten equivalents of ARS-1620 with KRAS$^{G12C}$ loaded with either GTP or GDP shows complete reactivity when the enzyme is bound to GDP and limited reactivity when GTP is bound (*see* Fig. 8). These results agree with the specificity of ARS-1620 toward the inactive, GDP-bound state of

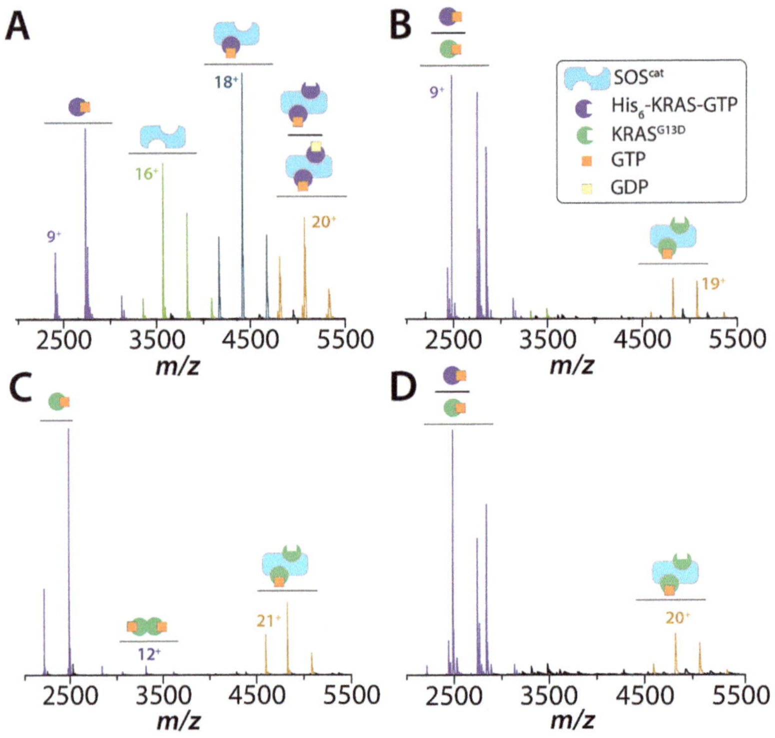


**Fig. 7** Competition of KRAS$^{G13D}$. (**a**) Native mass spectrum of 1 μM SOS$^{cat}$ in complex with 3 μM His6-KRAS-GTP. (**b**) Native mass spectrum after adding KRAS$^{G13D}$-GTP to the mixture in (**a**) to a final concentration of 3 μM. (**c**) Native mass spectra of 1 μM SOS$^{cat}$ mixed with three equivalents of KRAS$^{G13D}$-GTP. (**d**) The addition of His6-KRAS-GTP (3 μM final concentration) to preincubated mixture of 1 μM SOScat and 3 μM KRAS$^{G13D}$-GTP. (Adapted from [30])

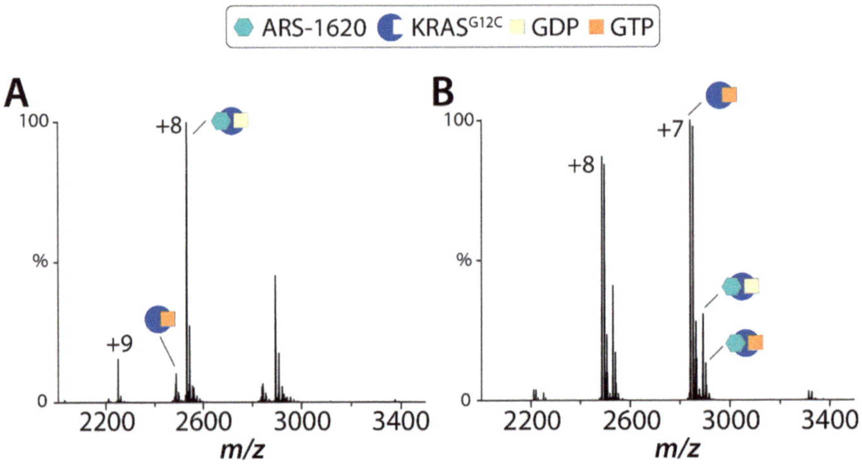


**Fig. 8** The specificity of ARS-1620 toward the inactive state of KRAS$^{G12C}$. The native mass spectra recorded for 10 μM ARS-1620 mixed with 3 μM (**a**) KRAS$^{G12C}$-GDP or (**b**) KRAS$^{G12C}$-GTP

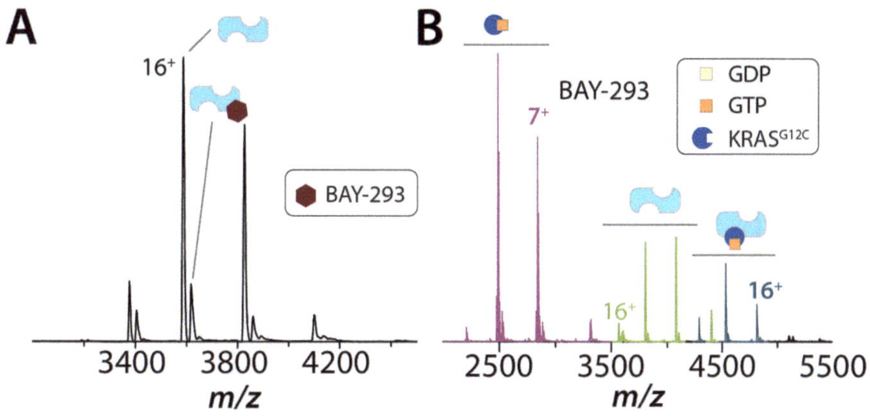


**Fig. 9** Disruption the complexes of SOS$^{cat}$ with KRAS$^{G12C}$ by BAY-293. (**a**) Native mass spectrum of 2 μM SOS$^{cat}$ and 10 μM of BAY-293. (**b**) The addition of BAY-293 at final concentration of 2.5 μM to a preincubated mixture of 1 μM SOS$^{cat}$ and 3 μM KRAS$^{G12C}$-GTP

KRAS$^{G12C}$ [27]. Moreover, the addition of ARS-1620 to an incubated mixture of SOS$^{cat}$ and KRAS$^{G12C}$-GDP resulted in complete disruption of complexes. However, ARS-1620 added to a preincubated mixture of SOS$^{cat}$ and KRAS$^{G12C}$-GTP had no appreciable disruption of binary and ternary complexes.

8. BAY-293 has shown to efficiently disrupt the interaction between KRAS$^{G12C}$ and SOS$^{cat}$ leading to antiproliferative activity [28]. BAY-293 displays direct binding to SOS$^{cat}$ and disrupts the formation of the ternary complexes for KRAS$^{WT}$ and KRAS$^{G12C}$ (*see* Figs. 3c and 9b). However, drug compound is unable to disrupt the complex formed between KRAS$^{G13D}$ and SOS$^{cat}$.
9. Kobe0065 is reported to bind directly to RAS-GTP and alters the binding at the allosteric site of SOS [29]. The addition of 2.5 equivalence of Kobe0065 to SOS$^{cat}$ in complex with KRAS-GTP or KRAS$^{G13D}$-GTP displayed no disruption, respectively (*see* Fig. 3b, e). Increasing the concentration of Kobe0065 to 200 μM, ten times the reported $IC_{50}$ value, does not disrupt binary and ternary complexes [30].
10. BI-3406 is reported to disrupt SOS1 and KRAS interaction with high selectivity and potency [31]. Although this drug compound is able to effectively disrupt KRAS$^{WT}$ from forming a ternary complex with SOS$^{cat}$, it is unable to disrupt the ternary complex formed with KRAS$^{G13D}$ (*see* Fig. 3d, h). Overnight incubation does not impact the KRAS$^{G13D}$●SOS complex (*see* Fig. 10b). At higher concentrations of the compound (4000-fold the reported $IC_{50}$) incubated overnight, the ternary complex between KRAS$^{G13D}$ and SOS$^{cat}$ is almost completely disrupted (*see* Fig. 10c).

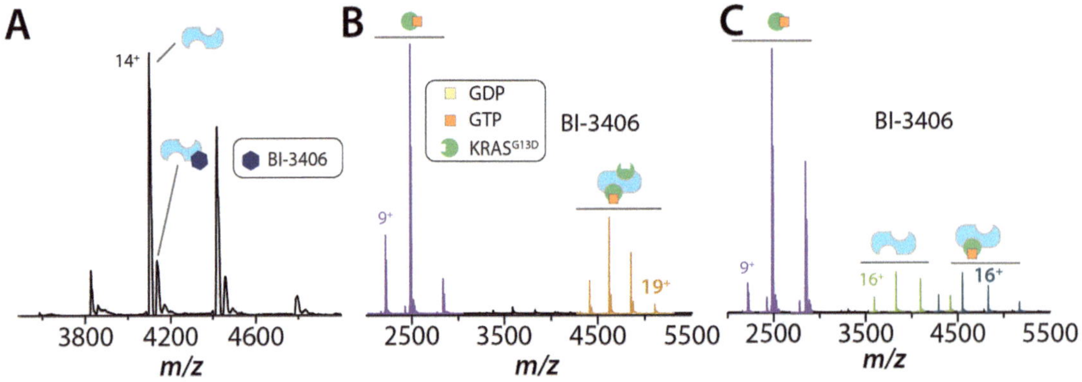


**Fig. 10** BI-3406 interaction with $SOS^{cat}$ and $SOS^{W729E}$:$KRAS^{G13D}$-GDP and $SOS^{cat}$:$KRAS^{G13D}$-GTP complexes. (**a**) Native mass spectrum of 1 μM $SOS^{cat}$ and 5 μM of BI-3406. The mass spectra of (**b**) 2.5 μM and (**c**) 20 μM BI-3406 added to a complex of 1 μM $SOS^{cat}$ and 3 μM $KRAS^{G13D}$-GTP and incubated overnight at 4 °C

## References


1. Vetter IR, Wittinghofer A (2001) The guanine nucleotide-binding switch in three dimensions. Science 294:1299–1304
2. Cox AD, Fesik SW, Kimmelman AC et al (2014) Drugging the undruggable RAS: mission possible? Nat Rev Drug Discov 13:828–851
3. Milburn MV, Tong L, deVos AM et al (1990) Molecular switch for signal transduction: structural differences between active and inactive forms of protooncogenic Ras proteins. Science 247:939–945
4. Goitre L, Trapani E, Trabalzini L et al (2014) The Ras superfamily of small GTPases: the unlocked secrets. Methods Mol Biol 1120:1–18
5. Vigil D, Cherfils J, Rossman KL et al (2010) Ras superfamily GEFs and GAPs: validated and tractable targets for cancer therapy? Nat Rev Cancer 10:842–857
6. Margarit SM, Sondermann H, Hall BE et al (2003) Structural evidence for feedback activation by Ras.GTP of the Ras-specific nucleotide exchange factor SOS. Cell 112:685–695
7. Sondermann H, Soisson SM, Boykevisch S et al (2004) Structural analysis of autoinhibition in the Ras activator son of sevenless. Cell 119: 393–405
8. Vo U, Vajpai N, Flavell L et al (2016) Monitoring Ras interactions with the nucleotide exchange factor son of sevenless (Sos) using site-specific NMR reporter signals and intrinsic fluorescence. J Biol Chem 291:1703–1718
9. Smith MJ, Neel BG, Ikura M (2013) NMR-based functional profiling of RASopathies and oncogenic RAS mutations. Proc Natl Acad Sci 110:4574
10. Freedman TS, Sondermann H, Kuchment O et al (2009) Differences in flexibility underlie functional differences in the Ras activators son of sevenless and Ras guanine nucleotide releasing factor 1. Structure 17:41–53
11. Lanucara F, Holman SW, Gray CJ et al (2014) The power of ion mobility-mass spectrometry for structural characterization and the study of conformational dynamics. Nat Chem 6:281–294
12. Allison TM, Bechara C (2019) Structural mass spectrometry comes of age: new insight into protein structure, function and interactions. Biochem Soc Trans 47:317–327
13. Bolla JR, Agasid MT, Mehmood S et al (2019) Membrane protein–lipid interactions probed using mass spectrometry. Annu Rev Biochem 88:85–111
14. Heck AJR (2008) Native mass spectrometry: a bridge between interactomics and structural biology. Nat Methods 5:927–933
15. Leney AC, Heck AJ (2017) Native mass spectrometry: what is in the name? J Am Soc Mass Spectrom 28:5–13
16. Cong X, Liu Y, Liu W et al (2016) Determining membrane protein–lipid binding thermodynamics using native mass spectrometry. J Am Chem Soc 138:4346–4349
17. Tamara S, den Boer MA, Heck AJR (2021) High-resolution native mass spectrometry. Chem Rev 122(8):7269–7326
18. Beveridge R, Migas LG, Payne KAP et al (2016) Mass spectrometry locates local and

allosteric conformational changes that occur on cofactor binding. Nat Commun 7:12163

19. Bennett JL, Nguyen GTH, Donald WA (2021) Protein-small molecule interactions in native mass spectrometry. Chem Rev 122(8): 7327–7385
20. Liu M, Van Voorhis WC, Quinn RJ (2021) Development of a target identification approach using native mass spectrometry. Sci Rep 11:2387
21. Nguyen GTH, Bennett JL, Liu S et al (2021) Multiplexed screening of thousands of natural products for protein–ligand binding in native mass spectrometry. J Am Chem Soc 143: 21379–21387
22. Monti MC, Cohen SX, Fish A et al (2011) Native mass spectrometry provides direct evidence for DNA mismatch-induced regulation of asymmetric nucleotide binding in mismatch repair protein MutS. Nucleic Acids Res 39: 8052–8064
23. Aron AT, Petras D, Schmid R et al (2022) Native mass spectrometry-based metabolomics identifies metal-binding compounds. Nat Chem 14:100–109
24. Tokmina-Lukaszewska M, Patterson A, Berry L et al (2018) The role of mass spectrometry in structural studies of flavin-based electron bifurcating enzymes. Front Microbiol 9:1397–1397
25. Marty MT, Baldwin AJ, Marklund EG et al (2015) Bayesian deconvolution of mass and ion mobility spectra: from binary interactions to polydisperse ensembles. Anal Chem 87: 4370–4376
26. Laganowsky A, Reading E, Hopper JT et al (2013) Mass spectrometry of intact membrane protein complexes. Nat Protoc 8:639–651
27. Ostrem JM, Peters U, Sos ML et al (2013) K-Ras(G12C) inhibitors allosterically control GTP affinity and effector interactions. Nature 503:548–551
28. Hillig RC, Sautier B, Schroeder J et al (2019) Discovery of potent SOS1 inhibitors that block RAS activation via disruption of the RAS-SOS1 interaction. Proc Natl Acad Sci USA 116: 2551–2560
29. Shima F, Yoshikawa Y, Ye M et al (2013) *In silico* discovery of small-molecule Ras inhibitors that display antitumor activity by blocking the Ras–effector interaction. Proc Natl Acad Sci 110:8182
30. Moghadamchargari Z, Shirzadeh M, Liu C et al (2021) Molecular assemblies of the catalytic domain of SOS with KRas and oncogenic mutants. Proc Natl Acad Sci USA 118: e2022403118
31. Hofmann MH, Gmachl M, Ramharter J et al (2021) BI-3406, a potent and selective SOS1–KRAS interaction inhibitor, is effective in KRAS-driven cancers through combined MEK inhibition. Cancer Discov 11:142

# Chapter 14

# Profiling Complex RAS-Effector Interactions Using NMR Spectroscopy

**Regina Strakhova and Matthew J. Smith**


## Abstract

Knowledge of how effectors interact with RAS GTPases is key to understanding how these switch-like proteins function in cells. Effectors bind specifically to GTP-loaded RAS using RAS association (RA) or RAS binding domains (RBDs) that show wide-ranging affinities and thermodynamic characteristics. Both normal development and RAS-induced tumorigenesis depend on multiple distinct effector proteins that are frequently co-expressed and co-localized, suggesting an antagonistic nature to signaling whereby multiple proteins compete for a limited pool of activated GTPase. NMR spectroscopy offers a powerful approach to multiplex effectors and/or regulatory enzymes and quantifies their interaction with RAS, expanding our biophysical and systems-level understanding of RAS signaling in a more integrated and physiologically relevant setting. Here we describe a method to directly quantitate GTPase binding to competing effectors, using wild-type KRAS complex with ARAF and PLCε1 as a model. Unlabeled RBD/RA domains are added simultaneously to isotopically labeled RAS, and peak intensities at chemical shifts characteristic of individually bound domains provide quantitation. Similar competition-based assays can be run with small molecule interactors, GEF/GAP domains, or regulatory enzymes that drive posttranslational modifications. Such efforts bring in vitro interaction experiments in line with more complex cellular environments.


**Key words** NMR spectroscopy, RAS GTPases, Effector binding, RBD domain, RAF, PLCε-1, Competition

## 1 Introduction

Small GTPases of the RAS superfamily are fundamental signaling enzymes able to transmit extracellular signals to multiple intracellular pathways. RAS GTPases function as molecular switches, cycling between GTP-bound active and GDP-bound inactive states [1]. When activated, RAS undergoes a conformational change that enables binding to numerous effector proteins that stimulate diverse downstream pathways. Notably, GTP-bound RAS activates the MAPK pathway via complex with RAF kinases and the PI3K/AKT pathway through engagement with PI3K p110α/γ subunits [2]. These interactions are responsible for triggering cell

Andrew G. Stephen and Dominic Esposito (eds.), *KRAS: Methods and Protocols*, Methods in Molecular Biology, vol. 2797, https://doi.org/10.1007/978-1-0716-3822-4_14,

proliferation and survival in both normal and cancer cells. Effector proteins such as RAF and PI3K, as well as numerous alternative effectors, bind to RAS-GTP using *RAS binding domains* [RBDs; also called *RAS association* (RA) domains] [3]. Nuclear magnetic resonance (NMR) spectroscopy has proven a valuable tool to determine how RAS GTPase variants cycle between active and inactive states [4]. Indeed, binding to GDP or GTP induces marked structural alterations in RAS that can be easily observed by NMR. To enable detection, GTPases are enriched with isotopic labels ($^{15}$N or $^{13}$C) and purified. NMR acts as a visualization filter that permits analysis of the labeled protein in even highly complex mixtures of unlabeled biomolecules. This includes measurements of activity from *guanine* nucleotide *exchange factors* (GEFs) and *GTPase activating proteins* (GAPs) as regulators, as well as the study of direct binding events including RAS interactions with small molecules or its complex with partner proteins [5]. Further, while conventional techniques to explore protein-protein interactions such as isothermal titration calorimetry (ITC) or surface plasmon resonance (SPR) are generally limited to systems of only two biomolecules, NMR can be used to study binding complexes of multiple proteins concurrently and in a quantitative manner [6]. This helps address the question of binding competition between multiple RAS effectors and regulators, as in the cellular environment small GTPases must engage simultaneously with various partners that are in direct competition. Complex of RAS with RBD domains from unique effectors induces distinct chemical shift perturbations (CSPs) in NMR spectra, enabling differentiation of several individual binding complexes in solution [7]. Additionally, NMR offers atomic-level resolution of the labeled protein, and the data provide highly accurate quantitation of individual complexes as effector RBDs compete for RAS.

Labeling of RAS GTPases for 2D protein NMR experiments can be uniform or specific. Uniform $^{15}$N labeling enables detection of all residues (except prolines), providing a comprehensive map of potential conformational changes or binding events across the entire protein [8]. This is particularly useful when mapping a new interaction based on a set of CSPs. Specific isotopic labeling is done on a limited set of amino acids which significantly reduces spectral complexity, as the experiments comprise only a few resonances. This is useful when CSPs are difficult to discriminate in spectral areas where peaks are crowded and permits the direct detection of multiple highly related GTPases in parallel [9]. In this chapter, we describe how to perform both uniform $^{15}$N and specific $^{13}$C-isoleucine ($^{13}$C-Ile) labeling of RAS GTPase proteins. We further detail a protocol for NMR-based competition assays that permit quantitative analyses of RAS interactions with multiple effector RBDs in parallel, including purification conditions for the required proteins and setup of NMR experiments.

## 2 Materials

Ensure all waste treatment and removal recommendations are followed when with working with cultured *E. coli*.

### 2.1 Bacterial Culture

1. Plasmids for protein expression and purification: KRAS (amino acids 1–171) in pET-15b; ARAF RBD (aa 17–94) and PLCε1 RBD (aa 150–233) in pGEX-4T-2. Plasmid selection is done with 100 μg/mL of ampicillin.
2. *E. coli*: BL21 (DE3 codon+) cells for protein expression and purification, maintained in culture with 34 μg/mL of chloramphenicol. Cultures are grown at 37 °C before induction; temperature for protein expression after induction must be optimized for each protein.
3. Agar plates are prepared by autoclaving LB agar in distilled water. LB agar mixture is prepared at 37 g/L. Desired antibiotic(s) are added once the autoclaved mixture is cool enough to touch.
4. Lysogeny broth (LB) medium: autoclave LB Miller's broth (contains yeast extract at 5 g/L, tryptone at 10 g/L, and NaCl at 10 g/L) in distilled water. LB is prepared at 25 g/L. Desired antibiotic(s) are added once autoclaved medium is cool enough to touch. M9 medium (for 1 L of bacterial cell growth): for M9 salts, weigh out 6.7 g of $Na_2HPO4$, 3 g of $KH_2PO_4$, and 0.5 g of NaCl. Dissolve in distilled water and autoclave. For $^{15}N$-uniformly labeled protein samples, add 1 g of $^{15}N$-$NH_4Cl$. Medium also includes the following reagents which should be filter sterilized: 20 mL of 20% glucose, 1 mL of 1 M $MgSO_4$, 1 mL of 0.1 M $CaCl_2$, 10 mg of thiamine, 10 mg of biotin and desired antibiotics. Protein expression is induced upon the addition of isopropyl-β-D-thiogalactopyranoside (IPTG) (200–500 μM final) (*see* **Note 1**).
5. Preparation of metabolic $^{13}C$-Ile precursor (for 1 L of bacterial growth): convert (3-$^{2}H$) $^{13}C$ 2-ketobutyric acid to (3,3-$^{2}H_2$) $^{13}C$ 2-ketobutyric acid by proton/deuterium exchange at C3: resuspend 50 mg of 2-ketobutyric acid in 5 mL of $D_2O$, adjust to pH to 10.45, and incubate for 13–16 h at 45 °C (*see* **Note 2**).

### 2.2 Laboratory Equipment

1. Benchtop centrifuge.
2. Floor centrifuge.
3. Size exclusion chromatography (SEC).
   (a) HiLoad 26/600 Superdex 75 pg size exclusion column is used to separate purified protein from impurities. To equilibrate this column, prepare 1 L of filtered (0.22 μm

filter) and degassed NMR buffer: 20 mM Tris–HCl pH 7.5 (20 mL), 150 mM NaCl (30 mL), 5 mM $MgCl_2$ (5 mL), 2 mM of fresh DTT (2 mL); equilibrate with 350 mL. Leftover buffer will be used for the run.

(b) Superdex 75 10/300 GL size exclusion column is used after KRAS loading with GMPPNP to remove excess $MgCl_2$ and GMPPNP. To equilibrate this column, prepare 250 mL of filtered (0.22 μm filter) and degassed NMR buffer: 20 mM Tris–HCl pH 7.5 (5 mL), 150 mM NaCl (7.5 mL), 5 mM $MgCl_2$ (1.25 mL), 2 mM of fresh DTT (500 μL); equilibrate with 45 mL. Leftover buffer will be used for the run.

### 2.3 Protein Purification

The following stock solutions will be used for preparation of all buffers listed in this chapter:

#### 2.3.1 Stock Solutions

1. 1 M Tris–HCl pH 7.5: dissolve 121.1 g of Tris in 800 mL of $H_2O$. Adjust the pH to 7.5 with HCl. Adjust the final volume to 1 L with $H_2O$.
2. 5 M NaCl: dissolve 292.2 g of NaCl in 800 mL of H2O. After complete dissolvement adjust the final volume to 1 L with $H_2O$.
3. 50% glycerol: mix 100 mL of 100% glycerol and 100 mL of $H_2O$.
4. 20% NP-40: measure 10 mL of 100% NP-40 and dilute by adding 40 mL of $H_2O$.
5. 1 M $MgCl_2$: dissolve 203.3 g of MgCl2·6$H_2O$ in 800 mL. After complete dissolvement adjust the final volume to 1 L with $H_2O$.
6. 2 M imidazole pH 7.5: dissolve 68.077 g of imidazole in 400 mL. Adjust the pH to 7.5 using HCl. Adjust the final volume to 500 mL with $H_2O$.
7. We use a non-diluted β-mercaptoethanol stock solution of 14.3 M.
8. 1 M dithiothreitol (DTT): dissolve 3 g of DTT in 16 mL of $H_2O$. After complete dissolvement adjust the final volume to 20 mL with $H_2O$. Store 1 mL aliquots at −20 °C.
9. 1 M IPTG: dissolve 2.38 g of IPTG in 8 mL of $H_2O$. After complete dissolvement adjust the final volume to 10 mL with $H_2O$. Store 1 mL aliquots at −20 °C.

#### 2.3.2 Purification Material

1. Sonicator: we use ½ inch probe tip for pellets from 1 to 2 L of bacterial culture.
2. We use Ni Sepharose™ 6 Fast Flow for purification of 6xHIS-tagged proteins.

3. We use Glutathione Sepharose® 4B beads for purification of GST-tagged proteins.
4. Gravity columns: 100 mL columns are used for protein purification; 50 mL columns are used for cleavage of affinity tag while the protein still binds affinity beads.
5. 3500 Da MWCO dialysis bag is used for buffer exchange during cleavage of 6xHIS tag to remove imidazole excess.
6. We use centrifugal tubes of 15 mL to concentrate purified proteins prior and post size exclusion chromatography. Membrane size must be chosen according to size of the purified protein: we use 10 kDa tubes for GTPases and 3 kDa tubes for RBDs.

#### 2.3.3 Buffers

The following buffers are prepared with stock solutions listed in Subheading 2.3.1. Stock solution volume is indicated between braces.

For 6xHIS-tagged KRAS:

1. Lysis buffer: 20 mM Tris–HCl pH 7.5 (1 mL), 150 mM NaCl (1.5 mL), 10% glycerol (10 mL), 0.4% NP-40 (1 mL), 5 mM $MgCl_2$ (250 μL), 10 mM imidazole pH 7.5 (250 μL). Add the following fresh reagents: 5 mM β-mercaptoethanol (17.5 μL), 10 μg/mL DNase I, and EDTA-free protease inhibitor cocktail. Adjust final volume to 50 mL with $H_2O$.
2. High salt buffer: 20 mM Tris–HCl pH 7.5 (6 mL), 500 mM NaCl (30 mL), 5 mM $MgCl_2$ (1.5 mL), 10 mM imidazole pH 7.5 (3 mL), 5 mM β-mercaptoethanol (105 μL). Adjust final volume to 300 mL with $H_2O$.
3. Low salt buffer: 20 mM Tris–HCl pH 7.5 (5 mL), 150 mM NaCl (7.5 mL), 5 mM $MgCl_2$ (1.25 mL), 5 mM β-mercaptoethanol (90 μL). Adjust final volume to 250 mL with $H_2O$.
4. Elution buffer: 20 mM Tris–HCl pH 7.5 (500 μL), 150 mM NaCl (750 μL), 5 mM $MgCl_2$ (125 μL), 250 mM imidazole pH 7.5 (3.125 mL), 5 mM β-mercaptoethanol (9 μL). Add fresh thrombin before transferring the protein into a dialysis membrane (amount of thrombin determined based on where the enzyme was obtained). Adjust final volume to 25 mL with $H_2O$.
5. NMR buffer: 20 mM Tris–HCl pH 7.5 (20 mL), 150 mM NaCl (30 mL), 5 mM $MgCl_2$ (5 mL), 2 mM of fresh DTT (2 mL). Adjust final volume to 1 L with $H_2O$.

For GST-tagged RBDs:

1. Lysis buffer: 20 mM Tris–HCl pH 7.5 (1 mL), 150 mM NaCl (1.5 mL), 10% glycerol (10 mL), 0.4% NP-40 (1 mL). Add the

following fresh reagents: 2 mM DTT (100 μL), 10 μg/mL DNase I, and EDTA-free protease inhibitor cocktail. Adjust final volume to 50 mL with $H_2O$.

2. High salt buffer: 20 mM Tris–HCl pH 7.5 (6 mL), 500 mM NaCl (30 mL), 2 mM of fresh DTT (600 μL). Adjust final volume to 300 mL with $H_2O$.
3. Low salt buffer: 20 mM Tris–HCl pH 7.5 (5 mL), 150 mM NaCl (7.5 mL), 2 mM of fresh DTT (500 μL). Adjust final volume to 250 mL with $H_2O$.
4. Cleavage buffer: 20 mM Tris–HCl pH 7.5 (500 μL), 150 mM NaCl (750 μL), 2 mM of fresh DTT (50 μL). Add fresh thrombin. Adjust final volume to 25 mL with $H_2O$.

#### 2.3.4 Nucleotides

Lyophilized GDP and GMPPNP are reconstituted in 20 mM Tris–HCl pH 7.5 at an approximate concentration of 50–100 mM. To determine the exact concentration, measure absorbance at 253 nm, and use the molar absorption coefficient for guanine ($\varepsilon 253 = 13{,}700\ cm^{-1}$). Stock solutions can be aliquoted and stored at −80 °C.

#### 2.3.5 Enzymes

1. Thrombin protease is used to remove affinity tags from purified KRAS GTPase, as well as the ARAF and PLCε1 RBDs. We use High Purity Bovine Thrombin (>2200 units/mg protein) from BioPharm Laboratories. For cleavage we use 20 units of active thrombin for purified protein from 1 to 2 L of bacterial culture.
2. Alkaline phosphatase is used to dephosphorylate GDP to guanosine during KRAS loading with GMPPNP.

#### 2.3.6 Material for Nuclear Magnetic Resonance

1. $D_2O$: deuterium oxide is used during preparation of NMR samples at a final concentration of 10%. In this chapter we also use $D_2O$ to exchange the $^{13}C$-Ile precursor. We use filtered $D_2O$ with a 99.8% isotopic purity rate.
2. Sample tubes: the following chapter describes the procedure to prepare an NMR sample in a 1.7 mm sample tube. Sample tubes vary in length and in size (e.g., 1.7 mm vs 3 mm vs 5 mm NMR tubes), and the choice ultimately depends on the NMR instrument and the amount of sample.

## 3 Methods

The following steps up to protein purification are described for 6xHIS-KRAS.

### 3.1 Bacterial Transformation

1. Transform BL21 (DE3 codon+) cells with the desired plasmid by performing a heat shock at 42 °C and then placing cells on ice for 2 min.
2. Recover cells in 1 mL of LB medium for 45 min at 37 °C.
3. Plate recovered BL21 (DE3 codon+) cells onto an agar plate containing ampicillin and chloramphenicol selection. Incubate the plate for 16 h at 37 °C.

### 3.2 Overnight Culture

1. Pick a colony of BL21 cells containing the KRAS expression plasmid for an overnight culture of 100 mL in LB medium containing desired antibiotics (ampicillin and chloramphenicol).
2. Incubate on shaker for 16 h (overnight) at 37 °C.

### 3.3 Inoculation, Growth, Induction (Different for $^{15}$N and $^{13}$C-Ile), and Storage

1. The following morning spin down the overnight culture in a benchtop centrifuge at 4000 × *g* to remove LB medium.
2. Resuspend the pellet using 5 mL of M9 minimal medium containing desired antibiotics and the labeling isotope ([$^{15}$N] $NH_4Cl$ or [$^{13}$C]2-ketobutyric acid).
3. Inoculate M9 medium with resuspended BL21 cells to have a final $OD_{600}$ of 0.2 (*see* **Note 2** for $^{13}$C-Ile labeling).
4. When $OD_{600}$ reaches 0.6, remove the culture from the incubator, and let it reach $OD_{600}$ 0.8 at room temperature. Meanwhile, decrease the temperature of the shaking incubator to 18 °C (*see* **Note 3**).
5. Induce the culture with 250 μM of IPTG at $OD_{600}$ 0.8, and grow for 16 h at 18 °C.
6. Pellet the bacterial culture the following morning in a floor centrifuge at 5000 × *g* for 20 min at 4 °C. Bacterial pellets can be stored short term at −20 °C and long term at −80 °C following a flash-freeze on dry ice.

### 3.4 Purification

The following steps are performed at 4 °C. Important, buffers contain β-mercaptoethanol (rather than DTT) to avoid oxidation of the nickel ($Ni^{2+}$)-coated resin.

#### 3.4.1 Purification of KRAS with an N-Terminal 6xHIS-Tag Followed by Tag Removal Using Thrombin Cleavage

1. If the bacterial pellet was frozen, thaw it at room temperature, and resuspend in 40 mL of ice-cold lysis buffer with fresh 5 mM β-mercaptoethanol, 10 μg/mL DNase I, and EDTA-free protease inhibitor cocktail.
2. Lyse resuspended cells using a sonicator. Adjust the sonication pulses according to the instrument used. Total sonication time is approximately 4 min at an amplitude of 30–35%, using 1 s "on" and 4 s "off" pulses.

3. Clear the lysate from insoluble cellular debris by centrifuging the suspension at 14,000 × *g* for 30 min at 4 °C.
4. During centrifugation, prepare a gravity-flow column by adding 10–15 mL of $Ni^{2+}$-bead slurry. Wash the slurry 2 times with 100 mL of filtered water, followed by 2 times with 100 mL of low salt buffer.
5. Drain the column containing the bead slurry and add the cleared lysate. Place the column with beads and the lysate on shaker, and allow gentle mix for 1 h. Do not exceed this time to avoid nonspecific binding.
6. After binding, drain the column to remove unbound proteins from the lysate, and wash the bead slurry 3 times with 100 mL of high salt buffer followed by 100 mL of low salt buffer.
7. Elute 6xHIS-KRAS by adding 7 mL of imidazole elution buffer: mix beads gently and incubate for 10 min. Repeat this **step 2** additional times.
8. Add GDP at a final concentration of 50 μM to eluted KRAS to avoid having GTPase protein without available nucleotide.
9. Add thrombin protease to the eluted KRAS (amount of the enzyme must be determined prior and will depend on the vendor and amount of KRAS being digested).
10. Perform an overnight dialysis using a 3500 Da MWCO dialysis bag in 1–2 L of NMR buffer containing 2 mM of fresh DTT with gentle stirring at 4 °C.

#### *3.4.2 Purification of GST-RBD Followed by Tag Removal Using Thrombin Cleavage*

Use glutathione agarose beads for purification of GST-tagged proteins. Purification steps are identical to those for 6xHIS-KRAS up to elution (Subheading 3.4.1, **step** 7). Refer to Subheading 2 for exact buffer composition for 6xHIS-tagged protein purification and GST-tagged protein purification.

1. After washing the bead slurry, resuspend it in 7 mL of thrombin cleavage buffer, and transfer it to a 50 mL gravity column. Repeat this **step 2** more times with 7 mL of cleavage buffer to transfer leftover beads from 100 mL gravity column to 50 mL gravity column (*see* **Note 4**).
2. Add thrombin protease to the resuspended beads and allow gentle mix overnight at 4 °C.
3. After 12 h, collect the cleaved RBD from the 50 mL gravity column.

#### *3.4.3 Size Exclusion Chromatography*

The following day perform Size exclusion chromatography (SEC).

1. Pre-equilibrate the column with NMR buffer containing 2 mM of fresh DTT (*see* **Note 5**).

2. Concentrate purified KRAS or RBD to a desired volume which will be injected, for example, 10 mL if using a 10 mL injection loop. Filter the protein with 0.2 μm filter prior to injection to remove precipitates.
3. After the SEC run, collect elution fractions (*see* **Note 6**).
4. Concentrate KRAS or RBD to a desired concentration using a centrifugal tube of 15 mL. To concentrate, pool SEC fractions together in the centrifugal tube, and centrifuge it in a benchtop centrifuge at 3500 × *g* for 10–15 min at 4 °C (*see* **Note 7**).

#### 3.4.4 KRAS Loading with Non-hydrolyzable GMPPNP

1. Concentrate KRAS in NMR buffer and add 10 mM of EDTA, 33 units of alkaline phosphatase, and a tenfold molar excess of GMPPNP to a final volume of 500 μL.
2. Incubate the sample at 37 °C for 10 min.
3. Add $MgCl_2$ at a final concentration of 20 mM.
4. Incubate the sample on ice for at least 10 min.
5. Spin the sample down in a benchtop centrifuge at 14,000 × *g* for 10 min at 4 °C to remove any precipitates.
6. Collect the supernatant in a fresh Eppendorf tube.
7. Pre-equilibrate a Superdex 75 10/300 GL column with NMR buffer, and perform SEC to remove excess nucleotide, $MgCl_2$, EDTA, and alkaline phosphatase.
8. Following the SEC run, collect elution fractions containing KRAS, and concentrate it to a desired concentration.

#### 3.4.5 NMR Sample Preparation and Experimental Setup

The following steps are described for a Bruker 600 MHz spectrometer equipped with 1.7 mm cryoprobe. Depending on the NMR instrument to be used, multiple parameters have to be optimized for data acquisition including the concentration of labeled protein, type of pulse program ($^{1}H$-$^{15}N$-HSQC, $^{1}H$-$^{15}N$-BEST-HSQC, $^{1}H$-$^{15}N$-HMQC, etc.), and sample size.

##### Acquisition of Control Spectra

Collect reference spectra for KRAS-GDP alone, KRAS-GMPPNP alone, and individual interactions with ARAF or PLCε-1 RBDs in NMR buffer (*see* **Note 8**).

Steps to follow to prepare KRAS alone samples:

1. Precisely measure concentrations of KRAS-GDP and KRAS-GMPPNP.
2. Prepare NMR samples in an Eppendorf tube containing 200 μM of KRAS-GDP or KRAS-GMPPNP in final volume of 35 μL with 10% $D_2O$. Use NMR buffer to dilute the protein if needed.
3. Spin samples down using a benchtop centrifuge at 14,000 × *g* for 10 min at 4 °C to remove any possible precipitates.

4. Transfer samples to a 1.7 mm NMR sample tube.
5. Insert the KRAS-GDP sample into the spectrometer.
6. Using the TopSpin software, create a new HSQC experiment using standard acquisition parameters (number of scans—NS, number of increments—TD, temperature). NS = 32 and TD = 128 are a good starting point for the experiment. Overall, the acquisition time should be within 30–45 min.
7. Prepare the magnet for data acquisition: lock the resonance frequencies of the magnet (command lock), tune and match the magnet (command atma for automatic tuning and matching or command atmm for manual tuning and matching), perform shimming (command topshim), and adjust the $^{1}$H pulse width (command pulsecal).
8. Start the acquisition with command zg (*see* **Note 9**).
9. Repeat **steps 5–8** for KRAS-GMPPNP.
10. After acquisition NMR samples can be stored at −80 °C.

#### Acquisition of Spectra for Individual Interactions

Collect spectra for individual interactions of KRAS with both ARAF and PLCε1 RBDs. NMR samples are prepared by mixing KRAS-GMPPNP with either ARAF or PLCε1 respecting a 1: 2 ratio of KRAS to RBD, ensuring saturation of KRAS binding. The high KRAS concentration (e.g., 200–250 μM) for these assays provides more intense and well-resolved peaks, compensating for a reduction in intensity due to interaction with binding partners. Reduction of peak intensities is more obvious when KRAS is in the presence of multiple RBDs (Table 1) (*see* **Note 10**). Generally, to compare numerous individual binding reactions, it is recommended to keep a consistent concentration of protein throughout.

**Table 1**
**Change of intensities for HRAS residue Ala18 in different conditions**

| | Peak ppm (HRAS-GMPPNP alone) | | HRAS-GMPPNP alone | With ARAF RBD | With PLCε1 RBD | With ARAF and PLCε1 RBDs |
|---|---|---|---|---|---|---|
| | $^{1}$H | $^{15}$N | Intensity | Intensity | Intensity | Intensity |
| Ala18 | 9.37 | 125.2 | 13.4 | 5.41 | 7.39 | ARAF RBD: 4.15<br>PLCε1 RBD: 4.33 |

Intensities were extracted using NMRView from the following NMR spectra: HRAS-GMPPNP alone, HRAS-GMPPNP in presence of ARAF RBD (1:2 ratio), HRAS-GMPPNP in presence of PLCε1 RBD (1:2 ratio), and HRAS-GMPPNP in presence of ARAF and PLCε1 RBDs (1:2 ratio to each RBD)

1. Precisely measure KRAS, ARAF, and PLCε1 RBD concentrations.
2. Prepare KRAS:ARAF RBD NMR sample in an Eppendorf tube by mixing 200 μM of KRAS and 400 μM of ARAF RBD in final volume of 35 μL with 10% of $D_2O$. Use NMR buffer to adjust the final volume if needed.
3. Repeat **step 2** with PLCε1 RBD. Final amount of protein in a sample with KRAS and a single RBD is 600 μM.
4. Allow equilibration of the binding reaction for 1 h at room temperature. This will ensure that all KRAS in the sample is in complex with RBD.
5. Spin the sample down using a benchtop centrifuge at 14,000 × *g* for 10 min at room temperature to remove any possible precipitates.
6. Transfer the sample into a 1.7 mm NMR sample tube.
7. Follow **steps 5–10** from Subheading 3.4.5.1 to step up the NMR experiment.

Acquisition of Spectra for Multiplexed Binding Reactions

1. Precisely measure KRAS, ARAF, and PLCε1 RBD concentrations.
2. Prepare the NMR sample in an Eppendorf tube by mixing 200 μM of KRAS, 400 μM of ARAF RBD, and 400 μM of PLCε1 RBD in a final volume of 35 μL with 10% of $D_2O$. Use NMR buffer to adjust final volume if needed. Final amount of protein in the sample is 1 mM.
3. Allow equilibration of the binding reaction for 1 h at room temperature. This will ensure that all KRAS in the sample is in complex with either ARAF or PLCε1.
4. Spin the sample down using a benchtop centrifuge at 14,000 × *g* for 10 min at room temperature to remove any possible precipitates.
5. Transfer the sample into a 1.7 mm NMR sample tube.
6. Follow **steps 5–10** from Subheading 3.4.5.1 to set up the NMR experiment.

### *3.4.6 Data Processing and Analysis*

1. Process your data with the software of your choice (TopSpin, NMRPipe, NMRFx Analyst). In our laboratory we use TopSpin to visualize data during and after acquisition and NMRFx Analyst to process data for analysis. By the end of the processing steps, each of your runs should be in the ".nv" type of files.
2. We perform data analysis using NMRViewJ: upload processed files (datasets), and compare each individual KRAS:RBD interaction with the corresponding competition assay. Figure 1

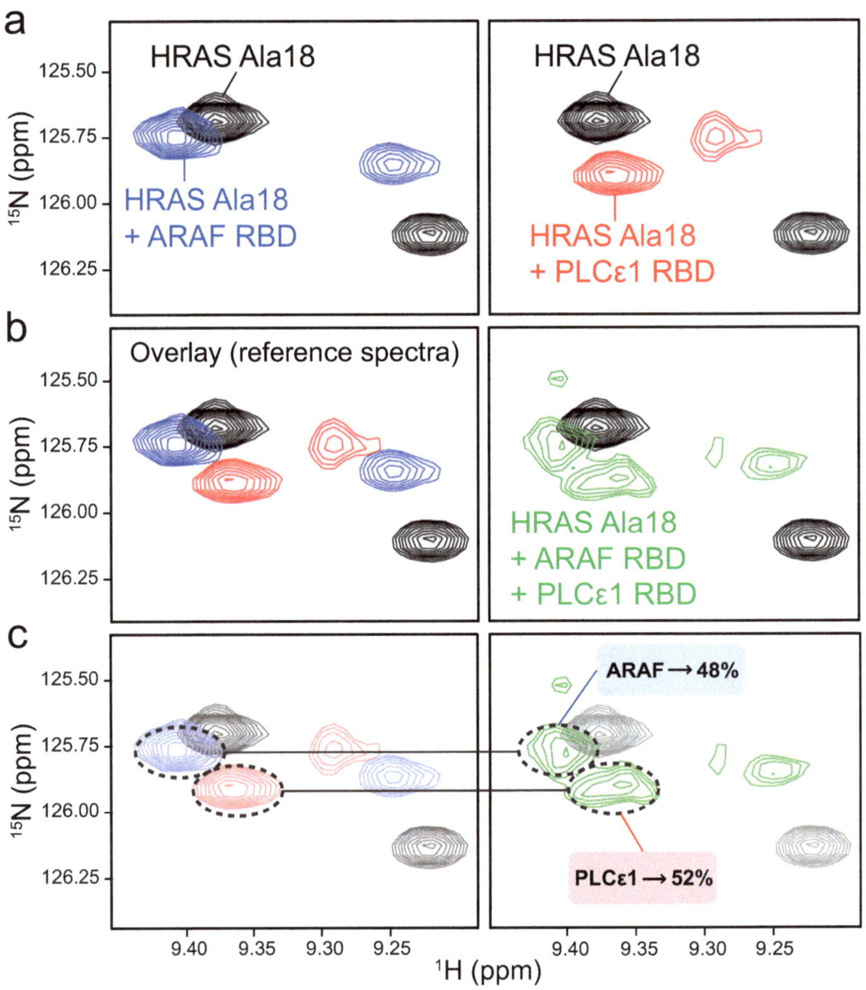


**Fig. 1** Quantification of multiplexed RAS-effector binding by NMR. (**a**) Characteristic CSPs of purified [$^{15}$N] HRAS induced by binding to purified ARAF RBD (left) or PLCε1 RBD (right). Panels show a zoomed area of $^{1}$H-$^{15}$N-HSQC spectra around the resonance from Ala-18 of HRAS. (**b**) Overlay of reference spectra with characteristic CSPs of HRAS in the presence of ARAF or PLCε1 RBDs added individually (left) and the spectrum from competitive binding between the RBDs in the same NMR sample tube (right). For the competition assay, 200 μM of [$^{15}$N]HRAS was mixed with twofold excess of each RBD (1 mM total protein in the sample tube). (**c**) Quantified intensities for CSPs characteristic of binding to ARAF or PLCε1 RBDs. Intensities are determined using NMRView. The percentage represents the ratio of HRAS bound to one RBD vs the other

shows results of the described experiment performed with HRAS, ARAF RBD, and PLCε1 RBD.

3. Extract peak intensities for residues with characteristic CSPs for each RBD bound to KRAS. Not all peaks will have a CSP; it is recommended to choose shifted peaks which remain as close as possible to the peaks seen on KRAS-GMPPNP alone spectra (*see* **Note 11**). The measured intensities for CSPs represent the ratio of binding between two RBDs (Fig. 1). Repeat this

**Table 2**
**Peak intensities for characteristic CSPs of HRAS-GMPPNP in the presence of ARAF and PLCε1 RBDs (1:2 ratio to each RBD)**

| | Peak ppm (HRAS-GMPPNP alone) | | ARAF RBD | | PLCε1 RBD | | |
|---|---|---|---|---|---|---|---|
| | $^{1}H$ | $^{15}N$ | Intensity | Percentage | Intensity | Percentage | Total intensity |
| Ser17 | 9.46 | 120.5 | 3.5 | 44 | 4.4 | 56 | 7.9 |
| Ala18 | 9.37 | 125.2 | 4.15 | 49 | 4.33 | 51 | 8.48 |
| Val44 | 9.04 | 121.8 | 4.16 | 53 | 3.56 | 47 | 7.72 |
| Ile46 | 8.33 | 126.2 | 4.4 | 58 | 3.07 | 42 | 7.47 |
| Leu79 | 9.24 | 127.5 | 3.27 | 59 | 2.22 | 41 | 5.49 |
| Leu6 | 9.22 | 126.2 | 3.36 | 61 | 2.17 | 39 | 5.53 |
| | | | | Average: 54 | | Average: 45 | |

Intensities were extracted using NMRView from single NMR spectra. Listed peaks show small and distinct shift with well-resolved peaks which allow measurement of respective intensities. The binding ratio is calculated using the following formula: Binding ratio/percentage $= \frac{\text{RBD intensity}}{\text{Total intensity}}$

measurement for five to ten distinct resonances (Table 2) (*see* **Note 12**).

4. Calculate the average intensities across measured peaks to determine the percentage of binding of each RBD to KRAS in the multiplexed sample (Table 2).

## 4 Notes

1. IPTG concentration should be determined experimentally to have optimal protein expression. Insufficient concentration of IPTG will result in lower protein yield after purification, while excessive concentration of IPTG will result in formation of inclusion bodies in *E. coli* cells, which contain aggregated over-expressed protein.
2. For $^{13}C$-Ile labeling: prepare unlabeled M9 medium, use 5 mL to resuspend the BL21 cell pellet, and inoculate as for $^{15}N$ labeling. Introduce the exchanged $^{13}C$-Ile precursor 1 h prior to induction, and keep the culture on shaker at 37 °C [9].

   You can explore alternative $^{13}C$-specific labeling strategies by using different metabolic precursors for other amino acids [10]. It is also possible to have specific $^{15}N$ labeling by introducing into M9 media labeled vs unlabeled amino acids [8], though we have found this strategy to be inefficient for many amino acids. Finally, uniform $^{13}C$ labeling can be achieved by using $^{13}$C-glucose [10]. However, this type of labeling is costly and is mostly used for assignments.

3. The temperature of the overnight culture for protein expression must be determined experimentally to have optimal protein expression. Lower temperature will decrease the speed of transcription and protein translation which may help with protein solubility. In our laboratory, we generally use 18 °C for proteins expressed in pET-15b and pGEX-4T-2 backbones.
4. In our laboratory we use two different sizes of gravity columns during protein purification. We use 100 mL column for the binding and washing steps. We also use 50 mL gravity columns for the cleavage of the GST affinity tag. As the cleavage takes place in a lower volume compared to the binding volume (20 mL vs 40 mL, respectively), we transfer bead slurry into a smaller column to have an optimal ratio of total volume of the column compared to the volume of the solution (beads with cleavage buffer).
5. Use NMR buffer containing $MgCl_2$ to perform SEC for both KRAS and RBD. This will ensure that $MgCl_2$ concentration will not be decreased during NMR sample preparation, when KRAS will be mixed with an RBD.
6. For HiLoad 26/600 Superdex 75 pg runs, we set the flow rate to 2.5 mL/min; we collect 4 mL fractions starting at 5 mAU in 5 mL fraction collection tubes. For Superdex 75 10/300 GL runs, we set the flow rate to 0.7 mL/min; we collect 1 mL fractions starting at 5 mAU in 5 mL fraction collection tubes. We recommend to run your SEC fractions on SDS-PAGE before pooling them to verify the purity of your protein.
7. Our laboratory uses 10 kDa Amicon® tubes for KRAS and 3 kDa Amicon® tubes for RBDs. Other methods can be used to reconcentrate your protein, such as dialysis (protein placed in a dialysis bag in the presence which will absorb the excess of the buffer), vacuum centrifugation (water from the buffer is removed by a vacuum pump leaving in the sample tube buffer components and the protein of interest), and others.
8. Run [$^{15}$N]KRAS-GDP and [$^{15}$N]KRAS-GMPPNP spectra to verify if the GMPPNP loading was successful. The spectra for [$^{15}$N]KRAS-GMPPNP alone will also serve you as reference to measure CSPs when compared to [$^{15}$N]KRAS:RBD spectra. In our laboratory we typically use a $^{1}$H-$^{15}$N-BEST-HSQC pulse program.
9. Running the 1D proton experiment before running a 2D experiment is not compulsory but can be useful to quickly verify if the protein is folded correctly, in which case peaks between 7 and 11 ppm will be sharp and well dispersed.

10. When multiple RBDs are mixed with KRAS, some of the latter will interact with one RBD and some of it with the other. As a result, signal for one residue will be split between two populations. If you work with an RBD with limited stability or which aggregates easily and you cannot increase its concentration, you can run longer experiments by increasing number of scans in your run.
11. CSPs created by ARAF and PLC RBDs are highly distinct, with some peaks observed for only one RBD or the other. Some peaks shift considerably, and identification of the original peak just by comparison with the spectra of KRAS-GMPPNP alone is not possible. For the purposes of binding quantification, it is crucial to compare intensities of the same residue. Choosing incorrect peaks will lead to incorrect binding ratios following quantification.
12. Peak intensities for each RBD might be different from the ones from the mixed sample. What is important is to compare intensities for one given residue in the mixed sample, and not intensities of CSPs for the mixed sample to the sample where each RBD is alone.

## References


1. Gimple RC, Wang X (2019) RAS: striking at the core of the oncogenic circuitry. Front Oncol 9:965. https://doi.org/10.3389/fonc.2019.00965
2. De Luca A, Maiello MR, D'Alessio A et al (2012) The RAS/RAF/MEK/ERK and the PI3K/AKT signalling pathways: role in cancer pathogenesis and implications for therapeutic approaches. Expert Opin Ther Targets 16:S17–S27. https://doi.org/10.1517/14728222.2011.639361
3. Wohlgemuth S, Kiel C, Krämer A et al (2005) Recognizing and defining true Ras binding domains I: biochemical analysis. J Mol Biol 348:741–758. https://doi.org/10.1016/j.jmb.2005.02.048
4. Smith MJ, Neel BG, Ikura M (2013) NMR-based functional profiling of RASopathies and oncogenic RAS mutations. Proc Natl Acad Sci 110:4574–4579. https://doi.org/10.1073/pnas.1218173110
5. Marshall CB, Meiri D, Smith MJ et al (2012) Probing the GTPase cycle with real-time NMR: GAP and GEF activities in cell extracts. Methods 57:473–485. https://doi.org/10.1016/j.ymeth.2012.06.014
6. Smith MJ, Ikura M (2014) Integrated RAS signaling defined by parallel NMR detection of effectors and regulators. Nat Chem Biol 10:223–230. https://doi.org/10.1038/nchembio.1435
7. Smith MJ, Marshall CB, Theillet F-X et al (2015) Real-time NMR monitoring of biological activities in complex physiological environments. Curr Opin Struct Biol 32:39–47. https://doi.org/10.1016/j.sbi.2015.02.003
8. Killoran RC, Smith MJ (2019) Conformational resolution of nucleotide cycling and effector interactions for multiple small GTPases determined in parallel. J Biol Chem 294:9937–9948. https://doi.org/10.1074/jbc.RA119.008653
9. Gardner KH, Kay LE (1997) Production and incorporation of $^{15}$N, $^{13}$C, $^{2}$H ($^{1}$H-δ1 methyl) isoleucine into proteins for multidimensional NMR studies. J Am Chem Soc 119:7599–7600. https://doi.org/10.1021/ja9706514
10. Rowlinson B, Crublet E, Kerfah R, Plevin MJ (2022) Specific isotopic labelling and reverse labelling for protein NMR spectroscopy: using metabolic precursors in sample preparation. Biochem Soc Trans 50:1555–1567. https://doi.org/10.1042/BST20210586

# Chapter 15

# Probing RAS Function Using Monobody and NanoBiT Technologies


## Michael Whaby, Rakesh Sathish Nair, and John P. O'Bryan



## Abstract

Missense mutations in the *RAS* family of oncogenes (*HRAS*, *KRAS*, and *NRAS*) are present in approximately 20% of human cancers, making RAS a valuable therapeutic target (Prior et al., Cancer Res 80: 2969–2974, 2020). Although decades of research efforts to develop therapeutic inhibitors of RAS were unsuccessful, there has been success in recent years with the entrance of FDA-approved $KRAS^{G12C}$-specific inhibitors to the clinic (Skoulidis et al., N Engl J Med 384:2371–2381, 2021; Jänne et al., N Engl J Med 387:120–131, 2022). Additionally, $KRAS^{G12D}$-specific inhibitors are presently undergoing clinical trials (Wang et al., J Med Chem 65:3123–3133, 2022). The advent of these allele specific inhibitors has disproved the previous notion that RAS is undruggable. Despite these advancements in RAS-targeted therapeutics, several RAS mutants that frequently arise in cancers remain without tractable drugs. Thus, it is critical to further understand the function and biology of RAS in cells and to develop tools to identify novel therapeutic vulnerabilities for development of anti-RAS therapeutics. To do this, we have exploited the use of monobody (Mb) technology to develop specific protein-based inhibitors of selected RAS isoforms and mutants (Spencer-Smith et al., Nat Chem Biol 13:62–68, 2017; Khan et al., Cell Rep 38:110322, 2022; Wallon et al., Proc Natl Acad Sci USA 119:e2204481119, 2022; Khan et al., Small GTPases 13:114–127, 2021; Khan et al., Oncogene 38:2984–2993, 2019). Herein, we describe our combined use of Mbs and NanoLuc Binary Technology (NanoBiT) to analyze RAS protein–protein interactions and to screen for RAS-binding small molecules in live-cell, high-throughput assays.



**Key words** Transfection, Monobody, RAS, Protein–protein interaction (PPI), PEI, Luminescence, NanoBiT


## 1 Introduction

Monobodies (Mbs) are small (~10 kDa) synthetic binding proteins based on the human fibronectin type III repeat [4]. Mbs are genetically encodable, are stable when expressed in cells, and display high affinity for their targets, often having inhibitory activity upon target engagement [4, 5]. We have previously described the screening for and use of Mbs to probe RAS function [6]. As an example, the NS1 Mb inhibits RAS signaling and biological activity in cells by


Andrew G. Stephen and Dominic Esposito (eds.), *KRAS: Methods and Protocols*, Methods in Molecular Biology, vol. 2797, https://doi.org/10.1007/978-1-0716-3822-4_15,

disrupting RAS–RAS interactions at the plasma membrane [1]. In addition to NS1, the R15 Mb identified the nucleotide-free state of RAS (apo RAS) as a targetable state that represents a vulnerability for certain oncogenic RAS mutants [2]. These Mbs, as well as others [3, 7, 8], exemplify the utility of Mbs to identify novel mechanisms and therapeutic vulnerabilities of RAS. Because common methods to analyze protein–protein interactions, such as co-immunoprecipitations, are time-consuming, lack quantitative power, and incompatible with live-cell, high-throughput assays, we turned to NanoLuc Binary Technology (NanoBiT) to develop a methodology to detect RAS–RAS and RAS–Mb interactions in live cells [9, 10]. NanoBiT relies on two peptide fragments of NanoLuc: large BiT (LgBiT; 17.6 kDa) and small BiT (SmBiT; 1.3 kDa). Alone, SmBiT and LgBiT do not generate luminescence. PPIs between two proteins, one tagged with SmBiT and the other with LgBiT, will allow for complementation of these tags to reconstitute NanoLuc luciferase activity. It is important to note that the SmBiT and LgBiT tags were designed to have low affinity for one another so as to not influence PPIs. Luminescence from NanoBiT can be detected in live cells (or cell lysates) allowing for real-time measurements of PPIs [9]. Not only have we utilized this system to analyze the engagement of protein–protein interactions (PPIs), we have also set up a methodology to screen for molecules that may disrupt RAS PPIs in live cells (Fig. 1a, b).

## 2 Materials

### 2.1 Reagents and Cell Lines

1. NanoBiT® PPI Starter Systems (MCS Starter System) (Promega):
   (a) pBiT1.1-N[TK/LgBiT] Vector.
   (b) pBiT2.1-N[TK/SmBiT] Vector.
   (c) LgBiT-PRKAR2A Control Vector.
   (d) SmBiT-PRKACA Control Vector.
   (e) NanoBiT® Negative Control Vector.
   (f) Nano-Glo Live-Cell Substrate (LCS).
   (g) Nano-Glo LCS Dilution Buffer.
2. HiBiT Control Protein (Promega).
3. 10% Triton X-100.
4. HA-tagged RAS expression vectors (pCGN-based)—available by request from authors, Addgene, and other sources.
5. pCW57-MCS1-2A-MCS2 (Cat # 71782, Addgene).
6. SmBiT-KRAS (G12D): Promega.

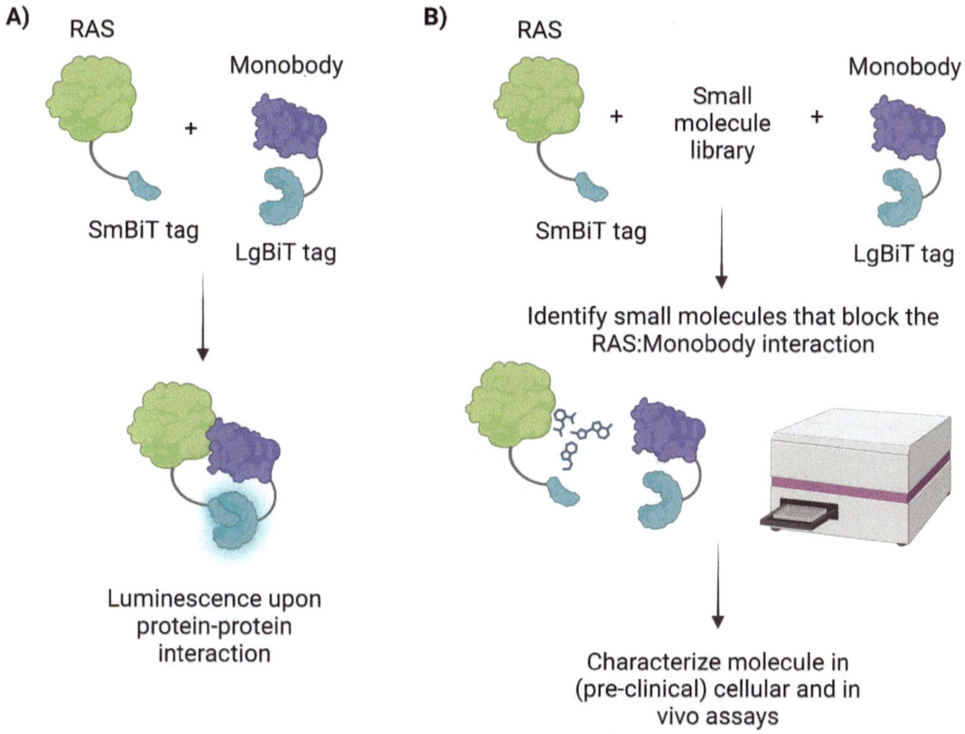


**Fig. 1** Using NanoBiT to detect gain and loss of PPIs. (**a**) Illustration of SmBiT-RAS and LgBiT-Mb interaction followed by luciferase-generated luminescence upon LgBiT–SmBiT complementation. (**b**) Illustration of workflow for using NanoBiT to screen for small molecule Mb mimetics

7. MYC-tagged ERK1 (pEFCL-based)—available by request from authors, Addgene, and other sources.
8. pMD2.G (VSV-G envelope) (Cat # 12259; Addgene).
9. pCMVR8.74 (packaging vector) (Cat # 22036; Addgene).
10. Dulbecco's Modified Eagle Medium (DMEM media).
11. Opti-MEM media.
12. Fetal bovine serum (FBS).
13. Blasticidin S HCL.
14. Polyethylenimine (PEI) linear, MW 25,000.
15. 96-Well White/Clear Bottom Plate, TC Surface.
16. 1X DPBS (without calcium or magnesium).
17. Trypsin-EDTA (0.25%), phenol red.
18. CloneAmp™ HiFi PCR Premix (Takara).
19. PCR tubes.
20. DNA Gel Extraction Kit.
21. 1 Kb Plus DNA ladder.

22. Gibson Assembly® Cloning Kit.
23. LB agar.
24. 1000X ampicillin (100 mg/mL).
25. Petri dishes.
26. Pierce™ BCA Protein Assay Kit.
27. Anti-pan-RAS Antibody (C-4) (Cat # sc-166691, Santa Cruz Biotechnology).
28. Anti-mouse MYC (Cat # 05-724, Clone A46, Millipore-Sigma).
29. Anti-HA monoclonal (Cat # 90154, Clone 16B12, Biolegend).
30. Anti-phospho-ERK (Cat # 9101, Thr202/Tyr204, CST).
31. Anti-total ERK (Cat # 9102, CST).
32. Vinculin antibody (Cat # sc-25336, Santa Cruz Biotechnology).
33. HRP-conjugated anti-mouse secondary antibody.
34. RASless mouse embryonic fibroblasts (MEFs) (BRAF V600E).
35. HEK293 cells.
36. HEK293-T cells.

### 2.2 Media, Solutions, and Buffers

1. PEI (1 mg/mL) is dissolved in 30% ethanol in sterile water and stored in 100 μL aliquots at −80 °C. It can be freeze-thawed up to three times without loss of transfection efficiency.
2. Complete media for RASless MEFs is DMEM supplemented with 10% FBS and 4 μg/mL Blasticidin.
3. Complete media for HEK293 cells is DMEM supplemented with 10% tetracycline-free FBS.
4. PLC lysis buffer (50 mM HEPES, pH 7.5, 150 mM NaCl, 10% glycerol, 1% Triton X-100, 1 mM EGTA, 1.5 mM $MgCl_2$, 100 mM sodium fluoride). Supplement with 1 mM sodium vanadate, 10 μg/mL leupeptin, and 10 μg/mL aprotinin immediately prior to use.
5. Polybrene (Hexadimethrine bromide): make stock solutions of 8 mg/mL (1000×) in deionized, autoclaved water. Working concentration of 8 μg/mL (1×) is used.

## 3 Methods

### 3.1 Analyzing RAS–RAS Interactions in Live Cells

The interactions of RAS monomers on the cell membrane to form "nanoclusters" is a necessary step for RAS signaling [11, 12]. While RAS dimerization has been proposed as a requisite step in RAS function [13–16], little evidence exists in support of a defined RAS dimerization interface. For instance, the NS1 Mb inhibited RAS by binding to the α4–α5 allosteric lobe and decreasing RAS multimers [1], but mutating amino acids within this region had no effect on RAS signaling or self-association [1, 10, 17]. Nevertheless, RAS–RAS interactions in cells clearly play a role in their function. Thus, a better understanding of these interactions will provide valuable insights into RAS biology and may lead to novel approaches to therapeutic targeting of RAS. Here, we describe a method to analyze RAS–RAS interactions in live cells using NanoBiT technology. We also describe a high-throughput assay for screening for small molecules that disrupt RAS–Mb interactions using NanoBiT.

#### 3.1.1 Cloning of RAS NanoBiT Vectors

For the NanoBiT system, each protein of interest must contain a LgBiT or SmBiT tag on either the N- or C-termini. For both protein partners, testing combinations of the LgBiT and SmBiT on both the N- and C-termini is usually recommended to determine the best orientation to detect PPIs [18]. There are eight possible combinations to test for each protein pair. For RAS, however, the C-terminus contains the hypervariable region (HVR) which undergoes posttranslational modifications necessary for membrane localization. For this reason, RAS should only be tagged on the N-terminus. For cloning these constructs, we used the Gibson cloning approach to insert PCR amplified RAS into pBiT1.1-N [TK/LgBiT] Vector and pBiT2.1-N [TK/SmBiT] Vector (*see* **Note 1**).

1. PCR amplifies KRAS from pCGN-HA-KRAS as a template using 3′KRAS-nBiT (EcoRI) (5′- GCCGCCCCGACTCTA GAAGATCTGCTAGCTTAGACTGAATTTTACAT AATTA-CACACTTTGTCT-3′) and 5′nBiT-KRAS (EcoRI) (5′ CGGGAGCGGAGGTGGAG GCTCGAGCGGTGGAGCT-CAGGGGAATATGACTGAATATAAACTTGTG-3′) using the following volumes and running conditions (Table 1):

   Add these reagents into a single PCR tube, and use the following thermal cycler running conditions: denature at 98 °C for 5 min; 4 cycles extension of 98 °C for 15 s, 44 °C for 15 s, and 72 °C for 10 s; 30 cycles extension of 98 °C for 15 s, 65 °C for 15 s, and 72 °C for 10 s; extend for 5 additional min at 72 °C, and hold at 4 °C.

**Table 1**
**Volumes of reagents to use for PCR reaction**

| CloneAmp HiFi PCR Premix | pCGN-HA-KRAS (25 ng/μL) | 5′ nBiT-KRAS (EcoRI) (10 μM) | 3′KRAS-nBiT (EcoRI) (10 μM) | Water |
|---|---|---|---|---|
| 12.5 μL | 1.0 μL | 0.5 μL | 0.5 μL | 10.5 μL |

2. Add 5 μL of 6X DNA loading dye to PCR product, and run full amount on a 0.8% agarose gel (120 V for 30 min).
3. Elute PCR fragment (651 bp) using Gel Extraction Kit per manufacturer protocol.
4. Clone PCR fragment into EcoRI digested pBiT1.1-N [TK/LgBiT] and pBiT2.1-N [TK/SmBiT] using Gibson Assembly Cloning Kit per manufacturer protocol to generate pTK-HSV-LgBiT-KRAS and pTK-HSV-SmBiT-KRAS.
5. Colonies on antibiotic (ampicillin) agar plates are screened by colony PCR using 5′HSV-TK (5′-GCGACCCTGCAGCG ACCCGCT-3′) and 3′KRAS-nBiT (EcoRI) (5′-GCCG CCCCGACTCTAGAAGATCTGCTAGCTTAGACTGAATT TTACATAATTACACACTTTGTCT-3′). DNA from positive clones is isolated and sequenced.

*3.1.2 Validation of Plasmid Expression*

Before performing the live-cell NanoBiT assay, it is recommended to confirm that the cloned NanoBiT vectors are expressing after transfection. To do this, a large-scale transfection (6-well or 60 mm TC plates) can be performed with SmBiT- and LgBiT-tagged RAS cDNA vectors, and lysates analyzed via Western blot to detect both tagged RAS proteins (Fig. 2) (*see* **Note 2**). Regardless of the cell line chosen, the transfection parameters should be optimized to obtain the optimal signal from the PPIs. We use polyethylenimine (PEI) transfection method (described previously [6]). Here, we use RASless MEFs (BRAF V600E) which have a triple knockout of H/K/NRAS and express the BRAF (V600E) protein to rescue proliferation [19, 20]. This eliminates the possibility of SmBiT- or LgBiT-tagged RAS from interacting with untagged, endogenous RAS, thereby maximizing NanoBiT RAS interactions.

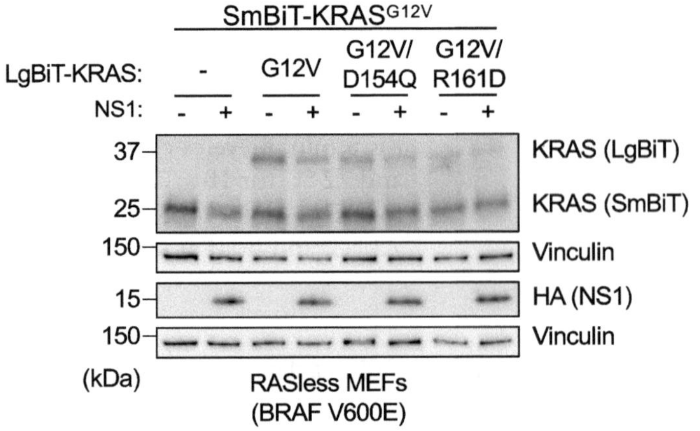


**Fig. 2** Validation of expression of NanoBiT constructs. Representative Western blot of cell lysates from RASless MEFs cotransfected with or without HA-tagged NS1, SmBiT-KRAS, and various LgBiT-KRAS mutants

**Table 2**
**Volumes of reagents to prepare for transfection of cells in 60 mm plates**

| Stock (ng/μL) | DNA | ng/plate | # plates | Total ng | Vol (μL) from stock | PEI (μL) | Opti-MEM (μL) |
|---|---|---|---|---|---|---|---|
| 500 | pTK-HSV-SmBiT-KRAS | 2000 | 1 | 2000 | 4 | 6 | 200 |
| 500 | pTK-HSV-LgBiT-KRAS | 2000 | 1 | 2000 | 4 | 6 | 200 |

1. Aspirate media and wash up to two 100 mm TC plates of RASless MEFs with DPBS, trypsinize, and count the cells.
2. Plate 1.0E6 cells per plate in two 60 mm TC plates (*see* **Note 3**).
3. Incubate overnight in a 37 °C/5% CO incubator.
4. The next day, prepare DNA for transfection in sterile 1.5 mL tubes using the following parameters (Table 2):

   Since both SmBiT- and LgBiT-KRAS are being transfected into the same plate, only one 1.5 mL tube is needed for the DNA/PEI/Opti-MEM incubation. Combine 400 μL Opti-MEM, 12 μL PEI, and 4 μL from each pTK-HSV-SmBiT-KRAS and pTK-HSV-LgBiT-KRAS (500 ng/μL stock) into a single 1.5 mL tube. Tap tube to mix and incubate for 30 min at room temperature.
5. After incubation, aspirate the complete media from the two 60 mm TC plates, and add 3.6 mL serum-free media to plates.
6. Add the DNA/PEI/Opti-MEM mix to the appropriate plate.
7. Incubate DNA/PEI/Opti-MEM/serum-free media with cells in 37 °C/5% $CO_2$ incubator for 5 h (*see* **Note 4**).
8. After incubation, aspirate media from cells and replace with 4 mL complete media.
9. Incubate cells for 24–72 h post transfection in 37 °C/5% $CO_2$ incubator.
10. Gently wash cells with DPBS, and, on ice, collect lysates in 250 μL PLC lysis buffer with a cell scraper.
11. Incubate lysates in 1.5 mL tubes on nutator at 4 °C for 30 min.
12. Centrifuge tubes at 12,000–14,000 RPM at 4 °C.
13. Transfer supernatant to new 1.5 mL tubes, careful not to disturb the pellet.
14. Quantify protein concentration in cell lysates using BCA assay per manufacturer protocol.
15. Run equal amounts of protein from each sample on SDS-PAGE gels and transfer to PVDF membranes for Western blot analysis with a pan-RAS antibody (*see* **Note 5**).

*3.1.3 NanoBiT Assay for RAS–RAS PPIs: Plating and Transfecting Cells*

Once expression is confirmed, the next step is to analyze the NanoBiT luminescence signal from PPIs in live cells. This can be done in a 96-well plate format. (It is recommended to use white-walled TC plates to reduce background luminescence and maximize signal.) Cells should be seeded overnight to be 50–75% confluent the next day (3.0E4 cells per well for RASless MEFs). The next day, cells will be transfected following the PEI transfection method mentioned in Subheading 3.1.2. A general procedure for performing a transfection of cells in 96 wells of a 96-well plate with SmBiT-KRAS and LgBiT-KRAS is outlined here:

1. Aspirate media and wash up to two 100 mm TC plates of RASless MEFs (~80–90% confluent) with DPBS, trypsinize, and count the cells.
2. Resuspend 3.0E6 cells in 20 mL media in a 25 mL sterile reservoir. Using a multichannel pipette (preferably), add 200 μL (3.0E4 cells) per well in 96 wells of a 96-well, white-walled TC plate. Incubate overnight in 37 °C/5% $CO_2$ incubator.
3. The next day, prepare DNA for transfection using the following parameters (Table 3):

   Since both SmBiT- and LgBiT-KRAS are being transfected into the same wells, only one 15 mL tube is needed for the DNA/PEI/Opti-MEM incubation. Combine 2 mL of Opti-MEM, 90 μL of PEI, 30 μL of pTK-HSV-SmBiT-KRAS (500 ng/μL), and 30 μL of pTK-HSV-LgBiT-KRAS (500 ng/μL) to a single 15 mL conical tube. Tap tube to mix and incubate for 30 min at room temperature.
4. In a 25 mL sterile reservoir, add 17.85 mL serum-free media, and then add the DNA/PEI/Opti-MEM mixture (2.15 mL). Mix slowly with 5 or 10 mL pipette.
5. Aspirate media from all wells in 96-well plate.
6. Add 200 μL per well of DNA mixture from reservoir using multichannel pipette.
7. Incubate in 37 °C/5% $CO_2$ incubator for 5 h.

**Table 3**
**Volumes of reagents to prepare for transfection of cells in a 96-well plate**

| Stock (ng/μL) | DNA | ng/well | # wells | Total ng | Vol (μL) from stock | PEI (μL) | Opti-MEM (μL) |
|---|---|---|---|---|---|---|---|
| 500 | pTK-HSV-SmBiT-KRAS | 150 | 100 | 15,000 | 30 | 45 | 1000 |
| 500 | pTK-HSV-LgBiT-KRAS | 150 | 100 | 15,000 | 30 | 45 | 1000 |

8. Aspirate media from all wells after incubation, and replace with 200 μL of complete media using a sterile reservoir and multichannel pipette.
9. Incubate cells in 37 °C/5% $CO_2$ incubator for 24–72 h.

#### 3.1.4 NanoBiT Assay for RAS–RAS PPIs: Detecting Luminescence from Live Cells

NanoBiT luminescence signal from PPIs can typically be detected 24–72 h after transfections. To prepare cells for NanoBiT luminescence reading, carry out the following steps:

1. Thaw Nano-Glo LCS Dilution Buffer at room temperature. Keep Nano-Glo LCS at −20 °C until use.
2. To prepare Nano-Glo substrate mixture, first make 4X Nano-Glo LCS by diluting Nano-Glo LCS 1:20 in Nano-Glo LCS Dilution Buffer (125 μL of Nano-Glo into 2.375 mL Nano-Glo LCS Dilution Buffer).
3. Aspirate media from all wells in 96-well plate.
4. Add 2.5 mL of 4X Nano-Glo LCS to 7.5 mL of pre-warmed Opti-MEM in a 25 mL sterile reservoir, mix, and add 100 μL per well to the cells in the 96-well plate.
5. Luminescence can be measured up to 30 min after Nano-Glo LCS addition with a luminometer without considerable loss of signal.

#### 3.1.5 NanoBiT Assay for RAS–RAS PPIs: Quantifying LgBiT Using HiBiT

Because protein expression levels can influence the number of PPIs—and therefore NanoBiT luminescence—it is important to normalize the live-cell luminescence to another variable such as cell viability or protein expression. Since cell viability does not always correlate with protein expression, we use the HiBiT peptide to detect the total level of LgBiT-tagged proteins. Here, the cells are lysed after the live-cell luminescence reading, and excess HiBiT peptide is added. HiBiT has high affinity ($K_D = 0.7$ nM) for the LgBiT peptide and will outcompete SmBiT binding to LgBiT [10, 21]. HiBiT is used to generate a luminescent signal that is representative of 100% LgBiT binding. This signal can be used as a normalization factor to the live-cell luminescent reading.

1. After live-cell luminescence is acquired, prepare the 2X HiBiT solution (Table 4) in a 25 mL sterile reservoir and mix by pipetting.
2. Add 100 μL per well to 96-well plate.

**Table 4**
**Volumes of reagents to use to prepare 2X HiBiT solution**

| HiBiT peptide (20 μM) | 10% Triton X | 4X Nano-Glo LCS | Opti-MEM |
|---|---|---|---|
| 0.1 mL | 2.0 mL | 2.5 mL | 5.4 mL |

3. Place plate on orbital shaker a 4 °C for 10 min.
4. Read luminescence using the same settings used for the live-cell luminescence reading.
5. The live-cell luminescence readings (NanoBiT PPIs) can be normalized to the HiBiT luminescence readings (LgBiT peptide quantification).

### 3.2 High-Throughput Screening of Small Molecule RAS–Mb Disruptors

Mbs and other biologics are powerful tools to study and perturb RAS function [22]. However, their limited cell permeability presents a challenge for translational efforts [22]. Therefore, we developed a high-throughput assay to screen for molecules that can disrupt the RAS–Mb interaction and potentially mimic Mb function. We generated stable cells that co-express a LgBiT-tagged Mb fused to a self-cleaving peptide (P2A) fused in-frame to SmBiT-tagged KRAS. This approach eliminates difficulties with expression of the two proteins from separate plasmids. In this system, both proteins are expressed under the control of a doxycycline (DOX)-inducible promoter. Interaction of Mb with KRAS is detected via complementation of the LgBiT and SmBiT which constitutes luciferase activity for quantification as described above. The following subheadings describe the development of the stable lines expressing the fusion proteins and use of high-throughput screening for small molecule Mb mimetics.

#### 3.2.1 Subcloning of Fusion Vector Containing SmBiT-RAS and LgBiT-Mb

1. Amplify monobody (R15 or NS1) with FLAG epitope from mammalian vector pCGN-ECFP-Mb (R15m10 or NS1) vector using 5′ LgBiT FLAG Mb (5′-GACGATGACGACAA GGGATCCGTTTCTTCTGTTCC-3′) and 3′ Mb-LgBiT (5′-CCTGAGCTCCACCGCCTAGGTACGGTAGTTAAT-3′). PCR amplification should be done using high-fidelity Taq polymerase.
2. Subclone PCR-amplified monobody fragments into the XhoI-digested N-LgBiT-TK vector using Gibson Assembly to produce the pTK-nLgBiT-FLAG Mb vector.
3. Amplify the SmBiT-KRAS (G12D) fragment using the 5′P2A-nSmBiT(5′-GCCACGAACTTCTCTCTGTTAAAGCAAGC AGGAGACGTGGAAGAAAACCCCGGTCCTATGGTGACC GGCTACCGG-3′) and 3′pCW-KRAS (G12D) (5′-AAAAGGC GCAA CCCCAACCCCGGATCCTTACATAATTACACAC TTTGTC-3′) primers from pTK-SmBiT-KRAS (G12D).
4. Amplify the LgBiT-Mb from the pTK LgBiT-FLAG Mb vector using the 5′pCW-nLgBiT (5′-GCCTGGAGAATTGGCTA GCGAATTCA TGGTCTTCACACTCGAAGATTTC-3′) and 3′P2A-Mb (5′-AGGACCGGGGTT TTCTTCCACGTCT CCTGCTTGCTTTA ACAGAGAGAAGTTCGTGGCGGT ACGGTAGTTAATCG-3′) primers.

5. Subclone the PCR-amplified and gel-purified fragments of LgBiT FLAG Mb and SmBiT-KRAS (G12D) into the EcoRI- and BamHI-digested pCW57-MCS1-2A-MCS2 using the Gibson Assembly Cloning Kit.
6. Confirm the final vector, pCW-LgBiT-FLAG-Mb (R15m10 or NS1)-P2A-nSmBiT-KRAS (G12D), using sequencing and restriction digest analysis.
7. Generate lentiviruses by transfecting pCW-LgBiT-FLAG-Mb (R15m10 or NS1)-P2A-nSmBit-KRAS (G12D) along with packaging vectors pMD2.G and pCMVR8.74 in HEK 293T cells using previously published protocols [6]. This purified virus is used for generating stable lines.

#### 3.2.2 Development of Primary (LgBiT-R15–SmBiT-KRAS (G12D)) and Counter (LgBiT-NS1–SmBiT-KRAS (G12D)) Assay Cell Lines

1. Plate HEK 293 cells in two different 6-well plates for generating two stable lines with a cell density of 0.3E6 per well the day before transduction.
2. After 24 h transduce the cells with pCW-SmBIT-R15–LgBiT-KRAS (G12D) and pCW-SmBiT–NS1-LgBIT-KRAS (G12D) in complete DMEM with 1X polybrene.
3. Twenty-four hours after transduction, add complete media to cells after removing the virus-containing media.
4. Place cells in puromycin (1 μg/mL) on day 2 for selection of infected clones.
5. Replate cells in 10 cm petri dishes further selecting for two passages.
6. Isolate single clones, and further expand in 100 mm petri dishes with another four passages in puromycin selection.
7. Freeze the stable lines for R15-KRAS (G12D) (for primary assay) and NS1-KRAS (G12D) (for counter assay) for future use.
8. These cell lines can be used for subsequent high-throughput screening of Mb mimetics (Fig. 1b).

#### 3.2.3 High-Throughput Screening of Small Molecule Mb Mimetics

1. On day 1, plate 4.0E4 cells per well in complete media in a white-walled 96-well plate in the morning.
2. After 6 h, treat the cells with or without DOX (2 μg/mL). In this plate, one lane of cells should remain as a control without DOX.
3. After 24 h, add a second dose of DOX (2 μg/mL) to the cells.
4. On day 3, treat the cells with compounds (from a compound library) or DMSO for 2 h (*see* **Note 6**).
5. Remove the compound-containing media, and dispense 60 μL of Nano-Glo LCS reagent mixed with Opti-MEM (20X Nano-Glo diluted in Nano-Glo LCS Buffer 4X final Nano-Glo LCS

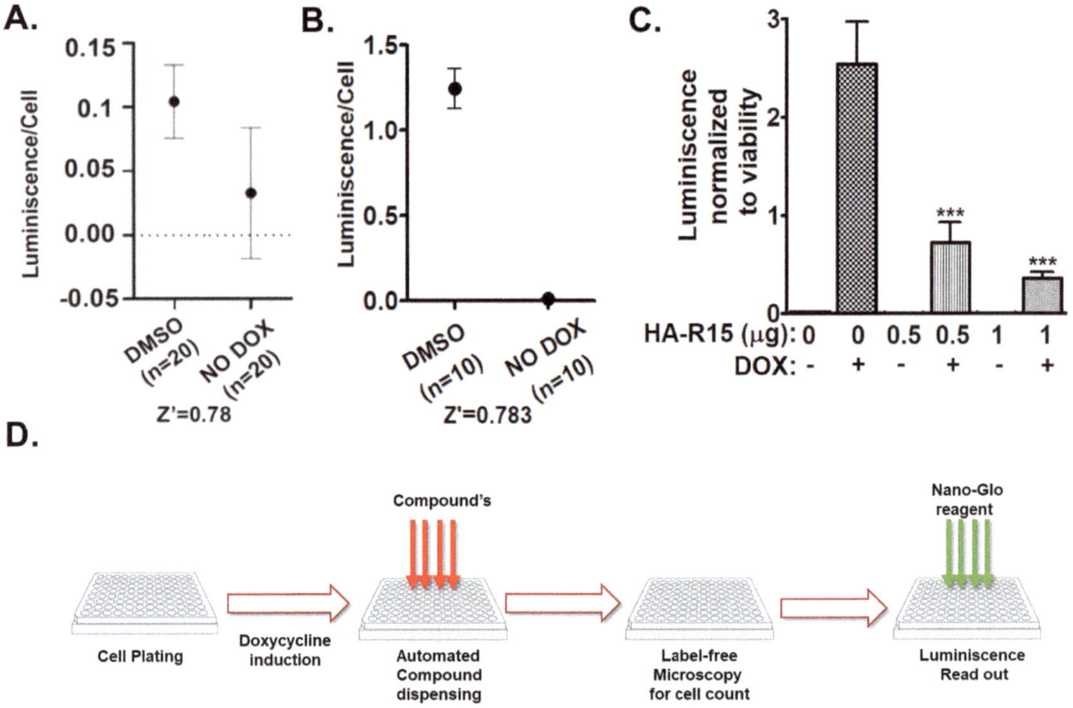


**Fig. 3** NanoBiT assay for screening small molecule Mb mimetics. The graph showing the maximum and minimum signal in the presence and absence of DOX and the *Z′* factor for the assay in the stable cell line (**a**) LgBiT-R15:SmBit-KRAS (G12D) and (**b**) LgBiT-NS1:SmBiT-KRAS (G12D). (**c**) HEK293 cells expressing LgBiT-R15 and SmBiT-KRAS (G12D) transfected with plasmid expressing HA-R15 (0.5 and 1 μg). The untagged R15 blocked the R15-KRAS (G12D) interaction with significantly reduced luminescence compared to untransfected control. The group without DOX was used as control. (**d**) A schematic diagram showing workflow for the screening of Mb mimetics using the stable cells expressing LgBiT-Mb and SmBiT-KRAS (G12D)

reagent) per well and immediately read using SpectraMax iD5 Multi-Mode Microplate Reader with data acquisition using the SoftMax Pro 7 software.

6. The maximum read is given by DMSO lane (MAX) and minimum read by cells without DOX (MIN). Compound-treated cells should be compared to these values to determine level of inhibition (Fig. 3a, b) (*see* **Note 7**).

7. As a positive control for inhibition of the Mb–RAS interaction, an untagged Mb vector can be transfected into the primary screening line. For example, expression of HA-R15 in the R15-KRAS (G12D) stable line treated with DOX is used to inhibit LgBiT-R15 and SmBiT-KRAS (G12D) interaction (Fig. 3c).

8. Normalize the NanoBiT luminescence reading to cell count determined using dye-free cell counting with a BioTek Cytation 5 Cell Imaging Multi-Mode Microplate Reader in a sterile workstation (Fig. 3d). Capture imaging data and analyze using Gen5™ software.
9. The quality of the HT assay is determined using the *Z′* factor for individual assay plates (*see* **Note 8**).

## 4 Notes

1. It is suggested to use a Halo-tagged SmBiT or LgBiT as negative controls for PPIs. As a negative control for RAS binding, we used LgBiT-tagged EGFR (donated from Promega) which did not generate a luminescent signal over background when co-expressed with SmBiT-RAS [10].
2. This approach allows for assessment of the expression of SmBiT- and LgBiT-tagged RAS protein expression compared to endogenous RAS levels (if using a pan-RAS antibody in cells with endogenous RAS).
3. One plate will be included as an untransfected control.
4. Equal amounts of PEI and Opti-MEM can be added to the untransfected control plate.
5. SmBiT-RAS should be detected at ~25 kDa and LgBiT-RAS at ~35 kDa.
6. Do not remove the media with DOX during compound treatment.
7. The counter assay line is used in parallel while testing compounds to make sure that Mb mimetics are selective for inhibiting interaction of the specific Mb with KRAS. For example, an R15 mimetic should reduce the R15-KRAS (G12D) interaction without blocking NS1-KRAS (G12D).
8. The *Z′* factor for the assay is determined from the DOX-treated and DOX-untreated controls according to the formula $Z' = 1 - (3\sigma_{c+} + 3\sigma_{c-})/(|\mu_{c+} - \mu_{c-}|)$ where ($\sigma_{c+}$) and ($\sigma_{c-}$) are the data standard deviation for the high reference control and low reference control, respectively, and $|\mu_{c+} - \mu_{c-}|$ is the absolute value of the difference of the two control signal means. An assay is considered excellent when the *Z′* value is between 0.5 and 1.0 [23].

## Acknowledgments

This work was supported in part by a Merit Review Award (1I01BX002095) from the United States (US) Department of Veterans Affairs Biomedical Laboratory Research and Development Service, NIH awards (CA212608, CA138313, and GM130457), and start-up funds from the Hollings Cancer Center at MUSC to JPO. M.W was supported in part by an NIH T32 (GM132055). The contents of this article do not represent the views of the US Department of Veterans Affairs or the US government.


### References


1. Spencer-Smith R et al (2017) Inhibition of RAS function through targeting an allosteric regulatory site. Nat Chem Biol 13:62–68
2. Khan I et al (2022) Identification of the nucleotide-free state as a therapeutic vulnerability for inhibition of selected oncogenic RAS mutants. Cell Rep 38:110322
3. Wallon L et al (2022) Inhibition of RAS-driven signaling and tumorigenesis with a pan-RAS monobody targeting the Switch I/II pocket. Proc Natl Acad Sci USA 119:e2204481119
4. Koide A, Wojcik J, Gilbreth RN, Hoey RJ, Koide S (2012) Teaching an old scaffold new tricks: monobodies constructed using alternative surfaces of the FN3 scaffold. J Mol Biol 415:393–405
5. Hantschel O, Biancalana M, Koide S (2020) Monobodies as enabling tools for structural and mechanistic biology. Curr Opin Struct Biol 60:167–174
6. Khan I, O'Bryan JP (2021) Probing RAS function with monobodies. Methods Mol Biol 2262:281–302
7. Teng KW et al (2021) Selective and noncovalent targeting of RAS mutants for inhibition and degradation. Nat Commun 12:2656
8. Akkapeddi P et al (2023) Exploring switch II pocket conformation of KRAS(G12D) with mutant-selective monobody inhibitors. Proc Natl Acad Sci U S A 120:11
9. Dixon AS et al (2016) NanoLuc complementation reporter optimized for accurate measurement of protein interactions in cells. ACS Chem Biol 11:400–408
10. Whaby M et al (2022) Mutations in the α4-α5 allosteric lobe of RAS do not significantly impair RAS signaling or self-association. J Biol Chem 298:102661
11. Nan X et al (2015) Ras-GTP dimers activate the mitogen-activated protein kinase (MAPK) pathway. Proc Natl Acad Sci USA 112:7996–8001
12. Santos E, Nebreda AR, Bryan T, Kempner ES (1988) Oligomeric structure of p21 ras proteins as determined by radiation inactivation. J Biol Chem 263:9853–9858
13. Ambrogio C et al (2018) KRAS dimerization impacts MEK inhibitor sensitivity and oncogenic activity of mutant KRAS. Cell 172:857–868.e815
14. Lee KY et al (2021) Oncogenic KRAS G12D mutation promotes dimerization through a second, phosphatidylserine-dependent interface: a model for KRAS oligomerization. Chem Sci 12:12827–12837
15. Muratcioglu S et al (2015) GTP-dependent K-Ras dimerization. Structure 23:1325–1335
16. Inouye K, Mizutani S, Koide H, Kaziro Y (2000) Formation of the Ras dimer is essential for Raf-1 activation. J Biol Chem 275:3737–3740
17. Grozavu I et al (2022) D154Q mutation does not alter KRAS dimerization. J Mol Biol 434:167392
18. Bodle CR, Hayes MP, O'Brien JB, Roman DL (2017) Development of a bimolecular luminescence complementation assay for RGS: G protein interactions in cells. Anal Biochem 522:10–17
19. Drosten M et al (2010) Genetic analysis of Ras signalling pathways in cell proliferation, migration and survival. EMBO J 29:1091–1104
20. Lechuga CG et al (2021) RASless MEFs as a tool to study RAS-dependent and RAS-independent functions. Methods Mol Biol 2262:335–346
21. Schwinn MK, Steffen LS, Zimmerman K, Wood KV, Machleidt T (2020) A simple and scalable strategy for analysis of endogenous protein dynamics. Sci Rep 10:8953

22. Whaby M, Khan I, O'Bryan JP (2022) Targeting the "undruggable" RAS with biologics. Adv Cancer Res 153:237–266
23. Iversen PW, Eastwood BJ, Sittampalam GS, Cox KL (2006) A comparison of assay performance measures in screening assays: signal window, Z′ factor, and assay variability ratio. J Biomol Screen 11:247–252

# Chapter 16

# Analysis of the Guanine Nucleotide-Bound State of KRAS by Ion-Pair Reversed-Phase High-Performance Liquid Chromatography

**Makoto Araki and Kenji Kontani**


## Abstract

Guanine nucleotides can be quantitatively analyzed by high-performance liquid chromatography (HPLC). Here we describe an ion-pair reversed-phase HPLC (IP-RP-HPLC)-based method, which enables analyzing GDP and GTP bound to small GTPases immunoprecipitated from cells. The activation status of FLAG-KRAS expressed in HEK293T cells can be investigated with the IP-RP-HPLC method. This method also can be adapted to determine the effects of compounds such as the KRAS/G12C inhibitor sotorasib on the activation status of FLAG-KRAS in the cells.


**Key words** High-performance liquid chromatography, Ion-pair reagents, Immunoprecipitation, Guanine nucleotides, KRAS

## 1 Introduction

Small GTPases undergo a conformational change between GDP-bound and GTP-bound states, thereby regulating cellular responses such as cell growth, differentiation, cytoskeletal organization, and intracellular logistics [1, 2]. Disruption of the regulation of small GTPase activities (e.g., an increased proportion of GTP-bound forms associated with KRAS mutations) can lead to diseases, including cancer. Clarifying the intracellular activation state of the small GTPases is thus crucial for understanding the consequences of disease-associated mutations and the mechanisms underlying their functional regulation. Several methods have been developed to analyze the intracellular activation state of small GTPases: metabolic labeling of GDP and GTP using [$^{32}$P]orthophosphate, followed by TLC analysis of the GDP and GTP bound to the immunoprecipitated small GTPases from metabolically labeled cells, enables to investigate the intracellular activation states of small GTPases [3]. GST-fusion proteins or antibodies that bind

Andrew G. Stephen and Dominic Esposito (eds.), *KRAS: Methods and Protocols*, Methods in Molecular Biology, vol. 2797,
https://doi.org/10.1007/978-1-0716-3822-4_16,

specifically to GTP-bound small GTPases are used to estimate the activation status of small GTPases [4, 5]. Spatiotemporal dynamics of the activation status of small GTPases in living cells can be analyzed by FRET assay [6, 7].

HPLC-based analysis of the GDP and GTP bound to small GTPases has also been performed, primarily to investigate the guanine nucleotide binding activities of purified small GTPases in vitro [8–17]. We recently reported an ion-pair reversed-phase HPLC (IP-RP-HPLC)-based method to analyze intracellular guanine nucleotide-bound forms of small GTPases [18]. IP-RP-HPLC is an analytical chromatography in which ions with opposite charges (counter ions) are added to ionic substances (such as nucleotides) in an aqueous solution to form ion pairs, which neutralize the overall charge and convert them to low-polarity substances, making them easier to be separated on a reversed-phase column. Here we show that IP-RP-HPLC can be used to investigate the activation status of FLAG-KRAS expressed in HEK293T cells. With IP-RP-HPLC, a decrease in the proportion of the GTP-bound form of KRAS/G12C was observed upon treatment with the KRAS/G12C inhibitor sotorasib. Therefore, IP-RP-HPLC analysis provides a versatile and valuable tool for analyzing the intracellular activation state of various small GTPases, including KRAS.

## 2 Materials

Prepare all solutions using ultrapure water (18 MΩ-cm ionic purity at 25 °C) and analytical-grade reagents.

### *2.1 Immuno-precipitation*

1. (1×) PBS: weigh 4 g NaCl, 0.1 g KCl, 1.45 g $Na_2HPO_4 \cdot 12H_2O$, and 0.1 g $KH_2PO_4$ into a glass beaker, dissolve in about 400 mL of water, and then make up to 500 mL with water using a graduated cylinder. Store at 4 °C.
2. 1 M Tris–HCl (pH 7.5): weigh 60.5 g tris(hydroxymethyl)-aminomethane into a glass beaker, dissolve in about 450 mL of water, and adjust pH with HCl. Make up to 500 mL with water using a graduated cylinder. Store at 4 °C.
3. 4 M NaCl: weigh 116.9 g NaCl into a glass beaker, dissolve in about 400 mL of water, and then make up to 500 mL with water using a graduated cylinder. Store at room temperature.
4. 1 M $MgCl_2$: weigh 102 g $MgCl_2 \cdot 6H_2O$ into a glass beaker, dissolve in about 400 mL of water, and then make up to 500 mL with water using a graduated cylinder. Store at room temperature.

5. 10% Lubrol-PX: weigh 50 g Lubrol-PX into a glass beaker, dissolve in about 400 mL of water, and then make up to 500 mL with water using a graduated cylinder. Store at 4 °C.
6. 1 M DTT: weigh 1.54 g DTT into a graduated cylinder, dissolve it in about 8 mL of water, and then make up to 10 mL with water using the graduated cylinder. Store 1 mL aliquots in a plastic tube at −30 °C.
7. 0.5 M AEBSF: weigh 60 mg AEBSF hydrochloride (4-(2-aminoethyl)benzenesulfonyl fluoride hydrochloride, Pefabloc) into a plastic tube, and dissolve in 0.5 mL of water. Store at −30 °C.
8. 0.5 M $KH_2PO_4$: weigh 34.02 g $KH_2PO_4$ into a glass beaker, dissolve it in about 400 mL of water, and then make up to 500 mL with water using a graduated cylinder. Store at room temperature.
9. 0.5 M $K_2HPO_4$: weigh 43.5 g $K_2HPO_4$ into a glass beaker, dissolve it in about 400 mL of water, and then make up to 500 mL with water using a graduated cylinder. Store at room temperature.
10. 30 mM potassium phosphate (pH 7.0): dilute 0.5 M $KH_2PO_4$ and 0.5 M $K_2HPO_4$ with water to make 1000 mL of 30 mM $KH_2PO_4$ and 30 mM $K_2HPO_4$, respectively. Add about 450 mL of 30 mM $KH_2PO_4$ to 600 mL of 30 mM $K_2HPO_4$ to make 30 mM potassium phosphate (pH 7.0) (*see* **Note 1**). Store at room temperature.
11. 100 mM potassium phosphate (pH 3.0): mix 1.5 mL of 0.5 M $KH_2PO_4$, 1 mL of 100 mM $H_3PO_4$, and 6 mL of water. Store at room temperature.
12. 1 M TBA-B (tetrabutylammonium bromide): weigh 24 g TBA-B into a graduated cylinder, dissolve in about 60 mL of water, and then make up to 75 mL with water using the graduated cylinder (*see* **Note 2**). Store at room temperature in a bottle wrapped with aluminum foil.
13. Extraction buffer: 40 mM Tris–HCl (pH 7.5), 100 mM NaCl, 5 mM $MgCl_2$, 1% (w/v) Lubrol-PX, 1 mM DTT, 0.5 mM AEBSF (protease inhibitor cocktail). For 10 mL, mix 400 μL of 1 M Tris–HCl (pH 7.5), 250 μL of 4 M NaCl, 50 μL of 1 M $MgCl_2$, 1 mL of 10% (w/v) Lubrol-PX, 10 μL of 1 M DTT, 10 μL of 0.5 M AEBSF, and 8.28 mL of water. Prepare just before use and keep ice-cold.
14. Wash buffer-1: 40 mM Tris–HCl (pH 7.5), 100 mM NaCl, 5 mM $MgCl_2$, 0.1% (w/v) Lubrol-PX, 1 mM DTT. For 10 mL, mix 400 μL of 1 M Tris–HCl (pH 7.5), 250 μL of 4 M NaCl, 50 μL of 1 M $MgCl_2$, 100 μL of 10% (w/v) Lubrol-PX, 10 μL

of 1 M DTT, and 9.19 mL of water. Prepare just before use and keep ice-cold.

15. Wash buffer-2: 28.9 mM potassium phosphate (pH 7.0), 100 mM NaCl, 1 mM $MgCl_2$, 0.1% (w/v) Lubrol-PX. For 5 mL, mix 4.82 mL of 30 mM potassium phosphate (pH 7.0), 125 μL of 4 M NaCl, 5 μL of 1 M $MgCl_2$, and 50 μL of 10% (w/v) Lubrol-PX. Prepare just before use and keep ice-cold.
16. Elution buffer: 29.7 mM potassium phosphate (pH 3.0), 0.1% (w/v) Lubrol-PX. For 1 mL, mix 990 μL of 100 mM potassium phosphate (pH 3.0) and 10 μL of 10% (w/v) Lubrol-PX. Prepare just before use and keep ice-cold.
17. Anti-FLAG antibody-conjugated beads (*see* **Note 3**).
18. HEK293T cells with doxycycline-inducible FLAG-tagged KRAS.
19. Amicon Ultra-0.5 Centrifugal Filter Device 30K (Merck Millipore).
20. Protein LoBind 1.5 mL tube (Eppendorf).

### 2.2 IP-RP-HPLC Analysis of Guanine Nucleotides

1. HPLC system equipped with a DGU-20A3 degasser, an LC-20AD pump, a SIL-20AC autosampler, a CTO-20AC column oven, an SPD-20A UV/VIS detector, and a CBM-20A Communication Bus Module (Shimadzu).
2. Instrumental control and data analysis software: LC Solution (Shimadzu).
3. Reversed-phase C18 column: Gemini 5 μm NX C18 110 Å LC Column (150 × 4.6 mm) (Phenomenex) with a Gemini NX C18 SecurityGuard Cartridge (4.0 × 3.0 mm) (Phenomenex).
4. Solvent A: 29.7 mM potassium phosphate (pH 7.0), 10 mM TBA-B. For 887.88 mL, mix 879 mL of 30 mM potassium phosphate (pH 7.0) and 8.88 mL of 1 M TBA-B. Vacuum filtrate the solvent using the TPP vacuum filtration "rapid"--Filtermax, and place it in a light-shielded brown bottle at room temperature. Prepare the solvent just before use.
5. Solvent B: 100% acetonitrile (HPLC grade).
6. 5 mM GDP and 5 mM GTP: dissolve each nucleotide in water to a concentration slightly higher than 5 mM and determine the concentration of each nucleotide based on its molecular absorption coefficient. Dilute the solution with water to make a 5 mM solution of each nucleotide. Store the nucleotide solutions at −30 °C.

## 3 Methods

### 3.1 Immunoprecipitation of FLAG-KRAS Proteins

All procedures of immunoprecipitation are performed as coldly as possible. The time required for cell solubilization and antigen-antibody reaction is kept as short as possible to minimize dissociation and hydrolysis of guanine nucleotides bound to KRAS.

1. Grow HEK293T cells, which express FLAG-tagged KRAS in doxycycline (DOX)-inducible manner (*see* **Note 4**), in 150 mm plates (*see* **Note 5**) until cells are ~50% confluent.
2. Remove the media from the plate and add the culture media containing 1 μg/mL DOX (*see* **Note 6**).
3. Incubate the cells for ~16 h in a $CO_2$ incubator.
4. Remove the media completely from the plates.
5. Wash the plates twice with ~5 mL of ice-cold (1×) PBS. Remove the residual (1×) PBS thoroughly by aspiration after the second wash.
6. Put the plates on ice and add ice-cold extraction buffer (550 μL/plate). Scrape the cells from the plates and transfer the lysate to a microcentrifuge tube on ice.
7. Vortex the tube for 5 min at 4 °C.
8. Centrifuge the tube at 15,000 rpm for 5 min at 4 °C, and transfer the supernatant to a new microcentrifuge tube on ice (the "Input" sample). Withdraw 60 μL of the Input sample for Western blotting analysis (Fig. 2c).
9. Wash anti-FLAG antibody-conjugated beads (~10 μL/sample) twice with wash buffer-1. Each washing step consists of resuspending the beads with 500 μL of wash buffer-1, centrifuging at 2300 × *g* for 1 min at 4 °C, and removing the supernatant. Resuspend the washed beads in ~30 μL of wash buffer-1.
10. Add the Input sample (~500 μL) to the washed beads.
11. Rotate the sample for 30 min at 4 °C in an orbital rotator.
12. Spin down the beads at 2300 × *g* for 1 min at 4 °C and remove the supernatant.
13. Wash the beads twice with wash buffer-1. Each washing step consists of resuspending the beads with 500 μL of wash buffer-1, centrifuging at 2300 × *g* for 1 min at 4 °C, and removing the supernatant.
14. Resuspend the beads with 250 μL of wash buffer-2, and transfer the resuspended beads to a Protein LoBind 1.5 mL tube (Eppendorf) on ice.
15. Spin down the beads at 2300 × *g* for 1 min at 4 °C and remove the supernatant. Remove the wash buffer-2 (supernatant) as

completely as possible by aspiration using 27G needle attached to 1 mL syringe (*see* **Note** 7).

16. Add 60 μL of elution buffer to the beads (*see* **Note 8**).
17. Vortex the sample for 5 min at 4 °C.
18. Spin down the beads at 15,000 rpm for 1 min at 4 °C, and transfer 50 μL of the supernatant to a new Protein LoBind 1.5 mL tube (Eppendorf) on ice.
19. Add 11.2 μL of 0.5 M $K_2HPO_4$ to the sample, and mix (the "Elution" sample) (*see* **Note 9**). Withdraw 15 μL of the Elution sample for Western blotting analysis.
20. Incubate the Elution sample for 3 min at 90 °C using a heat block (*see* **Note 10**).
21. Add 113.7 μL of water and 1.6 μL of 1 M TBA-B to the sample and mix.
22. Add the sample to the Ultra-0.5 Centrifugal Filter Device 30K (Merck Millipore).
23. Centrifuge the device at 14,000 × *g* for 20 min at 4 °C, and collect the filtrated fraction (the HPLC sample) in the centrifuge tube (*see* **Note 11**).

### 3.2 IP-RP-HPLC Analysis of Guanine Nucleotides Bound to FLAG-KRAS

This protocol has been optimized for analyzing guanine nucleotides using the HPLC system described in the Subheading 2. The isocratic mobile phase consisted of 20% acetonitrile and 80% 30 mM phosphate buffer (pH 7.0) containing 10 mM TBA-B. Isocratic elution was performed at 40 °C with a 0.8 mL/min flow rate.

#### 3.2.1 Generation of a Standard Curve

1. Dilute 5 mM stock solutions of GDP and GTP with solvent A to make standard solutions of the following concentrations: 100, 500, and 1000 nM.
2. Inject and run 100 μL of solvent A as a negative control.
3. Inject and run 100 μL of the 100 nM GDP or GTP standard solution.
4. Repeat Subheading 3.2.1, **step 3** for the 500 and 1000 nM GDP or GTP standard solutions.
5. Obtain the peak area for each standard solution using the data analysis software.
6. Prepare a standard curve by plotting the GDP or GTP amount in pmol (e.g., 10 pmol for the 100 μL of the 100 nM solution) vs the peak areas (Fig. 1).

#### 3.2.2 Analysis of Guanine Nucleotides Bound to FLAG-KRAS

1. Analyze 100 μL of the HPLC sample obtained in Subheading 3.1, **step 23** (Fig. 2a).

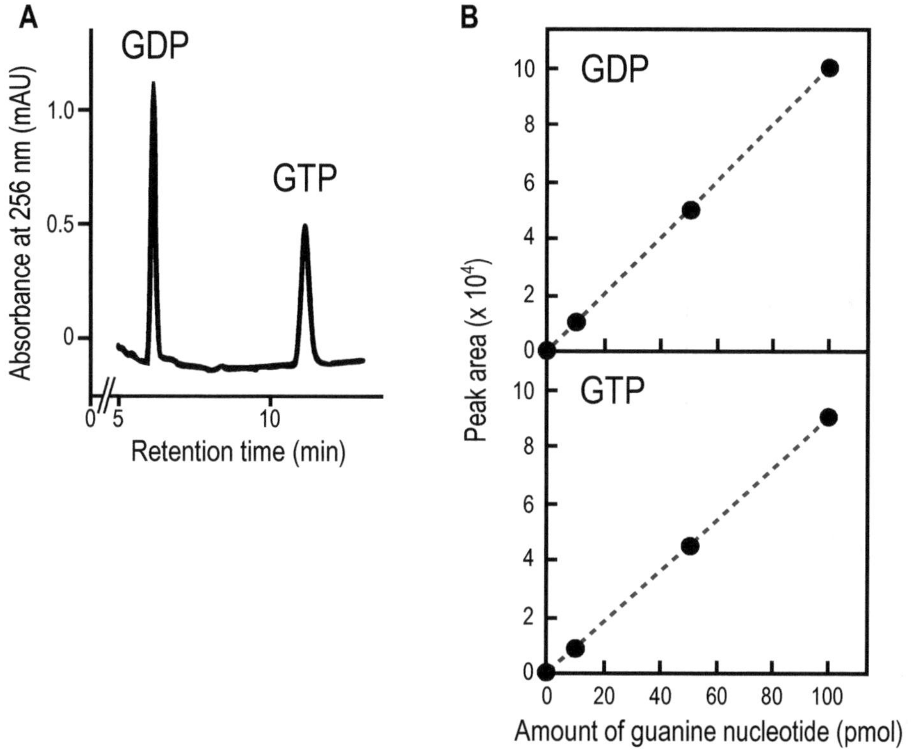


**Fig. 1** Separation of guanine nucleotides by IP-RP-HPLC. (**a**) Representative IP-RP-HPLC chromatogram of a standard mixture of guanine nucleotides (GDP and GTP, each 12.5 pmol). The guanine nucleotides were detected by a UV detector at a wavelength of 254 nm. The peaks of GDP and GTP were detected at 6.0 and 11.0 min, respectively, on a chromatogram. (**b**) Standard curve of GDP and GTP. The guanine nucleotides were determined by injecting 100 μL of each mixture with concentrations of 0, 100, 500, and 1000 nM of GDP and GTP by IP-RP-HPLC. A standard curve was plotted by measuring the area values of the peaks detected at each amount of nucleotide. The correlation coefficient ($r^2$) for each standard curve was ≥0.999. The data are the means from five independent experiments

2. Determine the peak area of GDP and GTP using the data analysis software.
3. Calculate the GDP and GTP amounts using the standard curve in Subheading 3.2.1, **step 6** (Fig. 2b).

## 4 Notes

1. Keep the pH of the mixed solution monitored while mixing 30 mM $K_2HPO_4$ and 30 mM $KH_2PO_4$ solutions.
2. Since TBA-B is highly hygroscopic, dissolve all of the contents in the reagent bottle in water to make a 1 M TBA-B solution after opening the bottle.
3. We use anti-DYKDDDDK-tag Antibody Beads (FUJIFILM Wako Pure Chemical Corporation).

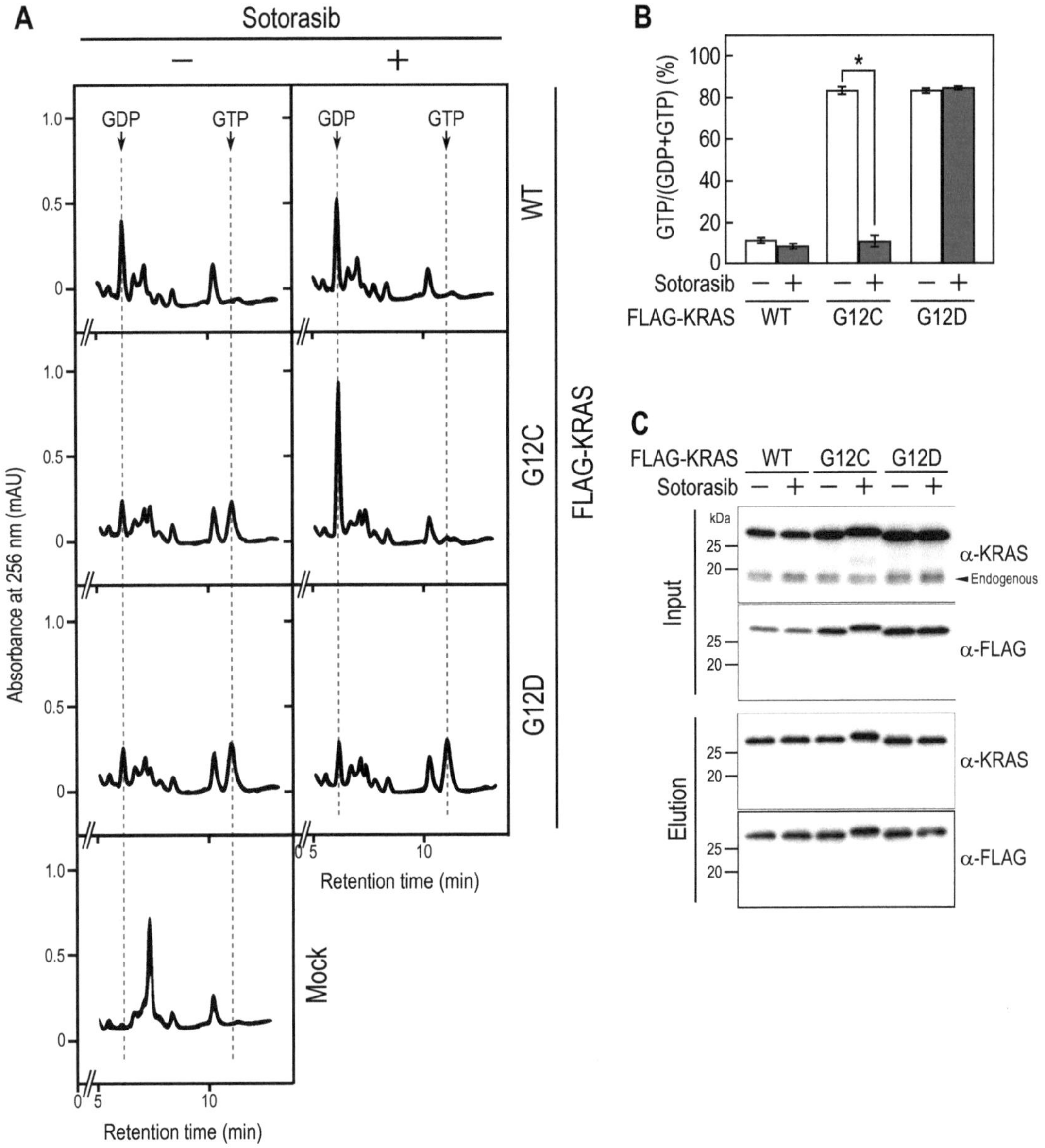


**Fig. 2** Analysis of the activation status of FLAG-KRAS in HEK293T cells. (**a**) Representative chromatogram of guanine nucleotides bound to immunoisolated FLAG-KRAS wild type (WT), G12C, and G12D from HEK293T cells treated with or without 1 μM sotorasib. (**b**) The relative amounts of guanine nucleotides bound to FLAG-KRAS wild type and mutants in HEK293T cells treated with (*closed bar*) or without (*open bar*) 1 μM of sotorasib. The data are the means ± SD from three independent experiments. $^*p < 0.05$, Student's *t* test. (**c**) The cell lysates (the "Input" sample) and the immunoprecipitates with anti-FLAG beads (the "Elution" sample) were subjected to Western blot analysis using the indicated antibodies

4. We established a HEK293T cell line expressing FLAG-KRAS in a DOX-inducible manner by inserting the transgene into the *AAVS1* locus (a human genomic "safe harbor" locus) via the

CRISPR-Cas9 genome engineering [19]. Parental HEK293T cells are also subjected to immunoprecipitation as a negative control.

5. Since HEK293T cells are easily detached from the plate, we use PEI-coated dishes [20].
6. A 1 μg/mL DOX concentration usually provides the maximum expression of FLAG-KRAS, although it depends on the cell line generated. When studying the effect of sotorasib, add sotorasib to the medium at the same time as DOX.
7. Careful aspiration with a 27G needle attached to a 1 mL syringe allows almost all of the wash buffer-2 to be removed without losing beads due to aspiration.
8. FLAG-KRAS can be dissociated from anti-FLAG antibodies under acidic conditions.
9. $K_2HPO_4$ is added to make the solution neutralized.
10. GDP and GTP bound to FLAG-KRAS are dissociated from FLAG-KRAS by heat denaturation.
11. The denatured proteins, which may interfere with the subsequent HPLC analysis, are removed from the sample by filtration using the device.

## Acknowledgements

This work was partly supported by the Japan Society for the Promotion of Science (JSPS) KAKENHI (grant numbers 23K05685 and 20K06565 to KK).

*Conflict of Interest*: The authors declare that they have no conflicts of interest with the contents of this article.

### References

1. Cherfils J, Zeghouf M (2013) Regulation of small GTPases by GEFs, GAPs, and GDIs. Physiol Rev 93:269–309
2. Colicelli J (2004) Human RAS superfamily proteins and related GTPases. Sci STKE 2004:RE13
3. Satoh T, Endo M, Nakamura S et al (1988) Analysis of guanine nucleotide bound to ras protein in PC12 cells. FEBS Lett 236:185–189
4. Taylor SJ, Resnick RJ, Shalloway D (2001) Nonradioactive determination of Ras-GTP levels using activated ras interaction assay. Methods Enzymol 333:333–342
5. Benard V, Bokoch GM (2002) Assay of Cdc42, Rac, and Rho GTPase activation by affinity methods. Methods Enzymol 345:349–359
6. Kiyokawa E, Aoki K, Nakamura T et al (2011) Spatiotemporal regulation of small GTPases as revealed by probes based on the principle of Förster resonance energy transfer (FRET): implications for signaling and pharmacology. Annu Rev Pharmacol Toxicol 51:337–358
7. Kim J, Lee S, Jung K et al (2019) Intensiometric biosensors visualize the activity of multiple small GTPases in vivo. Nat Commun 10: 211
8. Feuerstein J, Goody RS, Wittinghofer A (1987) Preparation and characterization of nucleotide-free and metal ion-free p21 "apoprotein". J Biol Chem 262:8455–8458
9. Weiss O, Holden J, Rulka C et al (1989) Nucleotide binding and cofactor activities of

purified bovine brain and bacterially expressed ADP-ribosylation factor. J Biol Chem 264: 21066–21072

10. John J, Sohmen R, Feuerstein J et al (1990) Kinetics of interaction of nucleotides with nucleotide-free H-ras p21. Biochemistry 29: 6058–6065
11. Bromberg Y, Shani E, Joseph G et al (1994) The GDP-bound form of the small G protein Rac1 p21 is a potent activator of the superoxide-forming NADPH oxidase of macrophages. J Biol Chem 269:7055–7058
12. Ahmadian M, Zor T, Vogt D et al (1999) Guanosine triphosphatase stimulation of oncogenic Ras mutants. Proc Natl Acad Sci U S A 96:7065–7070
13. Hanzal-Bayer M, Renault L, Roversi P et al (2002) The complex of Arl2-GTP and PDEδ: from structure to function. EMBO J 21:2095–2106
14. Rubio I, Pusch R, Wetzker R (2004) Quantification of absolute Ras-GDP/GTP levels by HPLC separation of Ras-bound [32P]-labelled nucleotides. J Biochem Biophys Methods 58: 111–117
15. Scrima A, Thomas C, Deaconescu D et al (2008) The Rap–RapGAP complex: GTP hydrolysis without catalytic glutamine and arginine residues. EMBO J 27:1145–1153
16. Sot B, Behrmann E, Raunser S et al (2013) Ras GTPase activating (RasGAP) activity of the dual specificity GAP protein Rasal requires colocalization and C2 domain binding to lipid membranes. Proc Natl Acad Sci 110:111–116
17. Hannan JP, Swisher GH, Martyr JG et al (2021) HPLC method to resolve, identify and quantify guanine nucleotides bound to recombinant ras GTPase. Anal Biochem 631:114338
18. Araki M, Yoshimoto K, Ohta M et al (2021) Development of a versatile HPLC-based method to evaluate the activation status of small GTPases. J Biol Chem 297:101428
19. Dalvai M, Loehr J, Jacquet K et al (2015) A scalable genome-editing-based approach for mapping multiprotein complexes in human cells. Cell Rep 13:621–633
20. Vancha AR, Govindaraju S, Parsa KV et al (2004) Use of polyethyleneimine polymer in cell culture as attachment factor and lipofection enhancer. BMC Biotechnol 4:23

# Chapter 17

# Real-Time Monitoring of RAS Activity Using In Vitro and In-Cell NMR Spectroscopy

Qingci Zhao, Ichio Shimada, and Noritaka Nishida


## Abstract

The activation level of RAS can be determined by GTP hydrolysis rate ($k_{hy}$) and GDP-GTP exchange rates ($k_{ex}$). Either impaired GTP hydrolysis or enhanced GDP-GTP exchange causes the aberrant activation of RAS in oncogenic mutants. Therefore, it is important to quantify the $k_{hy}$ and $k_{ex}$ for understanding the mechanisms of RAS oncogenesis and drug development. Conventional methods have individually measured the $k_{ex}$ and $k_{hy}$ of RAS. However, within the intracellular environment, GTP hydrolysis and GDP-GTP exchange reactions occur simultaneously under conditions where GTP concentration is kept constant. In addition, the intracellular activity of RAS is influenced by endogenous regulatory proteins, such as RAS GTPase activating proteins (GAPs) and the guanine-nucleotide exchange factors (GEFs). Here, we describe the in vitro and in-cell NMR methods to estimate the $k_{hy}$ and $k_{ex}$ simultaneously by measuring the time-dependent changes of the fraction of GTP-bound ratio under the condition of constant GTP concentration.


**Key words** Small GTPase, RAS, GTP hydrolysis rate, GDP-GTP exchange rate, GTP regeneration system, In-cell NMR, Real-time NMR

## 1 Introduction

RAS is a small GTPase protein that plays a crucial role in intracellular signaling pathways. RAS acts as a molecular switch, cycling between an active GTP-bound state and an inactive GDP-bound state. GTP-bound RAS transmits multiple signaling cascades, including the MAPK and PI3K pathways, thereby regulating cell growth, proliferation, and differentiation [1, 2]. Several mutations are known to constitutively activate RAS, leading to aberrant cell growth and contributing to the development of various cancers [3, 4]. Due to its significance in cancer and other diseases, RAS has been extensively studied as a potential therapeutic target [5, 6]. Therefore, understanding the mechanisms that regulate RAS activity is crucial for developing effective treatments to combat RAS-driven diseases.

Andrew G. Stephen and Dominic Esposito (eds.), *KRAS: Methods and Protocols*, Methods in Molecular Biology, vol. 2797,
https://doi.org/10.1007/978-1-0716-3822-4_17,

The activation levels of RAS are determined by two parameters (Fig. 1a). One is the GTP hydrolysis rate ($k_{hy}$), at which the GTP-bound form RAS is inactivated to the GDP-bound form by intrinsic GTPase activity. This reaction is also enhanced by the intracellular regulatory protein, called GTPase-activating proteins (GAPs) [7, 8]. The other parameter is the GDP-GTP exchange rate ($k_{ex}$), at which the inactive GDP-bound form RAS is converted to the GTP-bound form. This reaction proceeds in two steps: the dissociation of GDP from inactive RAS and the rebinding of the free GTP to apo-RAS. The dissociation of GDP is the rate-limiting step and facilitated by the guanine-nucleotide exchange factors (GEFs) [9]. Based on these two parameters, the fraction of GTP-bound (fGTP) in steady state can be determined as $k_{ex}/(k_{hy} + k_{ex})$.

Recently, the real-time NMR (RT-NMR) method has been utilized to individually measure $k_{hy}$ and $k_{ex}$ for the small GTPases, including RAS [10–12]. By conducting time-resolved NMR measurements on GTP-loaded RAS, the progress of GTP hydrolysis is monitored based on signal intensities of the GTP-bound and GDP-bound forms, and the rate constant for GTP hydrolysis is determined by single exponential fitting. Similarly, by adding a non-hydrolyzable GTP analog, GTPγS, to GDP-loaded RAS and continuously acquiring NMR spectra, the GDP-GTP exchange rate constant can be determined. The RT-NMR method can be performed without the need for fluorescent nucleotide analogs [13], $^{32}P$ radio isotope-labeled nucleotide [14], or any chemical modification of RAS. By using NMR, it is also possible to determine the activity of multiple GTPase proteins labeled with different stable isotopes in a single experiment [15, 16].

In the intracellular environment, RAS exists in an equilibrium where GTP hydrolysis and GDP-GTP exchange progress simultaneously. The balance between these bidirectional reactions determines the fGTP in the steady state. Since intracellular GTP concentration is maintained more than ten times higher than GDP [17], GTP preferentially binds to apo-RAS after GDP dissociation (Fig. 1a). On the other hand, when measuring $k_{ex}$ in vitro, GTP analogs are used instead of GTP because the concentration of free GTP cannot be maintained due to the GTP hydrolysis reaction [11]. If there is a difference in selectivity between GTP and the GTP analog, accurate determination of $k_{ex}$ may not be possible.

Furthermore, there exist various regulatory factors controlling RAS activity in the intracellular environment. The estimation of the steady-state fGTP for wild-type RAS based on individually measured $k_{hy}$ and $k_{ex}$ is about 30–40%, which contradicts its role as a molecular switch [18]. Therefore, the influence of intracellular regulatory factors needs to be considered. Although the influence of regulatory factors, such as GAPs or GEFs, has been examined by adding the recombinant proteins or cell lysates under in vitro

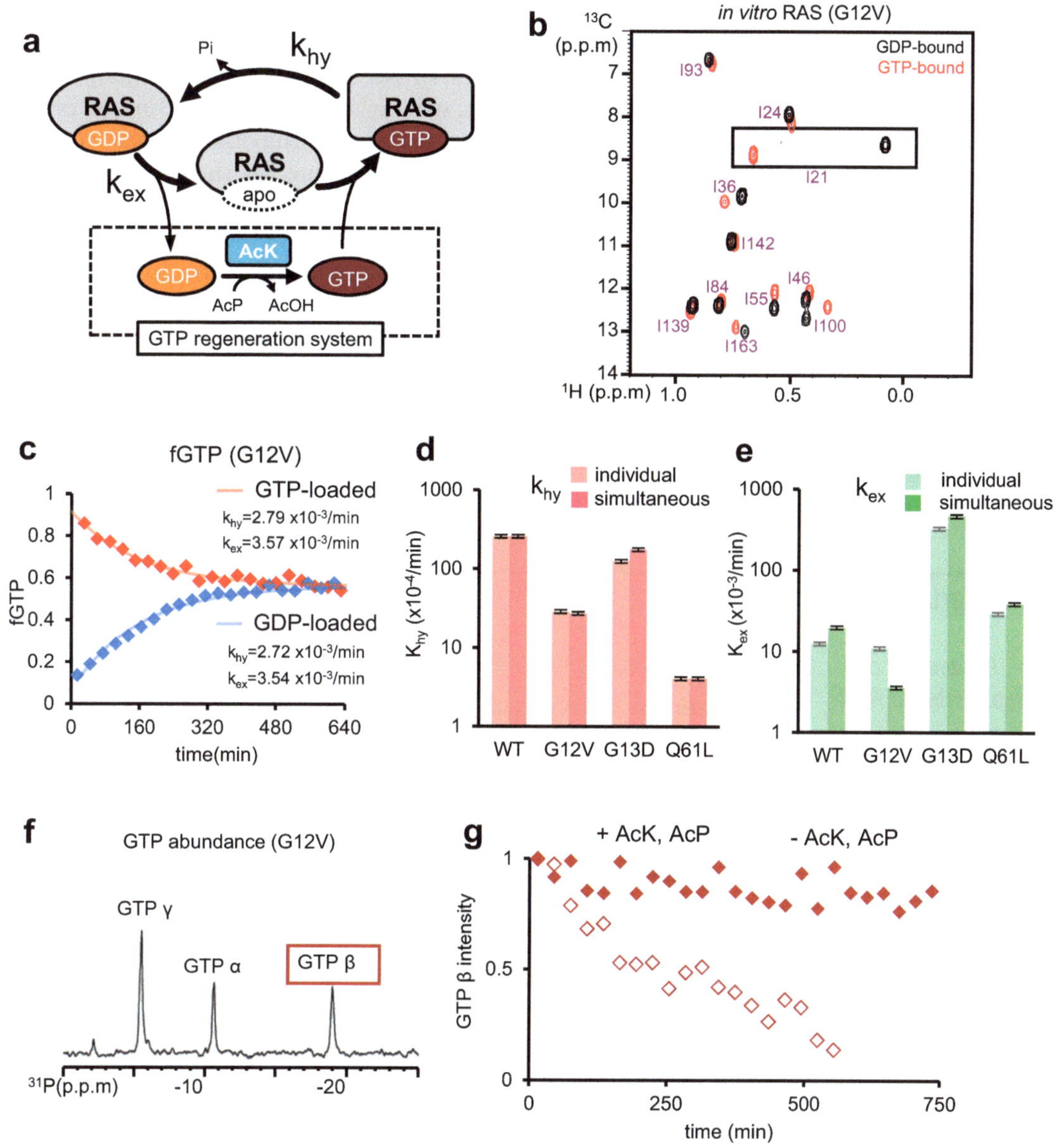


**Fig. 1** $k_{hy}$ and $k_{ex}$ measurements using GTP regeneration system. (**a**) Schematic illustration of the RAS cycling between active GTP-bound and inactive GDP-bound states. Under the GTP regeneration system, GDP released from RAS was regenerated to GTP by acetate kinase (AcK), which transfers the phosphate group of acetyl phosphate (AcP) to GDP. (**b**) Overlaid 2D $^{1}$H-$^{13}$C HSQC spectra of Ile δ1 methyl $^{13}$C-labeled RAS G12V bound to GDP (black) or GTP (red). (**c**) The time-course of fGTP starting from the GTP-loaded (red line) and GDP-loaded (blue line) RAS G12V in the GTP regeneration system. Estimated $k_{hy}$, $k_{ex}$, and fGTP in steady state are almost identical regardless of initial GTP-bound state. (**d**) In vitro GTP hydrolysis rate ($k_{hy}$) of RAS WT and mutants measured in the GTP regeneration system (magenta) or by individual method (light magenta). (**e**) In vitro GDP-GTP exchange rate ($k_{ex}$) of RAS WT and mutants measured in the GTP regeneration system (green) or by individual method (light green). (**f**) 1D $^{31}$P NMR spectrum to monitor the GTP abundance in the GTP regeneration system. (**g**) The time-dependent changes of free GTP abundance were estimated based on the signal intensity of β-phosphate of GTP as normalized by that of the first time point. The results are plotted for in the presence (filled diamond) and absence (open diamond) of AcK and AcP

conditions [11], it is impossible to fully reproduce the native intracellular environment. In such cases, the application of in-cell NMR, which allows the observation of NMR signals of the target proteins in the cellular environment, will be valuable. This technique proves to be effective for studying proteins within the cellular context, where the intracellular conditions influence their structure and activities [19, 20].

In this chapter, we first introduce a method to quantify RAS activity ($k_{hy}$ and $k_{ex}$) simultaneously under in vitro conditions utilizing a GTP regeneration system, which can maintain a constant GTP concentration. Additionally, we describe the in-cell NMR methods, which enable the direct observation of NMR signals of RAS in the physiological intracellular environment, to quantify the time-dependent changes of active GTP-bound form, thereby calculating the GTP hydrolysis and GDP-GTP exchange rates of RAS simultaneously [18].

## 2 Materials

### *2.1 Production of Isotope-Labeled RAS Protein*

1. Competent cells: BL21(DE3).
2. Expression vector: cDNA encoding HRAS G domain (residues 1–171) or KRAS G domain (residues 1–169) is cloned into the pET-15b vector (via Nde I-BamH I sites) which contains the 6× His-tag and thrombin cleavage sequence at N-terminus.
3. LB medium: 10 g/L Bacto tryptone, 5 g/L Bacto yeast extract, 5 g/L NaCl. Sterilize by autoclaving (120 °C 20 min).
4. M9 medium ($H_2O$): 17.7 g $Na_2HPO_4$ $12H_2O$, 3.0 g $KH_2PO_4$, 1.0 g $^{15}NH_4Cl$, 0.5 g NaCl. Dissolve with 1 L of Milli-Q water and sterilize by autoclaving (at 120 °C for 20 min). Add 4 g D-glucose, 20 mg biotin, 100 mg ampicillin, 2 M $MgSO_4$ solution 1 mL, solution M 1 mL, and solution V 1 mL.
5. Solution M: 100 mM $FeCl_3$, 100 mM $CaCl_2$, 50 mM $ZnCl_2$. Sterilize by 0.22 μm filter.
6. Solution V: 20 mg/mL thiamine, 10 mg/mL choline chloride, 10 mg/mL nicotine amide, 10 mg/mL pantothenic acid, 10 mg/mL pyridoxal, 0.1 mg/mL riboflavin, 20 mg/mL inositol. Sterilize by 0.22 μm filter.
7. M9 medium ($D_2O$): 12.9 g $Na_2HPO_4$ (anhydrate), 3.0 g $KH_2PO_4$, 1.0 g $^{15}NH_4Cl$, 0.5 g NaCl, and 3 g $^2H$ D-glucose. Dissolve in 1 L of $D_2O$. All other supplements are added either as powder or solution dissolved in $D_2O$. Sterilize by 0.22 μm filter Stericup (Millipore #SCGVU05RE) (*see* **Note 1**).

8. IPTG: 1.0 M solution dissolved either in $H_2O$ or $D_2O$. Sterilize by 0.22 μm filter.
9. Thrombin (Sigma-Aldrich #T4648).
10. Stable isotope-labeled compounds:
   [$^{15}$N] ammonium chloride (CIL #NLM-467).
   [Methyl-$^{13}$C, 2,2-D2] α-ketobutyric acid (CIL #CDLM-7318).
   [1,2,3,4,5,6,6-D7,97–98%] D-glucose (CIL #DLM-2062).
   $D_2O$ (CIL #DLM-4-1000).
11. NMR buffer: 20 mM HEPES-NaOH (pH 7.2), 150 mM NaCl, 5 mM $MgCl_2$, 1 mM TCEP.
12. GTP stock: 100 mM GTP (Sigma-Aldrich #G8877) dissolved in Milli-Q water. Stored at −80 °C.

### *2.2 GTP Regeneration System*

1. Acetate kinase (AcK): 100 unit/mL (Sigma-Aldrich #A7437) dissolved in Milli-Q water. Stored at −80 °C. Dilute this stock solution with the NMR buffer to 0.1 unit/mL for preparation of the NMR sample.
2. Acetyl phosphate (AcP): 500 mM AcP (Sigma-Aldrich #A0262) dissolved in Milli-Q water. Stored at −80 °C.
3. 10× NMR buffer: 200 mM HEPES-NaOH (pH 7.2), 1.5 M NaCl, 50 mM $MgCl_2$, 10 mM TCEP.

### *2.3 Protein Delivery into HeLa S3 Cells and Mebiol Gel Encapsulation*

1. HeLa S3 cells (JCRB#9010).
2. DMEM high glucose, no glutamine (Gibco #11960044) supplemented with 1% Pen-Strep (Gibco), 1% GlutaMAX (Gibco), 1 mM sodium pyruvate, and 10% (v/v) fetal bovine serum (FBS). Cells were cultured at 37 °C with an atmosphere containing 5% $CO_2$ using a spinner flask (at 90 rpm).
3. HBSS buffer: 30 mM HEPES-KOH (pH 7.2), 137 mM NaCl, 5.4 mM KCl, 0.25 mM $Na_2HPO_4$, 0.44 mM $KH_2PO_4$, 4.2 mM $NaHCO_3$, 0.1% w/v D-glucose.
4. HBSS (+) buffer: HBSS buffer containing 1 mM $CaCl_2$.
5. SLO stock: 5 μg/mL streptolysin O (1 mg/mL BioAcademia #01–531) dissolved in ice-cold PBS containing 50% glycerol and 10 mM DTT. 200 μL aliquots are stored in −80 °C.
6. FITC-labeled RAS (*see* **Note 2**).
7. Transport buffer: 25 mM HEPES-KOH (pH 7.4), 115 mM KOAc, 5 mM $MgCl_2$, 2 mM EDTA.
8. Propidium iodide (PI) stock: 500 μg/mL PI (Sigma-Aldrich, #P4170) dissolved in PBS buffer.
9. Percoll (Cytiva # 17089101).

10. Mebiol gel: lyophilized Mebiol gel (Mebiol) was dissolved with 10 mL of DMEM (20% $D_2O$) at 4 °C for 2 days according to the manufacturer's instructions. 1 mL aliquots are stored at −80 °C.

#### 2.4 In-Cell NMR Measurement Using Bioreactor System

1. DMEM (20% $D_2O$): DMEM powder (Thermo Fisher #12100046) dissolved with 25 mM HEPES-NaOH (pH 7.4), 20% $D_2O$. Supplemented with 1% Pen-Strep (Gibco), 1% GlutaMAX (Gibco), 1 mM sodium pyruvate, and 10% FBS. Sterilize with Stericup (Millipore #SCGVU05RE) (*see* **Note 3**).
2. Bioreactor equipment.

   The schematic diagram of the bioreactor setup is shown in Fig. 2a [21]. A PEEK tube (o.d. 1 mm, i.d. 0.5 mm) is attached to a glass pipette (o.d. 4 mm, i.d. 2.6 mm) using instant glue, while the other end of a PEEK tube is connected to a syringe pump for medium delivery. A PTFE tube (o.d. 2 mm, i.d. 1 mm) is connected to a vacuum pump through a waste reservoir. Both the PEEK tube and PTFE tube are inserted into a silicon tube (o.d. 6 mm, i.d. 4 mm). The silicon tube is connected to a Shigemi NMR tube (o.d. 5 mm, i.d. 4.2 mm). When the silicon tube is connected to the Shigemi NMR tube, the tip of the glass pipette should be precisely positioned at the bottom of the NMR sample tube. The position of the PTFE tube is set to be approximately 8 cm above the bottom of the NMR sample tube (*see* **Note 4**).
3. Syringe pump; PHD 2000 syringe pump (Harvard Apparatus).

## 3 Methods

#### 3.1 Preparation of Stable Isotope-Labeled RAS

1. Transform the expression plasmid into the *E. coli* competent cells.
2. Pre-culture in 20 mL of LB medium at 37 °C.
3. (For deuterium labeling) Collect the bacterial cells from the pre-culture by centrifugation, and transfer to 100 mL of M9 $D_2O$ medium, and cultivate for approximately 16 h at 37 °C (*see* **Note 5**).
4. Collect the bacterial cells from the pre-culture by centrifugation, and then transfer to the 1 L of M9 medium and cultivate at 37 °C.
5. When $OD_{600}$ reaches 0.5, add 40 mg [methyl-$^{13}$C, 2,2-D2] α-ketobutyric acid for isoleucine side chain δ1 methyl labeling, and set the temperature to 25 °C. After 1 h, add IPTG at a final concentration of 1 mM.
6. After 16 h of induction, harvest the cells by centrifugation.

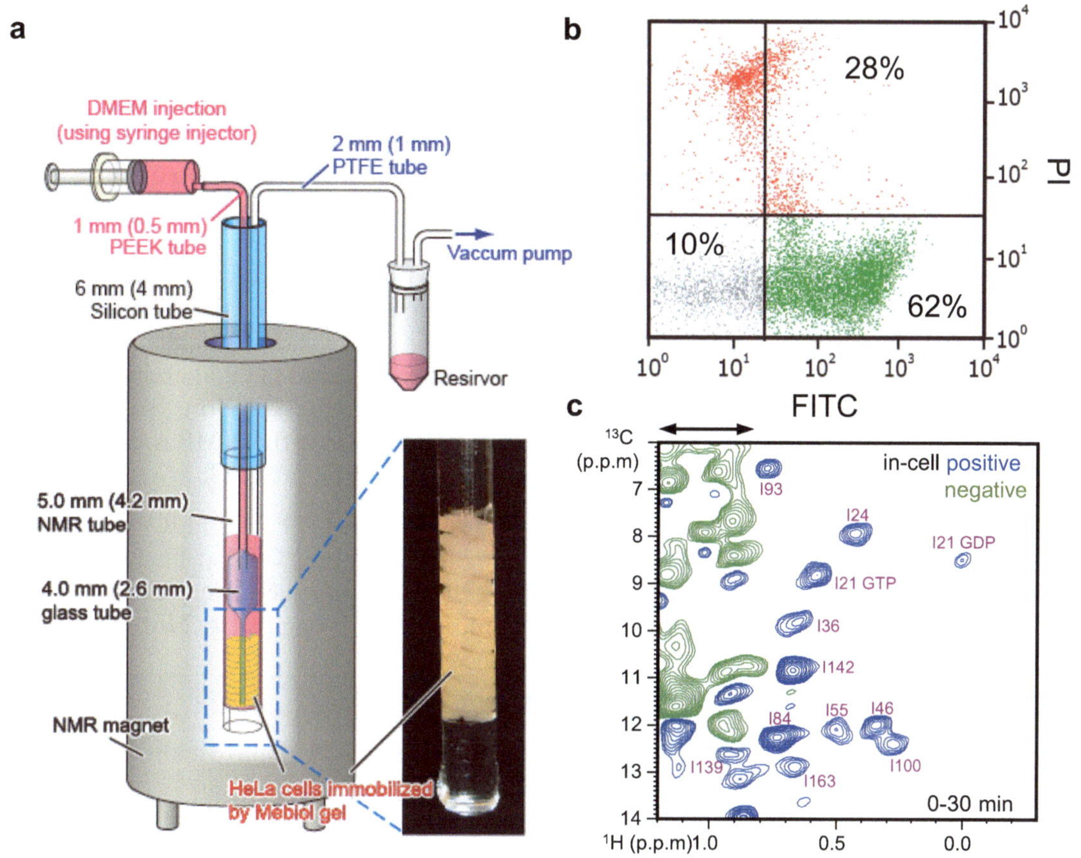


**Fig. 2** In-cell NMR measurement using the bioreactor system. (**a**) A schematic illustration of the gel-encapsulated bioreactor system. The outer diameter and inner diameter (in parenthesis) of each tube are indicated. The picture of the HeLa cells encapsulated within Mebiol gel in the NMR tube is shown. (**b**) A representative result of flow cytometric analysis of the SLO-treated HeLa S3 cells. The FITC positive and PI negative (lower right) region corresponds to the protein-introduced cells (green), while the PI positive (upper) region corresponds to the dead cells (red). (**c**) $^{1}$H-$^{13}$C SOFAST HMQC spectrum of the GTP-loaded RAS G12V introduced cells at the first 30 min. The positive signals are shown blue, while the aliased negative signals are colored green. The signals with $^{1}$H chemical shift more than 0.8 ppm (indicated by an arrow) are background signals derived from the medium or endogenous cellular proteins [21]. All 11 Ile signals were observed with sufficient sensitivity, and their chemical shift patterns were similar to those observed in vitro, without overlapping background signals except for I139

7. Lyse the cells by sonication, and purify RAS protein from the cell lysate using $Ni^{2+}$-affinity chromatography.
8. After tag cleavage with a protease, perform final purification by gel filtration.
9. To prepare GTP-loaded RAS, add 10 mM EDTA and 10 mM GTP to 10 mg/mL RAS solution (2 mL). Incubate at room temperature for 1 h, and then add 20 mM $MgCl_2$.

10. Add the RAS solution to a desalting column (Econo-Pac 10DG, BioRad) pre-equilibrated with NMR buffer, and collect eluates in 1 mL fractions.
11. Measure the UV absorbance spectrum of each fraction, and collect the fractions with a peak at 280 nm. Concentrate the fractions containing RAS to 500 μM for subsequent experiments (can be stored at −80 °C after flash freezing).

### 3.2 $k_{hy}$ and $k_{ex}$ Measurements in GTP Regeneration System

The GTP regeneration system utilizes the reaction in which acetate kinase (AcK) transfers the phosphate group of the substrate acetyl phosphate (AcP) to GDP [22]. Thus, GDP released from RAS following GTP hydrolysis is regenerated to GTP, thereby maintaining a concentration of free GTP in the solution. To avoid the progress of GTP hydrolysis of RAS before the NMR measurement, sample preparation should be carried out at 4 °C.

1. Add the following solutions to the GTP-bound RAS to prepare an NMR sample containing RAS and GTP at final concentrations of 200 μM and 400 μM, respectively. Keep the samples at 4 °C until the NMR measurement begins (*see* **Note 6**).

| | |
|---|---|
| 500 μM RAS(GTP-loaded) | 200 μL |
| $H_2O$ (Milli-Q) | 208.5 μL |
| $D_2O$ | 50 μL |
| 10× NMR buffer | 29.5 μL |
| 500 mM AcP | 5 μL |
| 100 mM GTP | 2 μL |
| 0.1 unit/mL acetate kinase | 5 μL |

2. Insert the samples into the NMR instrument (probe temperature is set to 37 °C), and perform lock, tuning, and shimming as quickly as possible. Continuously measure $^{1}H$-$^{13}C$ HSQC spectra every 30 min.
3. Based on the signal intensity of I21, which shows different chemical shifts between the GTP-bound and GDP-bound forms (Fig. 1b), calculate the fraction of the GTP-bound form at each time point.
4. Plot the time-dependent changes of the GTP-bound ratio (Fig. 1c), and calculate $k_{hy}$ and $k_{ex}$ by fitting the data to the theoretical equation shown below (Fig. 1d, e):

$$[fGTP]_t = \frac{k_{ex}\left(1 - \exp\left(-\left(k_{hy} + k_{ex}\right)t\right)\right) + \left(k_{hy} + k_{ex}\right)[fGTP]_0 \exp\left(-\left(k_{hy} + k_{ex}\right)t\right)}{k_{hy} + k_{ex}}$$

where $[fGTP]_t$ denotes the fGTP at the time $t$ (*see* **Note 7**).

5. Prepare the samples with the same composition as above, and measure 1D $^{31}$P NMR at 37 °C continuously every 30 min. Based on the $^{31}$P signals derived from GTP β phosphate (Fig. 1f at chemical shift value of −20 ppm), confirm that free GTP concentration is kept constant during the $^{1}$H $^{13}$C HSQC measurements (Fig. 1g) (*see* **Note 8**).

### 3.3 Protein Delivery into HeLa S3 Cells and Encapsulation Within Mebiol Gel

1. Harvest 1 × $10^8$ HeLa S3 cells from suspension cell culture (*see* **Note 9**).
2. Wash with 5 mL of HBSS buffer twice by centrifugation (400 × *g* 3 min), and resuspend with 10 mL of HBSS buffer to check the cell numbers and viability using a flow cytometer.
3. Transfer the cell suspension to a new 50 mL conical tube, and dilute with ice-cold HBSS buffer to a cell density of 2 × $10^6$ cells/mL.
4. Add SLO stock solution with a final concentration of 10–20 ng/mL, and incubate on ice for 10 min (*see* **Note 10**).
5. Remove the supernatant by centrifugation (400 × *g* 3 min), and then wash twice with 10 mL of ice-cold HBSS buffer (*see* **Note 11**).
6. Prepare a solution containing stable isotope-labeled RAS and FITC-labeled RAS at a concentration of 1–2 mM and $A_{495} = 0.3$, respectively, in transport buffer. The volume of the solution is typically 300–500 μL. Add this solution to the SLO-treated cell pellet, and gently pipette to obtain a homogeneous suspension (*see* **Note 12**).
7. Inoculate the cell suspension at 30 °C and 150 rpm using a bio-shaker. Take a small amount of cell suspension every 5–10 min for flow cytometer analysis. Estimate the protein-introduction efficiency and the percentage of dead cells based on FITC and PI fluorescence, respectively (*see* **Note 13**).
8. After confirming the protein uptake reached a maximum (or significant increase of dead cells), add 15 mL of HBSS (+) to facilitate membrane resealing.
9. Optional: in cases where significant cell aggregation is observed, perform filtration using a 100 μm mesh.
10. Estimate the cell numbers and percentage of dead cells using a flow cytometer. If the percentage of dead cells exceeds 30%, remove dead cells by density gradient centrifugation using 20% Percoll (*see* **Note 14**).
11. Collect 3 × $10^7$ cells and wash twice with HBSS (+) buffer. Completely remove the buffer.
12. Add 105 μL of 8% Mebiol gel to the cell pellet. Mix the cells gently (by pipetting or manually shaking), taking care to avoid making air bubbles. Perform this step on ice or in a cold room to prevent gelation of the Mebiol gel.

13. Aspirate the cell suspension using a precooled Pasteur pipette, and slowly dispense from the bottom of a 5 mm Shigemi tube filled with precooled DMEM. Ensure that the gel forms a coil-like shape inside the NMR tube. Gradually immerse the sample tube in a 37 °C water bath to gelate the dispensed Mebiol gel by warming (Fig. 2a).
14. Once fully filled, allow the Mebiol gel to completely gelate at 37 °C in a water bath. If necessary, use a centrifuge for NMR sample tubes to settle down evenly at the bottom of the sample tube.

### 3.4 Set Up of Bioreactor System and In-Cell NMR Measurement

1. Fill a 50 mL syringe with DMEM (20% $D_2O$) and connect it to the bioreactor device. Push the plunger manually to completely fill the flow path with DMEM (20% $D_2O$) up to the tip of the glass capillary.
2. Set the syringe to a syringe pump, and insert the tip of the glass capillary into an empty NMR sample tube. Allow it to flow at a rate of approximately 3 mL/h for about 1 h to reach the flow speed in equilibrium. The excess medium is aspirated through the PTFE tube using a vacuum pump.
3. Connect the silicon tube of the bioreactor system to the NMR sample tube containing the Mebiol gel-embedded cells prepared above. Ensure the connection point of the silicon tube and the NMR sample tube is tightly sealed.
4. Attach a spinner to the NMR sample tube and completely remove any water droplets on the surface. Set the probe temperature to 37 °C, and slowly lower the NMR sample tube connected to the bioreactor into the NMR magnet. Lock signal will be observed when the NMR tube reaches the designated position in the probe.
5. Perform standard NMR procedures including lock, tuning, and shim adjustments. Determine the 90-degree pulse for $^{1}$H. Complete this process as quickly as possible.
6. Measure $^{1}$H-$^{13}$C SOFAST HMQC spectrum [23]. Typically, a single measurement requires 30 min to obtain a sufficient signal-to-noise ratio (Fig. 2c). Based on the signal intensity of I21, estimate the fraction of the GTP-bound form. Repeat measurements until the fGTP reaches in a steady state. Measure $^{1}$H NMR spectra during the intervals of HMQC measurements (*see* **Note 15**).
7. Plot the time-dependent changes of the GTP-bound ratio, and calculate $k_{hy}$ and $k_{ex}$ by fitting the data to the theoretical equation shown (*see* Subheading 3.2, **step 4**) (Fig. 3, *see* **Note 16**).
8. After completing the measurements, lift the sample tube from the NMR magnet manually, and carefully disconnect the NMR sample tube from the bioreactor.

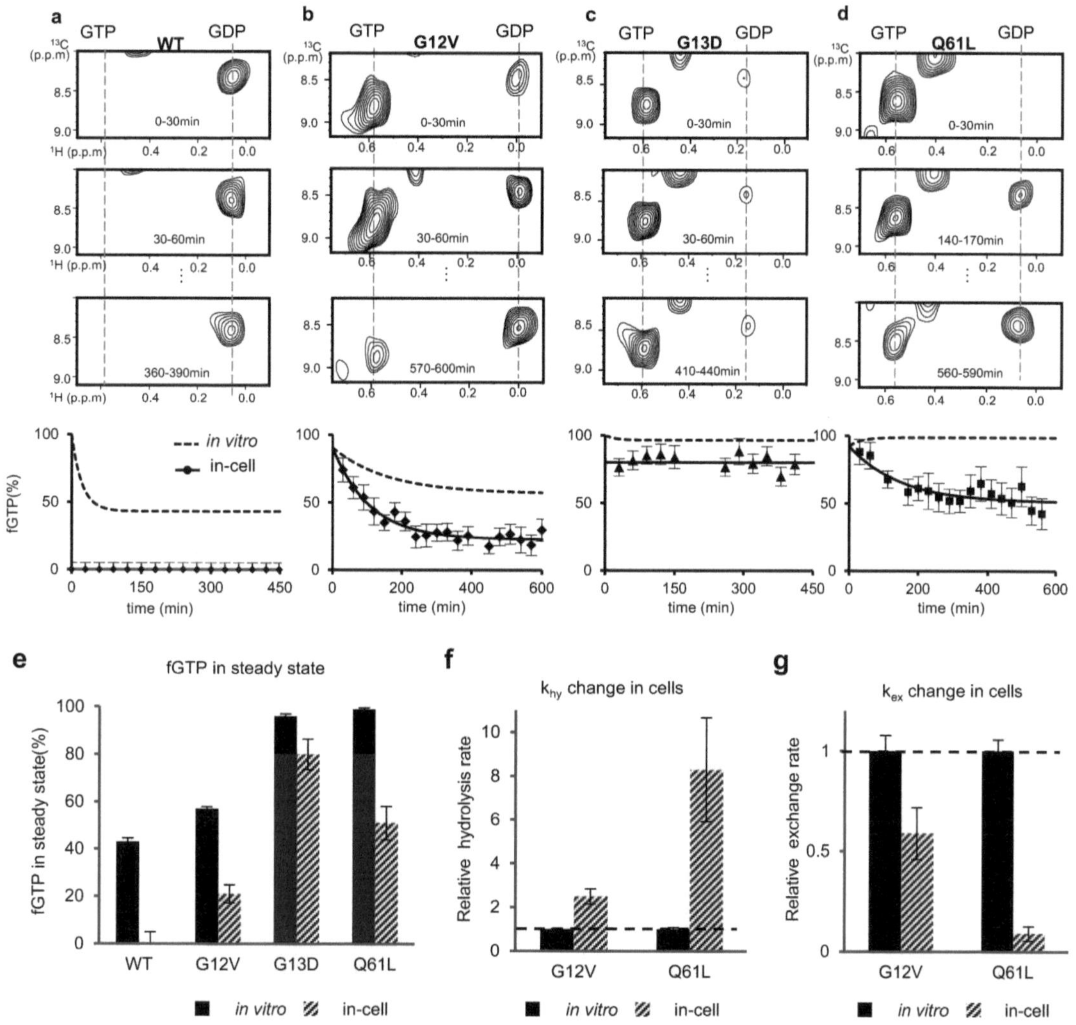


**Fig. 3** The estimation of fGTP, $k_{hy}$, and $k_{ex}$ under the intracellular condition. (**a–e**) The time-resolved $^{1}$H-$^{13}$C SOFAST HMQC spectra of GTP-loaded RAS WT (**a**), G12V (**b**), G13D (**c**), and Q61L (**d**) at representative time points. The position of I21 of GTP- and GDP-bound form is indicated by dotted line, and estimated fGTP at each time point was plotted at bottom with error bars calculated based on the signal-to-noise ratio of I21. For RAS WT (**a**) and G13D (**c**), the steady-state fGTP was estimated as an average of fGTP of all time points (indicated by solid lines). For RAS G12V (**c**) and Q61L (**e**), the $k_{hy}$, $k_{ex}$, and the steady-state fGTP were estimated by curve fitting of experimental data (indicated by solid lines). The dashed lines indicate the theoretical time-course of fGTP deduced from the in vitro $k_{hy}$ and $k_{ex}$ values. (**e**) Summary of the steady-state fGTP of RAS WT and mutants measured in vitro (solid) and in-cell (mesh). (**f**, **g**) The comparison of $k_{hy}$ (**f**) and $k_{ex}$ (**g**) of RAS G12V and Q61L in vitro (solid) and in-cell (mesh). $k_{hy}$ and $k_{ex}$ values were normalized to those in vitro (indicated by dashed lines)

9. After removing the excess medium from the NMR sample tube, cool the NMR sample tube in an ice bath to dissolve the Mebiol gel into a sol state, and then collect the cell suspension using a Pasteur pipette. Rinse the sample tube and Pasteur pipette with 300 μL of HBSS (+), and add it to the cell suspension.
10. Measure the cell viability using a flow cytometer. Measure the NMR spectrum of the lysate after sonication of the collected cells.
11. Clean the bioreactor device by flowing a sufficient amount of 0.5 M NaOH through it, followed by filling it with Milli-Q water.

## 4 Notes

1. The amount of $^{2}H$ glucose added to $D_2O$ M9 medium should be reduced, considering the cost of the isotope-labeled compound. However, reducing the amount of glucose in M9 medium may result in decrease of *E. coli* growth and expression levels. Therefore, it is necessary to determine the optimal amount of glucose using the non-isotope-labeled glucose.
2. Preparation of FITC-labeled RAS was carried out following the manufacturer's instructions. 5–10 equimolar FITC in DMSO was added to 10 mg/mL RAS solution dissolved in 100 mM carbonate buffer (pH 9.0) containing 5 mM $MgCl_2$ and incubated for 8 h at 4 °C under light-shielded condition. Excess ammonium chloride was added to stop the reaction, and the reaction mixture was applied to a desalting column (Econo-Pac 10DG, BioRad) equilibrated with transport buffer to separate RAS from unreacted FITC. The fraction containing RAS was collected and concentrated using Amicon Ultra centrifugal unit. The FITC labeling efficiency can be estimated based on the UV absorption at 280 and 495 nm. The FITC-labeled RAS was stored at −80 °C with 200 μL aliquots in light-blocking tubes.
3. In the standard protocol, sodium bicarbonate is used as a buffer for pH adjustment of DMEM. However, in this experiment, HEPES buffer is used to avoid the generation of air bubbles in the medium. The DMEM (20% $D_2O$) is degassed thoroughly and incubated under the 5% $CO_2$ atmosphere at 37 °C before use.
4. A similar bioreactor system is commercially available from Bruker as InsightCell™ [24] (https://www.bruker.com/en/products-and-solutions/mr/nmr-pharma-solutions/insight-cell.html). Additionally, various homemade bioreactor setups

are reported for in-cell NMR experiments for both *E. coli* [25] and mammalian cells [26, 27].

5. For the in vitro experiment using the GTP regeneration system, deuteration of RAS is not necessary. However, for in-cell NMR experiments, deuteration of RAS is essential to observe the NMR signals of intracellular RAS with sufficient sensitivity.
6. This condition is suitable for RAS, which exhibits a relatively slower GTP hydrolysis activity. On the other hand, for other small GTPases with higher GTPase activity, such as RAC1 [28], it is necessary to add a higher concentration of substrate and AcK to maintain the GTP concentration.
7. The results using regeneration system is summarized in Fig. 1. This experiment can be started using the GDP-loaded RAS. In the case of G12V mutant, calculated $k_{hy}$ and $k_{ex}$ are the same (Fig. 1c). If the steady-state fGTP is higher than 50%, it is advisable to start from the GDP-bound state. Simultaneously measured $k_{hy}$ (Fig. 1d) and $k_{ex}$ (Fig. 1e) using GTP regeneration system are compared with those measured individually using the conventional methods. The results are almost similar between two methods for RAS WT and all oncogenic mutant, except for $k_{ex}$ of G12V, because G12V has higher affinity for GDP than GTP (analog).
8. The concentration of GTP can be monitored based on the β-phosphate signal of GTP in $^{31}$P NMR spectrum (Fig. 1f). In the presence of AcK and AcP, the GTP concentration is maintained for 10 h, while it decreases gradually without AcK and AcP (Fig. 1g).
9. $3 \times 10^7$ cells are required for the final in-cell NMR sample. Therefore, it is advisable to start with a sufficiently large number of cells, considering the steps for the removal of dead cells.
10. The optimal SLO concentration may change depending on a lot of SLO and cell types. The optimal SLO concentration can be determined by conducting the pilot experiment using small number of cells.
11. At lower temperature, SLO accumulates on the cell membrane in a cholesterol-dependent manner and forms a ring-shaped pre-pore complex [29]. Pore formation is induced by structural changes in SLO to form a β-barrel structure at higher temperatures. To prevent pore formation, it is required to perform this step keeping the temperature as low as possible.
12. Protein influx through the pores formed by SLO proceeds by passive diffusion, utilizing the concentration gradient between intracellular and extracellular spaces. To achieve higher protein levels inside the cells, it is desirable to add the target protein to the extracellular environment at the highest possible concentration.

13. 2 μL of cell suspension is suspended in 200 μL of HBSS (+) buffer, and 2 μL of 500 μg/mL PI is added for flow cytometer analysis. The product of the percentage of FITC-positive cell population and the average fluorescence intensity of FITC-positive cells is defined as the introduction efficiency index, and the incubation is continued until the value reaches its maximum (typically 30 min incubation). Alternatively, if the percentage of dead cells from the fluorescence value of PI exceeds 30%, proceed to the next step at this point.
14. 2 mL of Percoll and 10 mL of HBSS buffer are mixed in a 15 mL tube and centrifuged at 2400 × *g* for more than 60 min to form a density gradient. The cell pellet is suspended with 500 μL HBSS (+) buffer and gently layered onto the top of the Percoll solution. The tube is centrifuged at 400 × *g* for 3 min. Live cells tend to be located at the bottom of the centrifuge tube, while dead cells will be at the top. By analyzing each fraction using flow cytometer, the fraction containing a greater number of live cells will be used for next steps.
15. If air bubbles are contaminated in the NMR sample tube during in-cell NMR measurements, it causes distortions in the line shape of the $^{1}$H NMR signals. In such cases, the measurement should be temporarily halted, and the sample tube should be lifted to remove the air bubbles before resuming the measurement.
16. The results of in-cell NMR observation of HRAS wild type and three oncogenic mutants (G12V, G13D, and Q61L) are summarized in Fig. 3. Based on the signal intensities of I21 signals on the SOFAST-HMQC spectra, fGTP is plotted every 30 min (Fig. 3a–d). The steady-state fGTP reflects the basal activation level of RAS in the cell. The intracellular steady-state fGTP was lower than in vitro for both the wild type and all oncogenic mutants (Fig. 1e). In particular, the wild-type RAS is present as predominantly the inactive GDP-bound form, in contrast to the in vitro results. This result demonstrates that within the cellular environment, wild-type RAS does not exhibit constitutive activation, and the result of in-cell NMR is consistent with the role of RAS as a molecular switch. In case of wild type and G13D mutant, the fGTP is already reached in steady state due to fast GTP hydrolysis and/or fast GDP-GTP exchange (Fig. 3b, d). For G12V and Q61L, $k_{hy}$ and $k_{ex}$ can be estimated by fitting the time-dependent decay of fGTP to the theoretical equation. Compared with in vitro data, the intracellular $k_{hy}$ is elevated in the cells (Fig. 3f), while $k_{ex}$ is significantly decreased (Fig. 3g). Therefore, both increase of $k_{hy}$ and decrease of $k_{ex}$ contributes to the lower intracellular fGTP of these mutants.

## Acknowledgments

This work is supported by the Japan Society for the Promotion of Science (JSPS) KAKENHI (Grant number 20H04693 and 21H05250 to NN and 17H06097A to IS, 21K20624, 22K15068 to QZ) and Funding Program for Core Research for Evolutional Science and Technology (CREST JPMJCR21E5 to NN) from the Japan Science and Technology Agency (JST) as well as Naito Foundation and Uehara Memorial Foundation (to NN).

### References

1. Vetter IR, Wittinghofer A (2001) The guanine nucleotide-binding switch in three dimensions. Science 294(5545):1299–1304. https://doi.org/10.1126/science.1062023
2. Simanshu DK, Nissley DV, McCormick F (2017) RAS proteins and their regulators in human disease. Cell 170(1):17–33. https://doi.org/10.1016/j.cell.2017.06.009
3. Schubbert S, Shannon K, Bollag G (2007) Hyperactive Ras in developmental disorders and cancer. Nat Rev Cancer 7(4):295–308. https://doi.org/10.1038/nrc2109
4. Prior IA, Hood FE, Hartley JL (2020) The frequency of Ras mutations in cancer. Cancer Res 80(14):2969–2974. https://doi.org/10.1158/0008-5472.CAN-19-3682
5. Ostrem JML, Shokat KM (2016) Direct small-molecule inhibitors of KRAS: from structural insights to mechanism-based design. Nat Rev Drug Discov 15(11):771–785. https://doi.org/10.1038/nrd.2016.139
6. Goebel L, Müller MP, Goody RS, Rauh D (2020) KRasG12C inhibitors in clinical trials: a short historical perspective. RSC Med Chem 11(7):760–770. https://doi.org/10.1039/D0MD00096E
7. Trahey M, McCormick F (1987) A cytoplasmic protein stimulates normal N-ras p21 GTPase, but does not affect oncogenic mutants. Science 238(4826):542–545. https://doi.org/10.1126/science.2821624
8. Hennig A, Markwart R, Esparza-Franco MA, Ladds G, Rubio I (2015) Ras activation revisited: role of GEF and GAP systems. Biol Chem 396(8):831–848. https://doi.org/10.1515/hsz-2014-0257
9. Bonfini L, Karlovich CA, Dasgupta C, Banerjee U (1992) The *Son of sevenless* gene product: a putative activator of Ras. Science 255(5044):603–606. https://doi.org/10.1126/science.1736363
10. Marshall CB et al (2012) Probing the GTPase cycle with real-time NMR: GAP and GEF activities in cell extracts. Methods 57(4):473–485. https://doi.org/10.1016/j.ymeth.2012.06.014
11. Smith MJ, Neel BG, Ikura M (2013) NMR-based functional profiling of RASopathies and oncogenic RAS mutations. Proc Natl Acad Sci U S A 110(12):4574–4579. https://doi.org/10.1073/pnas.1218173110
12. Marshall CB et al (2009) Characterization of the intrinsic and TSC2-GAP-regulated GTPase activity of Rheb by real-time NMR. Sci Signal 2(55):ra3. https://doi.org/10.1126/scisignal.2000029
13. Kanie T, Jackson PK (2018) Guanine nucleotide exchange assay using fluorescent MANT-GDP. Bio Protoc 8(7):e2795. https://doi.org/10.21769/BioProtoc.2795
14. Han JW, McCormick F, Macara IG (1991) Regulation of Ras-GAP and the neurofibromatosis-1 gene product by eicosanoids. Science 252(5005):576–579. https://doi.org/10.1126/science.1902323
15. Gebregiworgis T et al (2018) Multiplexed real-time NMR GTPase assay for simultaneous monitoring of multiple guanine nucleotide exchange factor activities from human cancer cells and organoids. J Am Chem Soc 140(13):4473–4476. https://doi.org/10.1021/jacs.7b13703
16. Killoran RC, Smith MJ (2019) Conformational resolution of nucleotide cycling and effector interactions for multiple small GTPases determined in parallel. J Biol Chem 294(25):9937–9948. https://doi.org/10.1074/jbc.RA119.008653
17. Traut TW (1994) Physiological concentrations of purines and pyrimidines. Mol Cell Biochem 140(1):1–22. https://doi.org/10.1007/BF00928361

18. Zhao Q et al (2020) Real-time in-cell NMR reveals the intracellular modulation of GTP-bound levels of RAS. Cell Rep 32(8): 108074. https://doi.org/10.1016/j.celrep.2020.108074
19. Nishida N, Ito Y, Shimada I (2020) In situ structural biology using in-cell NMR. Biochim Biophys Acta Gen Subj 1864(2):129364. https://doi.org/10.1016/j.bbagen.2019.05.007
20. Theillet F-X (2022) In-cell structural biology by NMR: the benefits of the atomic scale. Chem Rev 122:9497–9570. https://doi.org/10.1021/acs.chemrev.1c00937
21. Kubo S, Nishida N, Udagawa Y, Takarada O, Ogino S, Shimada I (2013) A gel-encapsulated bioreactor system for NMR studies of protein–protein interactions in living mammalian cells. Angew Chem Int Ed 52(4):1208–1211. https://doi.org/10.1002/anie.201207243
22. Small E, Addinall SG (2003) Dynamic FtsZ polymerization is sensitive to the GTP to GDP ratio and can be maintained at steady state using a GTP-regeneration system. Microbiology 149(8):2235–2242. https://doi.org/10.1099/mic.0.26126-0
23. Schanda P, Kupce E, Brutscher B (2005) SOFAST-HMQC experiments for recording two-dimensional heteronuclear correlation spectra of proteins within a few seconds. J Biomol NMR 33(4):199–211. https://doi.org/10.1007/s10858-005-4425-x
24. Luchinat E, Barbieri L, Campbell TF, Banci L (2020) Real-time quantitative in-cell NMR: ligand binding and protein oxidation monitored in human cells using multivariate curve resolution. Anal Chem 92(14):9997–10006. https://doi.org/10.1021/acs.analchem.0c01677
25. Sharaf NG, Barnes CO, Charlton LM, Young GB, Pielak GJ (2010) A bioreactor for in-cell protein NMR. J Magn Reson 202(2): 140–146. https://doi.org/10.1016/j.jmr.2009.10.008
26. Inomata K, Kamoshida H, Ikari M, Ito Y, Kigawa T (2017) Impact of cellular health conditions on the protein folding state in mammalian cells. Chem Commun 53:11245–11248. https://doi.org/10.1039/C7CC06004A
27. Breindel L, DeMott C, Burz DS, Shekhtman A (2018) Real-time in-cell nuclear magnetic resonance: ribosome-targeted antibiotics modulate quinary protein interactions. Biochemistry 57(5):540–546. https://doi.org/10.1021/acs.biochem.7b00938
28. Kawazu M et al (2013) Transforming mutations of RAC guanosine triphosphatases in human cancers. Proc Natl Acad Sci USA 110(8):3029–3034. https://doi.org/10.1073/pnas.1216141110
29. Dunstone MA, Tweten RK (2012) Packing a punch: the mechanism of pore formation by cholesterol dependent cytolysins and membrane attack complex/perforin-like proteins. Curr Opin Struct Biol 22(3):342–349. https://doi.org/10.1016/j.sbi.2012.04.008

# Chapter 18

# Studying RAS Interactions in Live Cells with BRET

**John Columbus and Thomas Turbyville**

## Abstract

Bioluminescence resonance energy transfer (BRET) is a valuable technique for studying protein-protein interactions (PPIs) within live cells (Pfleger and Eidne, Nat Methods 3:165–174, 2006). Among the various BRET methodologies, a recent addition called NanoBRET has emerged, leveraging advancements in donor and acceptor technologies (Machleidt and Woodroofe, ACS Chem Biol 10:1797–1804, 2015). In this study, we present a developed methodology designed to measure PPIs involving the RAS protein family and their effectors and interactors at the plasma membrane. By utilizing the NanoLuc and HaloTag BRET pair, we provide evidence of a saturable interaction between KRAS4b-G12D and full-length RAF1. Conversely, the RAF1 R89L mutant, known to impede RAF1 binding to active RAS, exhibits nonspecific interactions. The assay exhibits remarkable signal-to-background ratios and is highly suitable for investigating the interactions of RAS with effectors, as well as for high-throughput screening assays.

**Key words** Protein-protein interactions, BRET, RAS, MAP kinase, Signal transduction, High-throughput screening

## 1 Introduction

Resonance energy transfer is a natural process in which energy is transferred non-radiatively from an excited fluorophore (donor) to another fluorophore (acceptor) through long-range dipole-dipole coupling. Three conditions must be fulfilled for energy transfer to occur: first, the emission spectra of the donor must overlap with the excitation spectra of the acceptor; second, the distance between the fluorophores should be less than 10 nm; and third, the dipoles of the fluorophores must be aligned. In the context of BRET, the donor fluorophores can be bioluminescent proteins fused with the proteins of interest [1].

BRET encompasses multiple versions, each employing different donor-acceptor pairs and substrates. The selection of the donor-acceptor pair is primarily based on the emission spectrum of the donor and its overlap with the excitation spectrum of the acceptor [1, 2]. It is also crucial to consider factors such as luciferase and

Andrew G. Stephen and Dominic Esposito (eds.), *KRAS: Methods and Protocols*, Methods in Molecular Biology, vol. 2797,
https://doi.org/10.1007/978-1-0716-3822-4_18,

fluorescent protein brightness, spectral bleed through, expression levels, and signal-to-background ratio when conducting experiments.

BRET 1, the original method, uses Renilla luciferase (Rluc) from the sea pansy *Renilla reniformis* as the donor and enhanced yellow fluorescent protein (eYFP) as the acceptor. Coelenterazine serves as the substrate for Rluc. When Rluc catalyzes coelenterazine, it emits blue light at 482 nm, which closely matches the excitation spectra of YFP at 480 nm.

BRET 2 utilizes a chemical derivative called coelenterazine 400a as a substrate, which downshifts the maximum emission of Rluc to 395 nm. This modification allows the use of FP variants like GFP2 and GFP10 as acceptors, which have peak excitation at 400 nm and emission at 510 nm. The advantage of BRET 2 is the improved spectral separation between the donor and acceptor, reducing bleed through. However, the brightness of Rluc emission is significantly reduced when using coelenterazine 400a. To address this issue, the Rluc8 variant was developed, enhancing the quantum yield of coelenterazine 400a catalysis.

For developing BRET assays with a more red-shifted spectrum, firefly luciferase is employed as the donor (emitting at 565 nm), DsRed as the acceptor (with excitation at 558 nm), and luciferin as the substrate. Quantum dots, semiconductor nanocrystals, offer a broad excitation range from UV to 530 nm, and their emission can be tuned from blue to infrared based on the crystal size. Quantum dot-BRET (QD-BRET) typically involves Rluc or Rluc8 as the donor, quantum dots as the acceptor, and coelenterazine as the substrate. While quantum dots have the advantage of excellent spectral separation between donor and acceptor, they cannot be genetically encoded and expressed in living cells [1, 2].

NanoBRET employs a small, highly luminescent luciferase called NanoLuc (derived from a 19 kDa subunit of the deep-sea shrimp *Oplophorus gracilirostris* luciferase) as the donor, which luminesces in the presence of its engineered substrate, furimazine [3]. To harness the advantages of a bright, photostable fluorophore with a wide spectral separation and minimal background interference, the Halo enzyme was optimized to be expressed as a highly stable tag on the protein of interest. HaloTag is a modified bacterial enzyme mutated to form irreversible covalent bonds with molecules containing a chloroalkane group, allowing for the incorporation of a far-red dye, known as Halo618 ligand, as the acceptor molecule [4]. This approach reduces background signals, increases spectral separation, and increases sensitivity requiring less plasmid DNA, marking a significant improvement over other BRET techniques [4].

The RAS family of small GTPases interacts with numerous proteins in diverse signal transduction pathways [5]. NanoBRET is particularly suitable for investigating interactions within the RAS pathway in live cells. It enables the exploration of the interactions of RAS with effectors and inducers, quantitatively mapping and measuring binding affinities and preferences. Moreover, these assays exhibit excellent performance as high-throughput screening tools [6]. Here we describe a protocol to measure the interaction between KRAS4b-G12D and WT RAF1 using NanoBRET. As a control, we also include a RAF1 R98L mutant, which is defective in RAS binding.

## 2 Materials

1. HEK293T cells.
2. T75 or T150 flask, tissue culture treated.
3. 12-well tissue culture treated plates.
4. Benchtop centrifuge.
5. 1.5 mL microcentrifuge tube.
6. Promega NanoBRET™ Nano-Glo® Detection System which includes the furimazine substrate and the Halo618 ligand.
7. Plasmid DNA containing the CMV51p promotor and coding for the following three fusion proteins: RAF1:Nluc, RAF1 (R89L):Nluc, HaloTag7:KRAS4b-G12D (*see* **Note 1**).
8. Fugene-6.
9. Phenol red-free DMEM.
10. Complete media: phenol red-free DMEM, 10% FBS, 1× L-glutamine.
11. Assay media: phenol red-free DMEM, 1.0% FBS, 1× L-glutamine.
12. 0.25% trypsin.
13. 96-well polypropylene microplate U-shape.
14. OptiPlate-384, white opaque 384-well microplates.
15. Beckman MicroClime Environmental Lids.
16. Sterile water.
17. Perkin-Elmer Envision multimode plate reader. Reader needs to be equipped with the following to make it NanoBRET capable: emission filter 1, 460 nm, 50 nm bandwidth, chroma; emission filter 2, 600 nm long pass, chroma; filter holder, Perkin-Elmer; single luminescence dichroic mirror, Perkin-Elmer.

## 3 Methods

### *3.1 Transfection of Cells for a Saturation Curve*

1. From a flask of HEK239T cells that are 80–90% confluent, trypsinize the cells, count, and plate at a density of 125,000 cells/well in 12-well plates in 1.0 mL of complete media (*see* **Note 2**).
2. Dilute RAF1:Nluc and RAF1(R89L):Nluc to 5 ng/50 μL in phenol red-free DMEM, and add 50 μL to all wells of a 12-well tissue culture plate. This is the donor plasmid and so is kept constant to generate a saturation curve (*see* **Note 3**).
3. To generate a saturation curve as seen in Figs. 1 and 2, the acceptor is systematically varied across a large dilution range. Starting at 1000 ng/50 μL, make an 11-point, twofold serial dilution of the HaloTag7:KRAS4b-G12D plasmid in phenol red-free DMEM. Add 50 μL of each dilution to wells 2–12 of the 12-well tissue culture plate.
4. Add 50 μL of phenol red-free DMEM to well #1 of the 12-well tissue culture plate to bring the volume up to 100 μL (*see* **Note 4**).
5. Dilute enough Fugene-6 in phenol red-free DMEM for all wells to be tested, 8 μL of Fugene-6 and 42 μL of phenol red-free DMEM per reaction.
6. Incubate at room temperature for 5 min.
7. Add 50 μL of the Fugene-6 dilution to the plasmid tubes, and mix gently by pipetting several times.
8. Incubate at room temperature for 30 min.

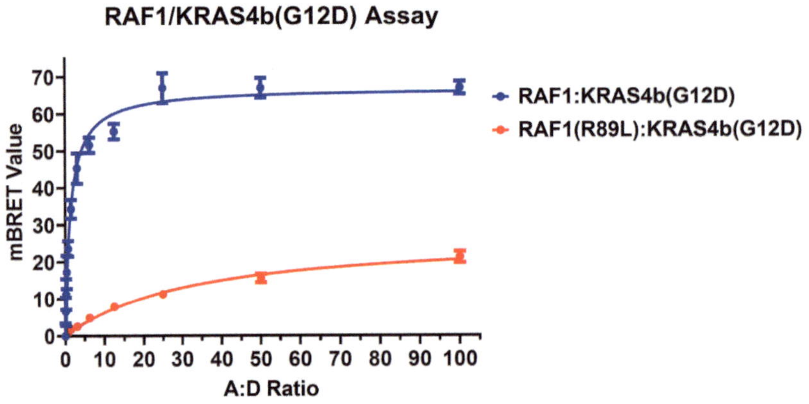


**Fig. 1** Saturation curves of the interaction between full-length RAF1 and the RAS binging domain mutant of RAF1, RAFD1(R89L), which disrupts binding to RAS and the G12D oncogenic mutant of KRAS4b. The error bars are the standard deviation between technical replicates ($n = 3$)

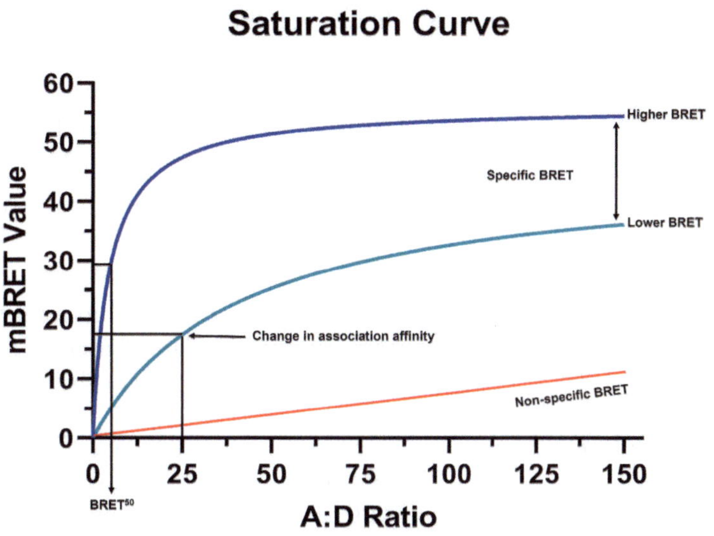


**Fig. 2** Saturation curves are generated by keeping the amount of the donor construct the same and increasing the ratio of the acceptor construct to the donor construct. The amplitude of the hyperbola denotes the amount of interaction occurring. Assessing the $BRET^{50}$ value is a way to compare the relative affinities of separate saturation curves. A nonspecific interaction, bystander BRET, will not show saturation at the higher acceptor to donor ratio, A:D ratio. (Figure adapted from Hamdan et al. [7])

9. Add plasmid/Fugene-6 mixture to individual wells of the 12-well plate.
10. Incubate in a tissue culture incubator for 24–48 h.

### 3.2 Setting Up Saturation Curve Assay

1. To harvest cells, thoroughly aspirate and discard the media and add 250 μL of trypsin.
2. Incubate at room temperature for 5 min, and then add 750 μL of complete media.
3. Mix with a pipet to break up clumps of cells and transfer to 1.5 mL microfuge tubes.
4. Spin cells down for 5 min at 300 g, and depending on pellet size, resuspend pellet in 100–500 μL of assay media (*see* **Note 5**).
5. Not all samples need to be counted, but count one or two samples to get a rough idea of how many cells per mL are in the samples.
6. In a 96-well round bottom plate, dilute cells to 12,000 cells/ 20 μL.
7. Add 20 μL to a 384-well plate in triplicate.

8. Dilute the Halo618 ligand from the NanoBRET kit from 100 μM stock to 0.32 μM in assay media, and add 20 μL to each well of the 384-well plate that contains cells.
9. Mix plate in a shaker, or tap by hand, and incubate 4–6 h in a tissue culture incubator (*see* **Note 6**).
10. After incubation, dilute the furimazine substrate from the NanoBRET kit from 5 mM stock to 30 μM in assay media, and add 20 μL to each well, 10 μM final concentration.
11. Incubate for 5–10 min, and then read in a Perkin-Elmer Envision multimode plate reader or equivalent plate reader set up with the appropriate NanoBRET emission filters (*see* **Note 7**).

### 3.3 Normalizing the Raw Data

1. To normalize the raw data, divide the acceptor channel results from the donor results. To obtain mBRET values, multiply by 1000. Next subtract the donor only control from every sample.

### 3.4 Data Analysis: Generation of Saturation Curves

1. In GraphPad Prism, or equivalent graphing software (*see* **Note 8**), use a nonlinear regression curve fit, hyperbola, where $x$ is concentration to generate saturation curves (*see* Fig. 1) (*see* **Note 9**).

## 4 Notes

1. Because the donor and acceptor pair distance and orientation will determine the efficiency of the resonance energy transfer, the N- or C-terminal orientation of the fusion partners needs to be determined empirically. Additionally, which protein has the donor NanoLuc or the acceptor HaloTag will need to be determined. Due to the requirement of the C-terminus of KRAS for membrane targeting, KRAS4b-G12D is tagged with an N-terminal HaloTag7. Optimal BRET signals were obtained using RAF1 that was tagged at the C-terminus with NanoLuc.
2. HEK293T cells are easy to transfect and express proteins very well. The ideal number of cells to seed and the amount of transfection reagent to use per reaction should be empirically determined. HeLa cells may also be amenable to NanoBRET.
3. The amount of donor plasmid to transfect is extremely important in maximizing the mBRET signal. This needs to be determined by doing multiple saturation curves, changing the amount of donor for each one until a curve with the highest $BRET^{max}$ is found. Too much or too little donor signal and the $BRET^{max}$ will be diminished.
4. There will be a BRET signal that is intrinsic to the system. This must be subtracted from each sample. Therefore, a control well

containing only the donor plasmid is required. The mBRET value derived from this well must be subtracted from each of the sample mBRET values.

5. Phenol red will interfere with the acceptor signal coming from the Halo618 ligand. Therefore, phenol red-free DMEM must be used.
6. The acceptor signal from the Halo618 ligand is a function of time. A 4-to-6-h incubation should be sufficient; the longer the better. Sufficient signal will be a function of several parameters: (1) a mBRET value in the blank well, well #1, that is less than 20, (2) a signal to background that is at least twofold, and (3) the $Z'$ values are between 0.5 and 1.0. If an overnight incubation is needed, a higher percentage of FBS in the assay media is required, up to 4%.
7. There are many plate readers capable of measuring BRET output. Finding which to use is a function of laboratory needs and available budget. Whether or not one comes BRET capable, additional filters may be required for NanoBRET donor and acceptor channels.
8. There are many graphing software packages available. Again, which one to use is a function of need and budget.
9. Saturation curves (*see* Fig. 2) should show a nice hyperbolic curve when graphed, but sometimes there will not be enough points to either saturate or have a slope. In those cases, having more points on the curve (higher or lower) is required. The amplitude of the curve describes how much PPI is happening. The shape and steepness of the curve describe the affinity of those interactions (*see* Fig. 2).

## Acknowledgement

Dominic Esposito and Vanessa Wall from the NCI RAS Initiative, Frederick National Laboratory, are gratefully acknowledged for the design and production of numerous NanoBRET expression plasmids, including the ones used in this study. This project was funded in part with federal funds from the National Cancer Institute, National Institutes of Health contract 75N91019D00024. The content of this publication does not necessarily reflect the views or policies of the Department of Health and Human Services, and the mention of trade names, commercial products, or organizations does not imply endorsement by the US government.

## References

1. Pfleger K, Seeber R et al (2006) Bioluminescence resonance energy transfer (BRET) for the real-time detection of protein-protein interactions. Nat Protoc 1:337–345
2. Bacart J, Corbel C et al (2008) The BRET technology and its application to screening assays. Biotechnol J 3:311–324
3. Hall M, Unch J et al (2012) Engineered luciferase reporter from a deep-sea shrimp utilizing a novel imidazopyrazinone substrate. ACS Chem Biol 7:1848–1857
4. Los G, Encell L et al (2008) HaloTag: a novel protein labeling technology for cell imaging and protein analysis. ACS Chem Biol 3:373–382
5. Stephen A, Esposito D et al (2014) Dragging RAS back in the ring. Cancer Cell 25:272–281
6. Durrant D, Smith E et al (2021) Development of a high-throughput nanoBRET screening platform to identify modulators of the RAS/RAF interaction. Mol Cancer Ther 9:1743–1754
7. Hamdan F, Percherancier Y et al (2006) Monitoring protein-protein interactions in living cells by bioluminescence resonance energy transfer (BRET). Curr Protoc Neurosci 34: 5.23.1–5.23.20

# Chapter 19

# FLIM-FRET Protein-Protein Interaction Assay

**Pedro Andrade Bonilla and Rebika Shrestha**

## Abstract

Fluorescence lifetime imaging performed under FRET conditions between two interacting molecules is a sensitive and robust way to quantify intermolecular interactions in cells. The fluorescence lifetime, an inherent property of the fluorophore, remains unaffected by factors such as concentration, laser intensity, and other photophysical artifacts. In the context of FLIM-FRET, the focus lies on measuring the fluorescence lifetime of the donor molecule, which diminishes upon interaction with a neighboring acceptor molecule. In this study, we present a step-by-step experimental protocol for applying FLIM-FRET to investigate protein-protein interactions involving various RAS isoforms and RAS effectors at the live cell's plasma membrane. By utilizing the FRET pair comprising enhanced green fluorescent protein (eGFP) and fluorescent mCherry, we demonstrate that the proximity and possible nanoclustering of eGFP-tagged KRAS4b G12D and mCherry-tagged KRAS4b WT led to a reduction in the donor eGFP's fluorescence lifetime. The donor lifetime of eGFP-tagged KRAS decreases even further when treated with a dimer-inducing small molecule, or in the presence of RAF proteins, suggesting a greater FRET efficiency, and thus less distance, between donor and acceptor.

**Key words** FLIM-FRET, FRET, RAS, RAF, TCSPC

## 1 Introduction

Förster radius energy transfer (FRET) is a photophysical phenomenon where energy is transferred non-radiatively from an excited donor molecule to a nearby acceptor molecule through dipole-dipole coupling [1]. FRET measurements are highly sensitive to the distance between the interacting molecules and are commonly used as a molecular ruler to measure fluorophore distances. The efficiency of FRET decreases as the sixth power of the distance between the fluorophores, with an effective range of <10 nm. Successful FRET requires spectral overlap between the donor emission and acceptor excitation, as well as alignment of their dipole orientations, allowing the acceptor to sense the electric field generated by the donor [1].

Andrew G. Stephen and Dominic Esposito (eds.), *KRAS: Methods and Protocols*, Methods in Molecular Biology, vol. 2797,
https://doi.org/10.1007/978-1-0716-3822-4_19,

Fluorescence lifetime imaging-FRET (FLIM-FRET) is a technique that measures FRET interactions between donor and acceptor molecules by examining changes in the fluorescence lifetime of the donor fluorophore. The fluorescence lifetime represents the average time a molecule remains in the excited state before emitting a photon and returning to the ground state. It is an inherent property of the fluorophore and is independent of its concentration but influenced by the local environment, including factors such as pH, ion concentration, viscosity, and interactions with other molecules. In a FRET process, the fluorescence lifetime of the donor noticeably decreases. As a result, FLIM-FRET is less susceptible to artifacts caused by photophysical processes like photobleaching, cross talk, variations in laser intensity, nonuniform illumination, or light pathlength. An important advantage of FLIM-FRET in cellular measurements is its ability to isolate FRET measurements within specific subcellular compartments, such as the plasma membrane or nucleus, based on fluorescent imaging.

FLIM-FRET has found widespread use in biology for various applications, including investigating molecular interactions between macromolecules such as protein-protein and protein-lipid interactions [2–5], assessing variations in cellular compartment viscosity [6], detecting temperature-related changes within living cells [7], monitoring conformational changes through glucose binding, and mapping pH or ion concentrations across cells [8, 9]. In this study, we employ FLIM-FRET to explore the interaction between RAS-RAS and RAS-effector proteins at the plasma membrane of live HeLa cells.

At the plasma membrane, GTP-bound RAS binds to the kinase effector protein RAF through the RAS binding domain (RBD), initiating a series of conformational changes that activate the kinase. Furthermore, RAS proteins form nanoclusters consisting of 6–8 molecules on the plasma membrane. These nanoclusters, facilitated by the interaction between the hypervariable region (HVR) of RAS and the diverse lipid microenvironment surrounding the RAS proteins, play a crucial role in their assembly. Previous studies have utilized FLIM-FRET to quantify the extent of isoform-specific RAS nanoclustering in mammalian cells [10, 11].

In this study, we present a detailed experimental protocol for measuring time-domain FLIM-FRET between KRAS G12D and the RAS binding domain cysteine-rich domain (RBDCRD) from RAF1 using a time-correlated single photon counting (TCSPC) technique, which enables the determination of fluorescence lifetime. In TCSPC, a high-speed stopwatch measures the time delay between pulse excitation and the arrival time of photons at the detector. These time measurements are repeated multiple times, and the photons are categorized into a histogram of time bins based on their arrival time [12].

For our experiments, we utilize HeLa cells expressing eGFP-tagged KRAS proteins as the donor and proteins tagged with mCherry as the acceptor fluorescent protein. The eGFP and mCherry pair has been previously validated to provide consistent and quantitative FRET measurements both in vivo and in vitro [13]. This donor-acceptor combination exhibits excellent spectral overlap, with the donor emission spectrum aligning well with the acceptor excitation spectrum. Moreover, it demonstrates minimal cross talk between the acceptor and donor channels, ensuring accurate measurements.

Our study demonstrates a reduction in the fluorescence lifetime of the donor molecule, eGFP KRAS G12D, at the plasma membrane when co-expressed with either mCherry KRAS WT or mCherry KRAS G12D, indicating the occurrence of RAS nano-clustering. This decrease in donor fluorescence lifetime is also observed when eGFP KRAS G12D interacts with mCherry-RBDCRD from RAF1 (Figs. 1 and 2). Conversely, the non-RAS binding mutant R89L RBDCRD and CRD from RAF1 do not lead to a decrease in the donor fluorescence lifetime of eGFP KRAS G12D.

To explore the potential application of this technique in assessing small molecule interactions with KRAS, we investigated the effect of a specific compound called BI-2852, known to induce nonfunctional dimers of KRAS [14]. In our FLIM-FRET experiment, treatment with 20 μM of BI-2852 resulted in a further decrease in the fluorescence lifetime of eGFP KRAS G12D when co-expressed with the FRET pair, mCherry KRAS WT, compared to the untreated conditions. These findings suggest the potential utility of FLIM-FRET in studying small molecule interactions with KRAS.

## 2 Materials

### 2.1 Cell Culture

1. HeLa cells.
2. T75 or T150 flask, tissue culture treated.
3. 6-well tissue culture-treated plates.
4. 1.5 mL microcentrifuge tube.
5. Plasmid DNA containing sequences of proteins of interest. The protein constructs used in this protocol are as follows:
   - eGFP KRAS4b G12D
   - mCherry-KRAS4b G12D
   - mCherry-KRAS4b
   - mCherry-RAF1(52–192)
   - mCherry-RAF1(52–188) R89L
   - mCherry-RAF1(136–188)

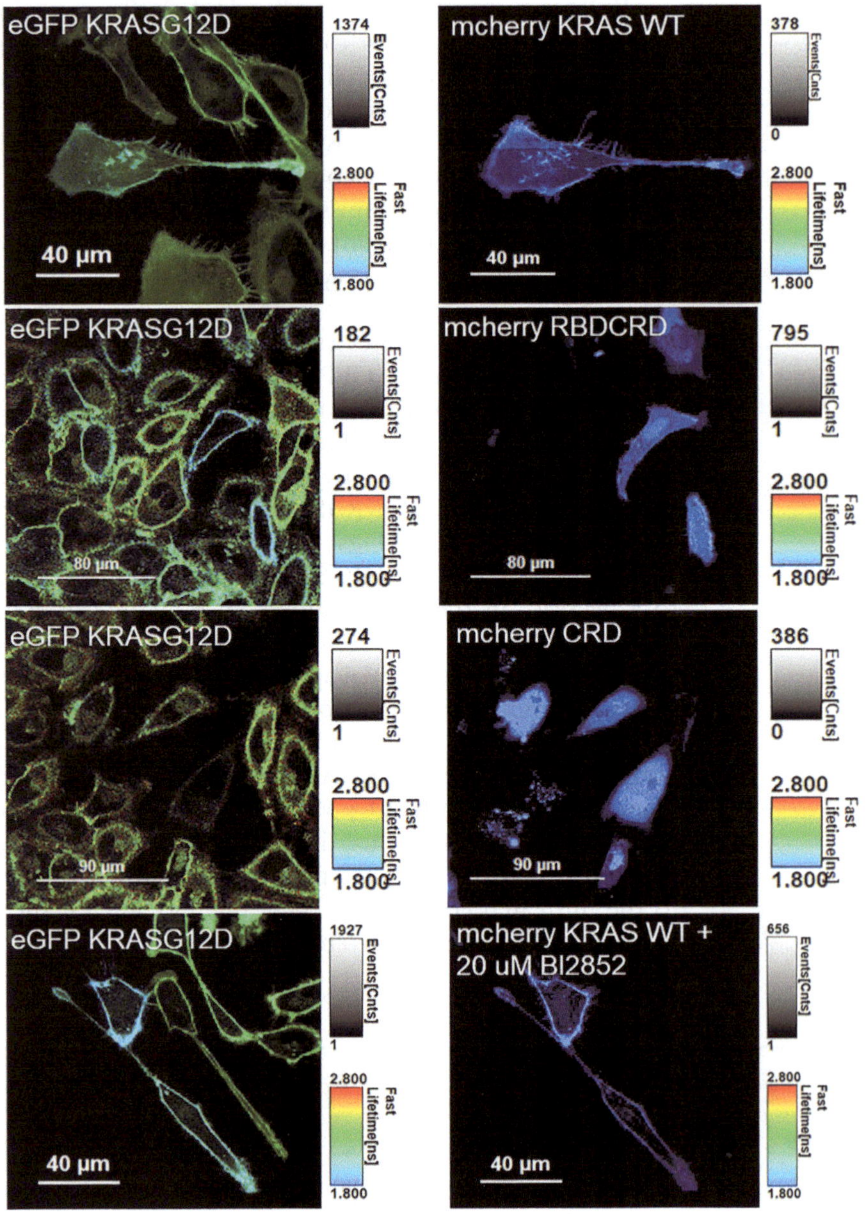


**Fig. 1** Left, fluorescence lifetime images of HeLa cells expressing donor, eGFP KRAS G12D (left) in the donor emission channel. Right, fluorescence lifetime images of cells expressing mCherry tagged to protein of interest, KRAS WT, RAF1 RBDCRD, RAF1 CRD, mCherry-KRAS WT treated with BI2852, respectively, imaged in the acceptor emission channel. The colors represent the fluorescence lifetime values of each pixel based on the fast lifetime scale bar on the side. Cells expressing both donor and acceptor show decreased donor lifetime represented by the teal color in the FLIM image for the donor channel

6. FuGENE 6.
7. Phenol red-free DMEM.

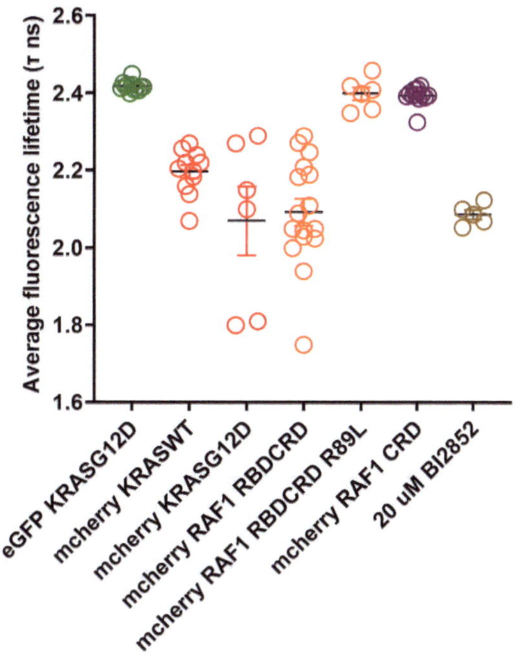


**Fig. 2** Average fluorescence lifetime calculated for eGFP KRAS G12D (donor only) under various co-expression conditions. A minimum of five FLIM images containing a minimum of two cells were analyzed for each donor-acceptor pair to get statistically significant data. Each circle represents the fluorescence lifetime calculated per image, and the horizontal black line represents the calculated mean, and vertical line represents the error bar

8. Complete media: phenol red-free DMEM, 10% FBS, 1× L-glutamine.
9. 0.25% trypsin.
10. Sterile phosphate buffer saline (PBS 1X).
11. 5 mM stock of BI2852 in DMSO.

### 2.2 Imaging

FLIM-FRET experiments are performed on Nikon Eclipse Ti2 equipped with MultiHarp 150 multichannel event timer system by PicoQuant GmbH which allows time-correlated single photon counting (TCSPC). In general, TCSPC instrumentation require the following key components.

1. Confocal or multiphoton microscope optics with environment-controlled stage (in these experiments we used a Nikon Eclipse Ti2 equipped with stage top incubator from Tokai Hit, but any other confocal or multiphoton microscopes can be used).
2. A pulsed laser source (in these experiments we used a PDL 828 Sepia II multichannel picosecond diode laser from PicoQuant; other options, a pulsed femtosecond laser such as titanium–sapphire crystal lasers used in two-photon excitation TPE can be used).

3. Single photon detection module with appropriate sensitivity and time resolution (in these experiments we used a PMA hybrid detector for single photon counting, PicoQuant; other options, photomultiplier tubes (PMTs), avalanche photodiodes, hybrid detectors, and single photon avalanche diodes (SPADs)).
4. Suited timing electronics for data registration (in these experiments we used a MultiHarp 150, PicoQuant; other options, PicoHarp 300 (PicoQuant), SPC modules (Becker and Hickl)).
5. Data acquisition and analysis software (in these experiments we used a SymPhoTime 64 software; other options, refer to [1]).

## 3 Methods

### 3.1 Cell Culture and Transfection

1. Trypsinize HeLa cells from a T-75 flask that are at 70–80% confluency. Count the cells and seed at a density of 125,000 cells/well in a 6-well plate in 2 mL of complete media per well.
2. On the second day, the cells should be approximately 50–60% confluent to be ready for transfection. Prepare a mixture of FuGENE 6 transfection reagent in serum-free DMEM by mixing 10 μL of FuGENE 6 and 190 μL of DMEM per reaction. Mix the mixture well and incubate at room temperature for 5 min (*see* **Notes 1**–**3**).
3. Dilute the plasmid DNA to a concentration of 1.1 μg/reaction in 200 μL DMEM (*see* **Notes 1**–**3**).
4. Add the 200 μL of the FuGENE 6 mix to each 200 μL plasmid dilution. Mix by pipetting and incubate for 20 min.
5. Add the FuGENE 6/DNA complexes to the appropriately labeled wells of the 6-well plate.
6. Gently swirl the mixtures in the plates and place in a 37 °C/5% $CO_2$ incubator.
7. One day after transfection, wash the cells with 2 mL of PBS once, and replace with 2 mL of complete media supplemented with doxycycline at 500 ng/mL concentration. Return the cells back into the 37 °C/5% $CO_2$ incubator, and plan to image after a further 24 h incubation (*see* **Notes 1** and **4**).
8. As a control for the donor-only fluorescence lifetime measurement (*see* **Note 5**), prepare another 6-well plate of the HeLa cells following Subheading 3.1, **step 1**. Skip Subheading 3.1, **steps 2**–**6**. Add doxycycline following the protocol explained in Subheading 3.1, **step 7**. Image after 24 h incubation at 37 °C/5% $CO_2$ incubator.

9. For drug treatment, wash each well with 2 mL of PBS once on the day of imaging.
10. Prepare 20 μM of drug (BI2852 in DMSO) in complete media. Add 1 mL of drug solution into each well, and incubate for 2 h or desired time (*see* **Notes 6** and 7).

### *3.2 Image Acquisition*

1. On the day of imaging, turn on the microscope mounted with the temperature controller and the humidification chamber, and set to 37 °C and 5% $CO_2$ for at least an hour prior to imaging. These conditions are required to maintain the cells in physiological condition during imaging experiment.
2. Mount the 6-well plate onto the appropriate stage and image using a Plan Apo IR 60× water immersion objective with numerical aperture (NA) of 1.27.
3. Focus the cells using the bright-field microscope, and then confirm the expression of the fluorophores by epifluorescence.
4. Select cells for high level of donor and acceptor fluorescence for FRET experiments and donor-only fluorescence for the control experiments by imaging in the confocal mode (*see* **Note 8**).
5. After selection of the appropriate cells, switch the acquisition mode to the FLIM mode. Adjust the pulsed laser intensity of both donor (eGFP) and acceptor (mCherry) excitation lasers to get photon counts between 200,000 and 500,000 counts/s, and set the pulse width/repetition rate to get a good fluorescence decay curve (the fluorescence signal must decay to zero within the pulse width) in each channel (*see* **Note 9**).
6. Acquire the fluorescence lifetime images for 30 frames or the brightest pixel reaches 1000 counts in both donor and acceptor channel by exciting with both donor and acceptor lasers for FRET experiments. For donor-only experiments, excite and collect images only in the donor channel.
7. Analyze the fluorescence lifetime images in the SymPhoTime 64 v1.6 software to calculate the fluorescence lifetime of donor on its own and when co-expressed with mCherry [15]. For a detailed description of the procedures for fluorescence lifetime data analysis in SymPhoTime software, refer to McClung and Walczak [2, 3] (*see* **Note 10**).
8. Fit the fluorescence lifetime decay curve to the n-exponential reconvolution fit to determine the fluorescence lifetime. In our experiments the decay curve for eGFP KRAS is fit with a two componential decay with fluorescence lifetimes of $t_1 = 2.8$ ns and $t_2 = 1.7$ ns (for donor only) with relative area of $A_1 = 63.8\%$ and $A_2 = 36.2\%$. The weighted average of the two is reported as the average fluorescence lifetime.

## 4 Notes

1. The protocol uses doxycycline-inducible system to achieve a maximum expression of the donor molecule eGFP KRAS protein, but one can use transient transfection to express donor and acceptor proteins of their choice. The production of the dox-inducible HeLa cell pool and DNA plasmid construction of transient transfection are explained in detail in [16].
2. The protocol uses a FRET pair of eGFP and mCherry proteins, but one can use other FRET pair of their choice such as CFP-YFP, GFP-RFP, and so on [4, 17] following the FRET guidelines discussed earlier.
3. The protocol uses HeLa cells for the ease of transfection, but the protocol could be extended to any cell lines. Seeding density and transfection conditions need to be optimized for each cell line.
4. Use phenol red-free media to avoid any background fluorescence from the media.
5. For the control experiment, one should measure the fluorescence lifetime of the donor only at different levels of expression. The donor lifetime should be independent of the donor concentration.
6. When using drugs, run a confocal experiment to test the effect of different drug concentration on cell health.
7. Keep the concentration of DMSO below 0.01–0.025% to prevent cytotoxicity in cells.
8. Use the correct filters and dichroic in the detector for the corresponding fluorophores.
9. The laser pulse width or the repetition rate must be optimized to allow the whole population to decay to baseline within the window. If the repetition rate is too low, not all molecules will have the chance to decay, and if the repetition rate is too high, the fluorescence intensity will not decay to the baseline.
10. For KRAS imaging, prior to calculating the fluorescence lifetime, select only the plasma membrane as the region of interest (ROI), either by thresholding or using several ROI tools like paint, free, or magic wand.

## Acknowledgments

We would like to thank Dom Esposito, Katie Powell, Vanessa Wall, and William Burgan (NCI RAS Initiative, Frederick National Laboratory for Cancer Research) for cell line production and plasmid construction. We would also like to acknowledge Anna Maciag for facilitating lead compound acquisition. This project was funded in part with federal funds from the National Cancer Institute, National Institutes of Health contract 75N91019D00024. The content of this publication does not necessarily reflect the views or policies of the Department of Health and Human Services, and the mention of trade names, commercial products, or organizations does not imply endorsement by the US government.

## References

1. Esposito A, Wouters FS (2004) Fluorescence lifetime imaging microscopy. Curr Protoc Cell Biol 25:4.14.11–14.14.30
2. Kaufmann T, Herbert S, Hackl B, Besold JM, Schramek C, Gotzmann J, Elsayad K, Slade D (2020) Direct measurement of protein–protein interactions by FLIM-FRET at UV laser-induced DNA damage sites in living cells. Nucleic Acids Res 48:e122
3. Ems-McClung SC, Walczak CE (2022) Using FLIM-FRET for characterizing spatial interactions in the spindle. Methods Mol Biol 2415: 221–243
4. Datta R, Heaster TM, Sharick JT, Gillette AA, Skala MC (2020) Fluorescence lifetime imaging microscopy: fundamentals and advances in instrumentation, analysis, and applications. J Biomed Opt 25:1–43
5. Stöckl MT, Herrmann A (2010) Detection of lipid domains in model and cell membranes by fluorescence lifetime imaging microscopy. Biochim Biophys Acta 1798:1444–1456
6. Kashirina AS, López-Duarte I, Kubánková M, Gulin AA, Dudenkova VV, Rodimova SA, Torgomyan HG, Zagaynova EV, Meleshina AV, Kuimova MK (2020) Monitoring membrane viscosity in differentiating stem cells using BODIPY-based molecular rotors and FLIM. Sci Rep 10:14063
7. Pliss A, Zhao L, Ohulchanskyy TY, Qu J, Prasad PN (2012) Fluorescence lifetime of fluorescent proteins as an intracellular environment probe sensing the cell cycle progression. ACS Chem Biol 7:1385–1392
8. Caron NS, Munsie LN, Keillor JW, Truant R (2012) Using FLIM-FRET to measure conformational changes of transglutaminase type 2 in live cells. PLoS One 7:e44159
9. Linders PTA, Ioannidis M, Ter Beest M, Van den Bogaart G (2022) Fluorescence lifetime imaging of pH along the secretory pathway. ACS Chem Biol 17:240–251
10. Parkkola H, Siddiqui FA, Oetken-Lindholm C, Abankwa D (2021) FLIM-FRET analysis of Ras nanoclustering and membrane-anchorage. Methods Mol Biol 2262:233–250
11. Abankwa D, Gorfe AA, Inder K, Hancock JF (2010) Ras membrane orientation and nanodomain localization generate isoform diversity. Proc Natl Acad Sci USA 107:1130–1135
12. Wahl M (2014) Technical note on time-correlated single photon counting. PicoQuant GmbH
13. Albertazzi L, Arosio D, Marchetti L, Ricci F, Beltram F (2009) Quantitative FRET analysis with the EGFP-mCherry fluorescent protein pair. Photochem Photobiol 85:287–297
14. Tran TH, Alexander P, Dharmaiah S, Agamasu C, Nissley DV, McCormick F, Esposito D, Simanshu DK, Stephen AG, Balius TE (2020) The small molecule BI-2852 induces a nonfunctional dimer of KRAS. Proc Natl Acad Sci USA 117:3363–3364
15. Picoquant Gmbh (2013) Tutorial: Lifetime-Fitting Using the FLIM Script. https://www.picoquant.com/images/uploads/downloads/lifetime-fitting_using_the_flim-script_step_by_step.pdf
16. Goswami D, Chen D, Yang Y, Gudla PR, Columbus J, Worthy K, Rigby M, Wheeler M, Mukhopadhyay S, Powell K, Burgan W, Wall V, Esposito D, Simanshu DK, Lightstone FC, Nissley DV, McCormick F, Turbyville T (2020) Membrane interactions of the globular domain and the hypervariable region of KRAS4b define its unique diffusion behavior. eLife 9:e47654
17. Bajar BT, Wang ES, Zhang S, Lin MZ, Chu J (2016) A guide to fluorescent protein FRET pairs. Sensors (Basel) 16:1488

# Chapter 20

# High-Throughput Cell-Based Screening of Small Molecule KRAS Signaling Inhibitors Using a Homogeneous Time-Resolved Fluorescence (HTRF) Assay

**Brian P. Smith, Megan Rigby, Roger Ma, and Anna E. Maciag**

## Abstract

With recent advances proving that effective inhibition of KRAS is possible, there have been significant efforts made to develop inhibitors of specific mutant alleles. Here we describe a detailed protocol that employs homogeneous time-resolved fluorescence (HTRF) to identify compounds acting on KRAS signaling in malignant cell lines. This method allows for high-throughput, cell-based screens of large compound libraries for the development of RAS-targeted therapeutics.

**Key words** HTRF, HTS, KRAS, Cell-based drug screen, KRAS signaling, KRAS inhibitors, MAPK, ERK, PI3K, AKT

## 1 Introduction

Homogenous time-resolved fluorescence (HTRF) is a widely used molecular assay that combines fluorescence resonance energy transfer (FRET) with a time-resolved measurement of fluorescence in a homogenous format that is ideal for high-throughput drug discovery efforts. This rapid, easily implemented, and scalable assay provides a sensitive and robust readout with reduced background, high reproducibility, and a low compound interference rate [1]. Since its inception, the technique has been used to study kinases, GPCRs, receptor-ligand interactions, and protein-protein interactions [1]. HTRF relies on FRET occurring between donor and acceptor fluorophores that are within close proximity of one another. Excitation of the donor fluorophore transfers energy to the nearby acceptor fluorophore, resulting in acceptor energy emission at a signature wavelength. Measurements of both donor and acceptor fluorescence are taken after a time delay of 50–150 μs from excitation, eliminating short-lived background fluorescence [1]. The ratio between donor and acceptor values is then calculated and

Andrew G. Stephen and Dominic Esposito (eds.), *KRAS: Methods and Protocols*, Methods in Molecular Biology, vol. 2797, https://doi.org/10.1007/978-1-0716-3822-4_20,

used as a quantitative readout. Donor and acceptor fluorophores can be conjugated to ligands, second messengers, or antibodies to proteins of interest [1]. In the protocol described below, we utilize antibodies conjugated to donor and acceptor fluorophores that detect both the phosphorylated and non-phosphorylated proteoforms of a downstream RAS effector kinase to determine its relative level of activation.

RAS oncoproteins are major drivers of human cancers that overactivate multiple signaling pathways [2, 3]. Wild-type RAS activity is driven by cycling from an inactive GDP-bound state to an active GTP-bound state—a process that is tightly regulated [4, 5]. However, most common mutations in oncogenic RAS occur at positions 12, 13, or 61 of the protein [6, 7], all of which impair the hydrolysis of GTP [8]. Thus, in cancers, mutant RAS is kept in a constitutively active GTP-bound state which drives tumor growth and progression.

The most important signaling pathways driven by mutant RAS are the RAF-MEK-ERK (MAPK) and PI3K-AKT pathways. Consequently, there has been great interest in targeting these pathways pharmacologically. Direct targeting of mutant RAS alleles has long been seen as an ideal yet elusive therapeutic strategy, as it could avoid the toxicity associated with broad pathway inhibition. Since the discovery of the cryptic switch II pocket in mutant KRAS by Shokat and colleagues [9], there has been much focus on directly targeting KRAS and its mutant alleles. Indeed, the development of the KRAS G12C-specific inhibitors AMG510-sotorasib (Amgen) [10] and MRTX849-adagrasib (Mirati Therapeutics) [11], and the KRAS G12D-specific inhibitor MRTX1133 (Mirati Therapeutics) [12], has proven that mutant KRAS is in fact a druggable target. Current strategies continue to exploit the cryptic switch II pocket to lock KRAS in its inactive GDP-bound conformation [9, 13–15]. Additionally, the unique biochemistry of different KRAS mutations can be exploited to expand the selection of allele-specific inhibitors. For instance, strategies to covalently target the G12S and G12R alleles were developed by Zhang, Shokat, and colleagues in 2022 [16, 17].

Here we describe a method that rapidly detects and quantifies the effectiveness of KRAS signaling inhibitors in malignant cell lines. HTRF detection of phospho-ERK (p-ERK) as a proportion of total ERK is utilized to measure active MAPK signaling. We show that KRAS mutant alleles can be targeted specifically in malignant cell lines by utilizing inhibitors of the KRAS G12C or KRAS G12D proteins. MAPK signaling was modified through KRAS inhibition, and the timing by which KRAS mutants respond to inhibitors was determined across several cell line models. Our assay allows for a rapid, high-throughput, and quantitative screening of libraries of KRAS inhibitors. Additionally, by modifying the antibodies, the effectiveness of inhibitors on AKT signaling can be determined using the same assay. Although we focus specifically on KRAS signaling, this protocol can be modified to study other signaling pathways or optimized for use in additional cell lines.

## 2 Materials

### 2.1 Cell Culture

Handle all materials under sterile conditions.

1. KRAS mutant cancer cell lines for G12C mutants (e.g., MIA PaCa-2, NCI-H358, SW1573).
2. KRAS mutant cancer cell lines for G12D mutants (e.g., ASPC-1, GP2d, HPAC).
3. KRAS mutant cancer cell lines for G12V mutants (e.g., HuP-T4, NCI-H441, SW620).
4. Cell culture medium (RPMI-1640, DMEM/F12, DMEM).
5. Tetracycline system-approved fetal bovine serum (FBS).
6. Complete cell culture medium: add 50 mL of FBS to a 500 mL bottle of cell culture media. Store at 4 °C.
7. 1× phosphate-buffered saline (PBS) at pH 7.4.
8. 0.25% trypsin EDTA.
9. 75 $cm^2$ cell culture flasks.
10. 150 $cm^2$ cell culture flasks.
11. 96-well clear cell culture plates.
12. 0.4% trypan blue solution.
13. Automated cell counter.

### 2.2 HTRF Assay

1. Labcyte 384-well low dead volume (LDV) plates.
2. ProxiPlate-384 Plus, white, tissue culture treated (PerkinElmer).
3. HTRF Adv.ERK p-T202/Y204 kit (PerkinElmer).
4. Dimethylsulfoxide (DMSO).
5. Compounds of interest. We optimized this protocol with the following: trametinib, AMG510-sotorasib (Amgen), MRTX849-adagrasib (Mirati Therapeutics), MRTX1133 (Mirati Therapeutics).
6. Labcyte Echo 555 liquid handler (Beckman Coulter).
7. Centrifuge with plate attachment.
8. Sterile reservoir.
9. Eppendorf repeater pipette and tips.
10. Multichannel 8-well vacuum manifold.
11. Orbital microplate shaker.
12. EnVision plate reader (PerkinElmer).

## 3 Methods

All procedures are carried out at room temperature unless otherwise noted.

### *3.1 Preparation of Cells*

Complete all procedures under sterile conditions in a biological safety cabinet.

1. Prepare each cell line to be tested in T150 flasks until 70–80% confluency is reached (*see* **Notes 1** and **2**).
2. Warm PBS and required complete cell culture media to between room temperature and 37 °C in a water bath or incubator.
3. Aspirate and discard media from flasks.
4. Pipette 10 mL of PBS gently into the flask, cap, and gently rock back and forth over cells to wash.
5. Aspirate and discard PBS.
6. Dissociate cells from the flasks by adding 3 mL of 0.25% trypsin EDTA and gently rocking the flask to cover the cells (*see* **Note 3**).
7. Watch for cell detachment from flask under a microscope. Cells may be incubated at 37 °C to speed up the process. This generally takes 3–5 min.
8. Once cells have detached, neutralize the trypsin by adding at least 1 equal volume of complete media. Rinse the bottom surface of the flask with this media to collect all the cells. Bring the volume up to 10 mL with complete media.
9. Take a 20 μL aliquot of the resuspended cells, and mix with 20 μL of 0.4% trypan blue solution in a 1.5 mL microcentrifuge tube.
10. Incubate for 2–3 min.
11. Take 10 μL of this solution and add to a cell counter chamber slide.
12. Count cells using a cell counter. Count twice and record the viability and concentration of viable cells. Take the average viable cells of both counts.
13. Using the concentration of viable cells, dilute with fresh media to obtain the desired cell densities for each cell line. Most cell lines work well in a range from 25k to 40k cells/well. Bring the total volume up to 8 mL of cells per plate.

| Common seeding densities | Concentration of viable cells |
|---|---|
| 25k cells/well | 3.13e5 cells/mL |
| 30k cells/well | 3.75e5 cells/mL |
| 35k cells/well | 4.38e5 cells/mL |
| 40k cells/well | 5.00e5 cells/mL |

14. Pour the cell suspension into a sterile reservoir, providing constant agitation with one hand to prevent cell settling, and dispense 80 μL/well to all wells using a multichannel pipette.
15. Centrifuge the plates at 230 × *g* for 1 min to evenly settle cells and prevent clumping.
16. Place plates in a 37 °C/5% $CO_2$ incubator for 18–24 h.
17. Cells should be at 70–80% confluency at the time of compound treatment. Cell seeding density can be optimized by performing dose-response curves (as described in Subheading 3.3) using a tool compound at various densities.

### 3.2 Preparation of Compound Stock Solutions

1. Obtain two Labcyte 384-well LDV plates.
2. Dilute compounds in DMSO. Depending on stock concentration and desired highest dose, the dilution made should be 1000-fold of the highest dose exposure on the cells.
3. Add 10 μL of each compound to one of the LDV plates in the layout below (Fig. 1). This will be the LDV intermediate plate.
4. Spin the LDV intermediate plate at 500 × *g* for 1 min.
5. Using the Labcyte Echo 555 and Echo Plate Reformat software to transfer appropriate volumes from the LDV intermediate plate to the second LDV plate (LDV source plate). This will be used to produce specific dose titrations of compounds.
6. Add 10 μL of DMSO backfill to all wells of the LDV source plate that had compounds transferred into them.
7. Transfer the compounds from the LDV intermediate plate to the same wells of the LDV source plate.
8. Add 11 μL DMSO to columns 17–24 of the LDV source plate (Fig. 2).
9. Use plate immediately or seal and place plate into −20 °C freezer overnight.

### 3.3 Source Plate Preparation

1. Thaw the LDV source plate (if frozen).
2. Spin at 500 × *g* for 1 min.
3. Prepare 96-well source plates of 5× compound dilutions. Use the Labcyte Echo 555 and Echo Dose-Response software to

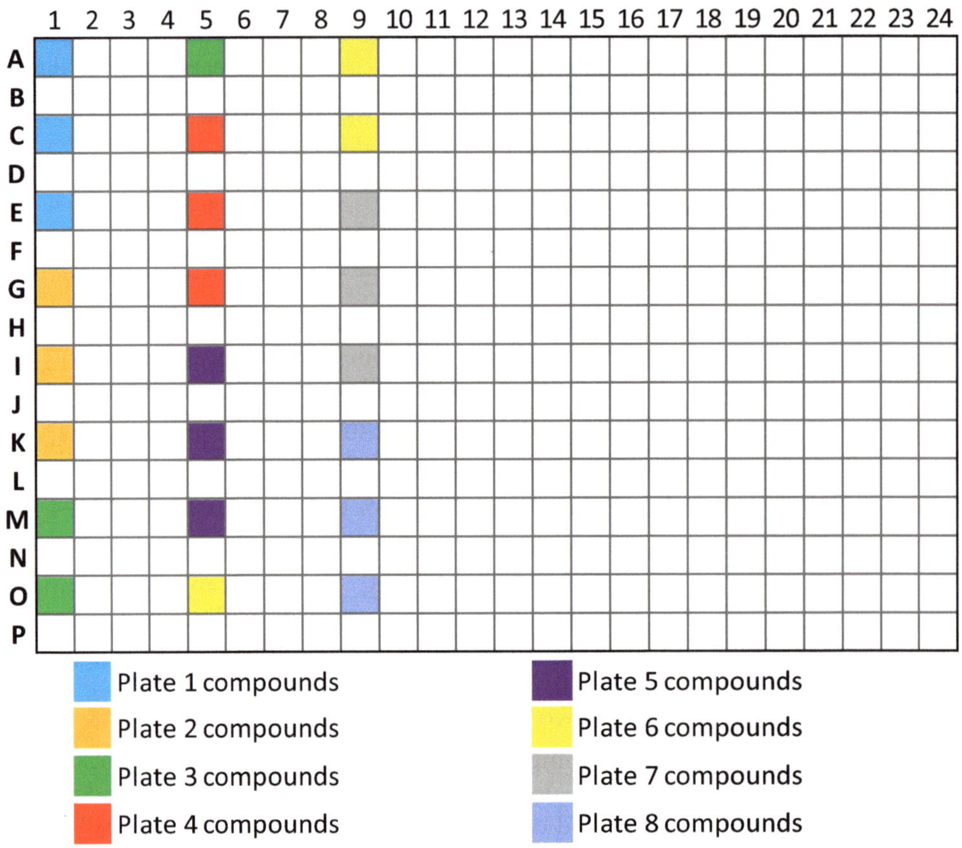


**Fig. 1** Plate layout for the intermediate LDV plate. Each LDV plate can hold up to eight 96-well plates of compounds. Each 96-well plate can hold three compounds

transfer the appropriate volumes from the LDV source plate to the 96-well source plates. Each compound dose-response curve is generated twice for a total of two replicates (Fig. 3).

4. Backfill the 96-well source plates with 185 μL of complete media using the Multidrop Combi, and briefly swirl by hand to mix.

### *3.4 Compound Treatment and Cell Lysis*

1. Add 20 μL from the wells of the source plate to the corresponding wells of the 96-well cell plate (*see* **Note 4**).
2. Swirl to mix the compounds and return to the 37 °C/5% $CO_2$ incubator.
3. Repeat **steps 1** and **2** for the remaining plates.
4. Incubate for the desired time.
5. During incubation, prepare fresh lysis buffer by diluting 4× kit lysis buffer in water and adding 100× blocking reagent to a final concentration of 1×. You will need 3 mL per plate plus extra for blank readings in HTRF analysis. We generally make an additional 2 mL (*see* **Note 5**).

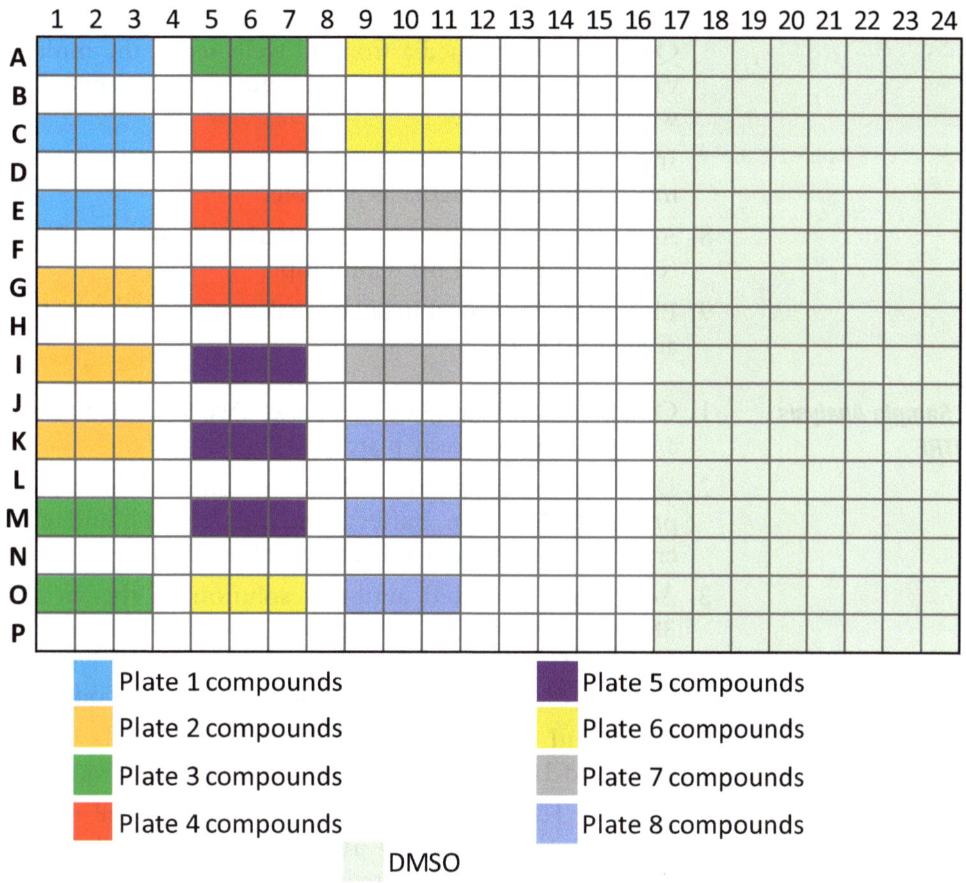


**Fig. 2** Layout of the LDV source plate. Each compound will have three concentrations on this source plate. The DMSO wells are used to backfill to ensure that all compound concentrations contain the same DMSO percentage

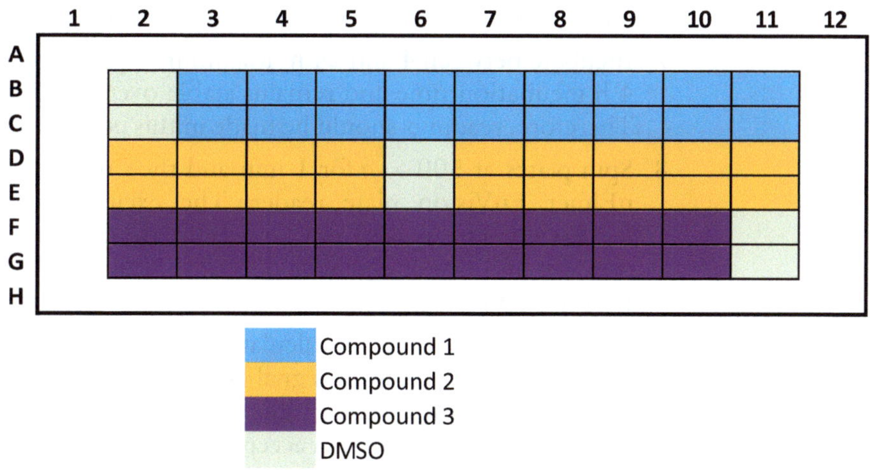


**Fig. 3** Layout of the 96-well source plate. Each compound is in a 9-point dose-response curve in duplicate wells. After addition of 185 μL media, all concentrations will be at 5× the final assay concentration. 20 μL from the wells of this plate will be added to the cell plates that were prepared the previous day

6. Remove the plates from the incubator at the desired time. Quickly aspirate media from all wells using the multichannel vacuum manifold, while being careful to not scrape cells off the wells.
7. Invert and smack the plate onto paper towels to remove as much remaining media as possible.
8. Add 50 μL lysis buffer to all treated wells using an Eppendorf repeater with a 2.5 mL Combitip.
9. Place plates on an orbital plate shaker at a speed of 850 rpm for at least 30 min.

### *3.5 Sample Analysis by HTRF*

1. Obtain a ProxiPlate-384 Plus plate. Two 96-well plates fit onto a single 384-well assay plate.
2. Dilute antibody aliquots 1:20 (29 μL antibody into 551 μL kit provided detection buffer), and mix the d2 and europium cryptate solution at a 1:1 ratio (*see* **Note 6**).
3. Add 4 μL of mixed antibody solution to the wells of the 384-well assay plate using an Eppendorf repeater with a 200 μL Combitip. All wells that contain cell lysates, blank samples, or positive control should receive antibody (Fig. 4).
4. Add 16 μL of the positive control sample to the appropriate well. Add 16 μL of lysis buffer to each of blank sample well.
5. Transfer 16 μL of cell lysates from the 96-well plate to the 384-well assay plate (20 μL total volume in all wells). Each well from the 96-well plate is transferred into duplicate wells to produce technical replicates. Before transfer, mix lysate thoroughly by pipetting several times to ensure lysate is homogenous.
6. Use parafilm to seal the edge of the plate where the lid and plate meet to ensure no evaporation occurs during incubation.
7. Incubate between 4 and 24 h. Maximum signal is reached after 4 h incubation time and remains stable over a period of 24 h. Therefore, readings should be made in this period (*see* **Note 7**).
8. Spin plates at 500 × *g* for 1 min and then read on the PerkinElmer EnVision plate reader. The excitation wavelength for Eu cryptate is 320 nm, and both the acceptor signal at 665 nm and donor signal at 620 nm should be captured (*see* **Notes 8** and **9**).
9. From the EnVision data, calculate the acceptor to donor ratio by dividing the acceptor signal by the donor signal for each well. Subtract the average value of blank wells (antibody and lysis buffer only) from each acceptor to donor ratio. Values can be normalized to percentage of DMSO control if desired.

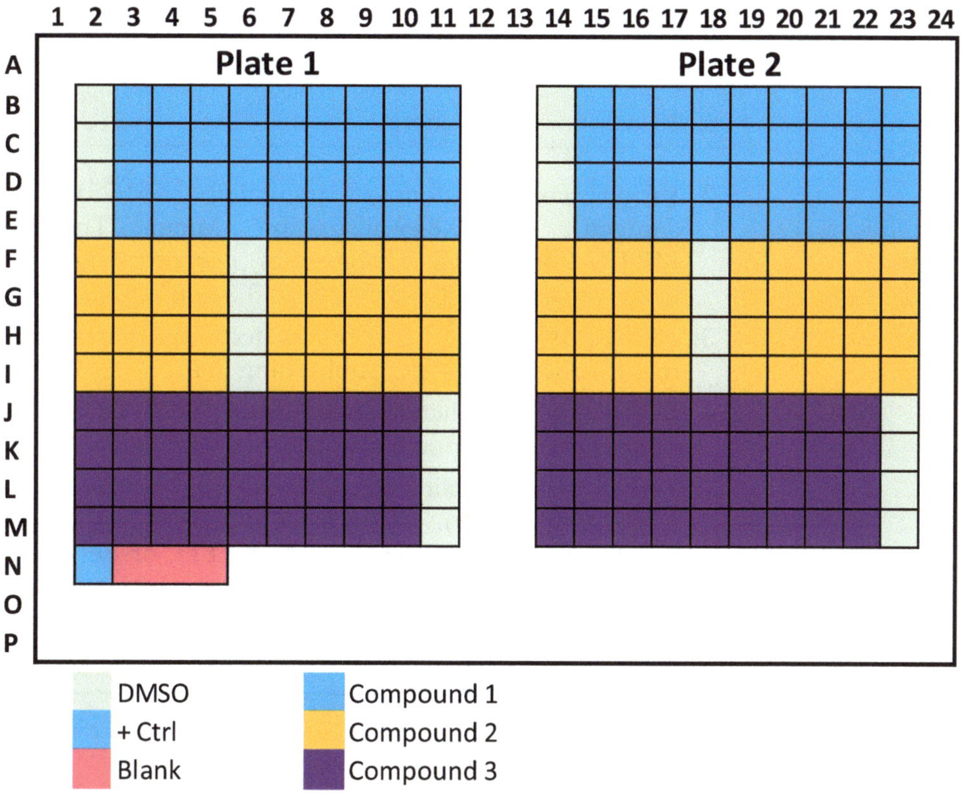


**Fig. 4** Layout of the assay plate. Two 96-well cell plates can fit on each assay plate. 16 μL of lysates are moved from the 96-well cell plate into duplicate wells of this plate to serve as technical replicates. Mix lysate thoroughly by pipetting several times before transfer to ensure lysate is homogenous

### *3.6 Inhibition of RAS Signaling Pathway: Analysis of p-ERK Inhibition in KRAS Mutant Cell Lines*

As an example of the usage of this assay to evaluate the effects of KRAS targeting compounds on MAPK signaling in cells, we determined the potency of several KRAS inhibitors across multiple KRAS mutant cell lines.

1. Seed cells from the following cell lines at specified densities, as per Subheading 3.1:

| Cell line | KRAS mutation | Seeding density |
|---|---|---|
| ASPC-1 | G12D | 3e4 cells/well |
| GP2d | G12D | 3e4 cells/well |
| HPAC | G12D | 3e4 cells/well |
| HuP-T4 | G12V | 3e4 cells/well |
| NCI-H441 | G12V | 3e4 cells/well |
| SW620 | G12V | 3e4 cells/well |
| SW1573 | G12C | 4e4 cells/well |

(continued)

| Cell line | KRAS mutation | Seeding density |
|---|---|---|
| MIA PaCa-2 | G12C | 4e4 cells/well |
| NCI-H358 | G12C | 4e4 cells/well |

2. Various concentrations of trametinib, AMG510-sotorasib (Amgen), MRTX849-adagrasib (Mirati Therapeutics), MRTX1133 (Mirati Therapeutics), or compound A were made, as indicated in Subheadings 3.2 and 3.3.
3. 18–24 h after seeding, cells are treated with compounds for up to 24 h and lysed, as per Subheading 3.4.
4. Cell lysates were then analyzed using HTRF to obtain IC50 values for the compounds, as per Subheading 3.5.
5. This assay can be validated with MEK inhibitors such as trametinib (*see* **Note 10**), can be used to analyze new compounds which target RAS directly (*see* **Note 11**), or can distinguish effects on different mutant KRAS alleles (*see* **Note 12**).

## 4 Notes

1. Refer to cell line information sheets for specific cell handling requirements. Consider handling only one cell line at a time to reduce the potential of cross contamination.
2. If cells were recovered from freezing, they should be passaged at least once after thawing before seeding.
3. Care should be taken to minimize cell exposure to trypsin.
4. We suggest delaying the compound treatment of additional plates by 1–2 min, as harvesting the plates takes longer than treating. At shorter times points, this can accumulate to significantly increase the compound incubation time.
5. HTRF lysis buffer should be made fresh every 24 h.
6. Do not vortex antibodies.
7. HTRF signal drops off rapidly after 24 h incubation at room temperature. We have found optimized signal quality between 4 and 24 h incubation at room temperature.
8. Instrument operations and settings depend on the manufacturer—check the specific instructions for your instrument for more details.
9. This protocol is also optimized for phospho-AKT analysis using a separate kit (PerkinElmer HTRF AKT p-S473 kit) substituting only the kit-specific antibody reagents.

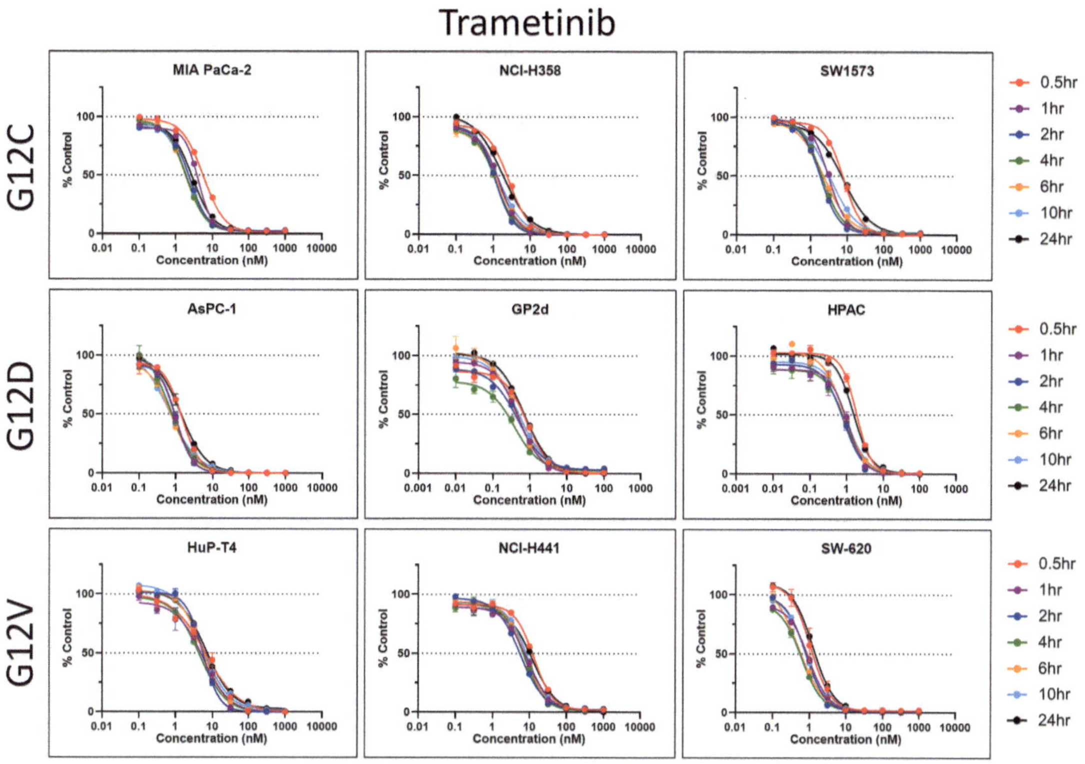


**Fig. 5** Dose-response and time-response analysis of p-ERK levels in KRAS mutant malignant cell lines carrying either KRAS G12C, G12D, or G12V mutations treated with trametinib. Quantification of cell line sensitivity was determined by $IC_{50}$ values for each time point and cell line (Table 1)

10. The commercially available MEK inhibitor, trametinib, was used to provide a reference for MAPK inhibition. MEK signals directly upstream of ERK, and therefore inhibition should be reflected in p-ERK levels (Fig. 5, Table 1).
11. We show data from a KRAS G12C inhibitor that targets both active (GTP-bound) and inactive (GDP-bound) KRAS G12C, compound A, developed in a collaboration with TheRas, Inc. We compared activity of this compound to the reduction in p-ERK levels seen in clinical KRAS inhibitors. Using this assay, we show that compound A is a more potent KRAS G12C inhibitor in malignant cell lines than current clinical inhibitors (Fig. 6, Table 2).
12. The assay can distinguish between KRAS mutant alleles effectively. As an example, MRTX1133, a KRAS G12D-specific inhibitor, reduces p-ERK levels in GP2d cells (KRAS G12D mutant) but not in SW620 cells (KRAS G12V mutant) (Fig. 7, Table 3).

**Table 1**
**$IC_{50}$ values (nM) for a panel of malignant cell lines treated with the MEK inhibitor trametinib**

| | KRAS G12C | | | KRAS G12D | | | KRAS G12V | | |
|---|---|---|---|---|---|---|---|---|---|
| **Time (h)** | **MIA PaCa-2** | **NCI-H358** | **SW1573** | **AsPC-1** | **GP2d** | **HPAC** | **HuP-T4** | **NCI-H441** | **SW-620** |
| 0.5 | 5.5 | 2.3 | 7.5 | 1.40 | 0.68 | 1.9 | 6.5 | 12.1 | 1.2 |
| 1 | 3.8 | 1.3 | 3.1 | 0.93 | 0.46 | 0.91 | 4.9 | 7.4 | 0.87 |
| 2 | 2.2 | 1.0 | 1.8 | 0.95 | 0.40 | 0.74 | 5.6 | 5.5 | 0.80 |
| 4 | 1.9 | 1.0 | 2.0 | 0.81 | 0.24 | 0.78 | 4.3 | 6.8 | 0.56 |
| 6 | 1.8 | 1.1 | 2.3 | 0.74 | 0.44 | 0.89 | 5.9 | 7.2 | 0.59 |
| 10 | 1.9 | 1.3 | 3.3 | 0.73 | 0.51 | 0.93 | 6.3 | 8.2 | 0.84 |
| 24 | 2.7 | 1.8 | 7.0 | 1.41 | 0.71 | 1.6 | 7.3 | 10.2 | 1.4 |

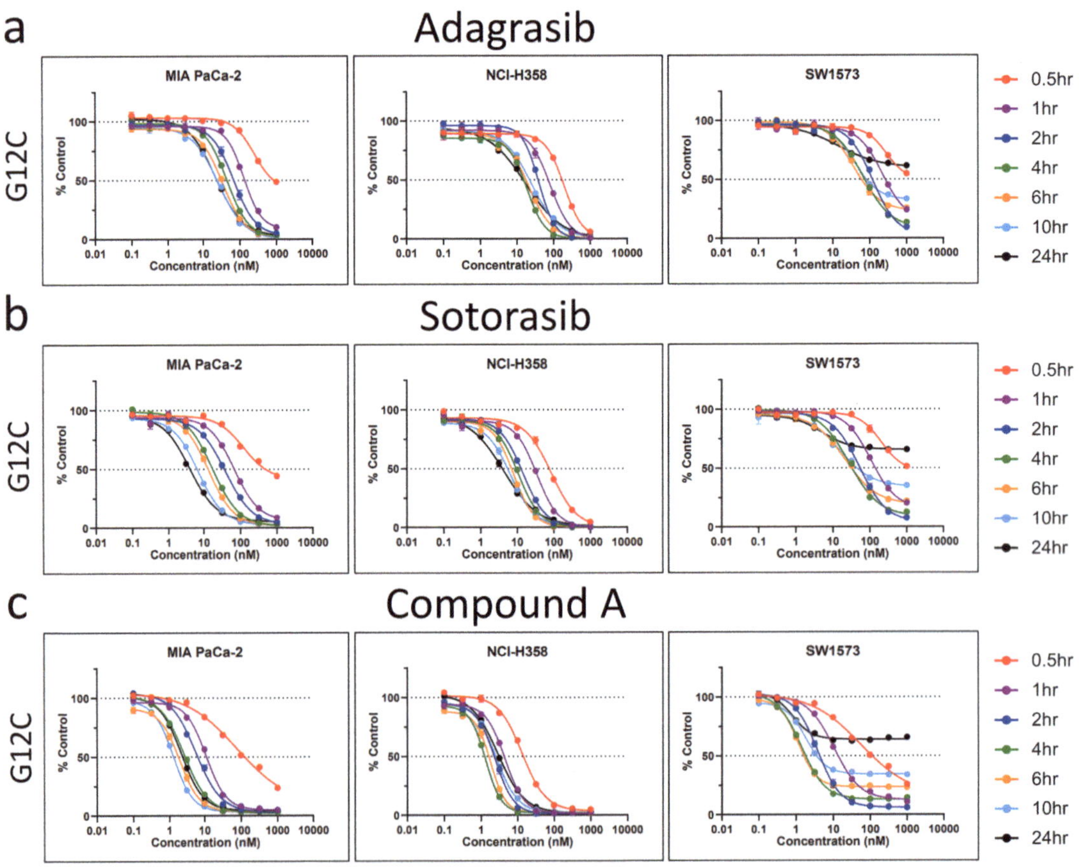


**Fig. 6** Dose-response and time-response analysis of p-ERK levels in MIA PaCa-2, NCI-H358, and SW1573 cells treated with KRAS G12C inhibitors. (**a**, **b**) Cells were treated with the two clinically approved KRAS G12C inhibitors, adagrasib (**a**) and sotorasib (**b**). (**c**) Cells were treated with the KRAS G12C inhibitor (compound A) developed in a collaboration with BridgeBio Inc. Quantification of cell line sensitivity toward each compound was determined by the $IC_{50}$ values (Table 2)

**Table 2**
**IC$_{50}$ values (nM) for KRAS G12C malignant cell lines treated with covalent G12C inhibitor compounds**

| | MIA PaCa-2 | | | NCI-H358 | | | SW1573 | | |
|---|---|---|---|---|---|---|---|---|---|
| Time (h) | Adagrasib | Sotorasib | Cpd A | Adagrasib | Sotorasib | Cpd A | Adagrasib | Sotorasib | Cpd A |
| 0.5 | 868 | 481 | 126 | 171 | 75.3 | 14.5 | >1000 | >1000 | 112 |
| 1 | 137 | 68.8 | 11.5 | 71.9 | 30.0 | 4.3 | 262 | 139 | 13.6 |
| 2 | 69.1 | 37.0 | 6.5 | 38.2 | 12.8 | 2.3 | 118 | 54.5 | 4.4 |
| 4 | 45.1 | 17.8 | 2.8 | 15.4 | 9.4 | 1.2 | 72.7 | 32.9 | 1.8 |
| 6 | 32.3 | 13.3 | 1.8 | 17.7 | 7.0 | 1.6 | 59.2 | 31.4 | 1.6 |
| 10 | 24.3 | 6.5 | 1.3 | 22.4 | 5.5 | 2.5 | 81.6 | 44.6 | 3.6 |
| 24 | 24.2 | 4.1 | 2.4 | 14.8 | 3.9 | 3.1 | >1000 | >1000 | >1000 |

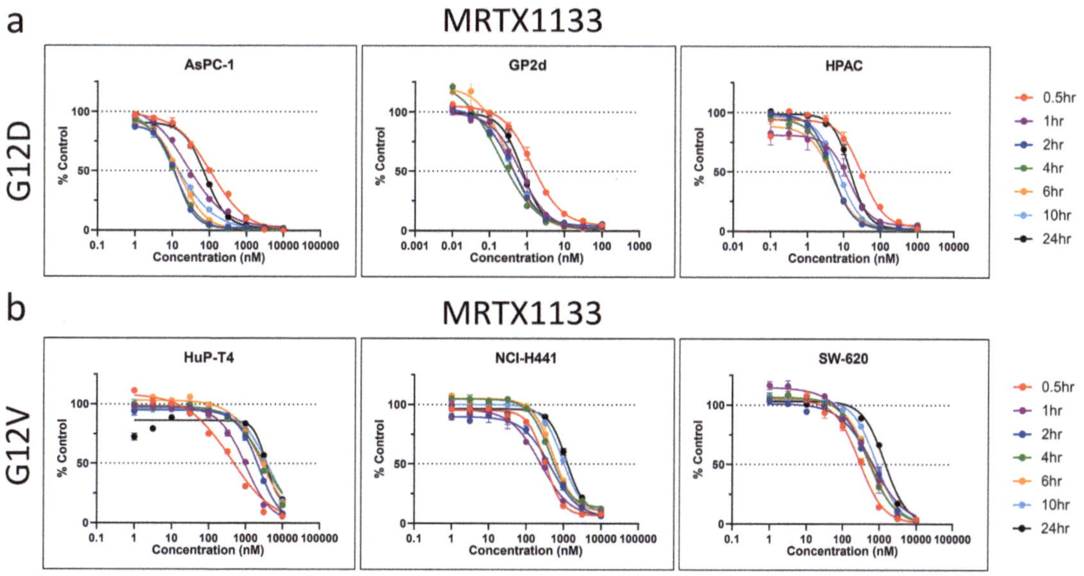


**Fig. 7** Dose-response and time-response analysis of p-ERK levels in KRAS G12D (**a**) and KRAS G12V (**b**) cell lines treated with MRTX1133. Quantification of cell line sensitivity was determined by the IC$_{50}$ values (Table 3)

**Table 3**
**$IC_{50}$ values (nM) for KRAS G12D and G12V cell lines treated with the KRAS G12D inhibitor MRTX1133**

| | KRAS G12D | | | KRAS G12V | | |
|---|---|---|---|---|---|---|
| Time (h) | AsPC-1 | GP2d | HPAC | HuP-T4 | NCI-H441 | SW-620 |
| 0.5 | 105 | 1.7 | 30.2 | 485 | 328 | 310 |
| 1 | 31.0 | 0.66 | 10.6 | 957 | 248 | 669 |
| 2 | 11.6 | 0.44 | 4.8 | 2120 | 411 | 682 |
| 4 | 11.0 | 0.30 | 4.5 | 3260 | 559 | 552 |
| 6 | 14.3 | 0.45 | 4.2 | 2800 | 650 | 597 |
| 10 | 15.1 | 0.45 | 7.0 | 3900 | 1040 | 876 |
| 24 | 69.4 | 0.76 | 15.3 | 3640 | 1280 | 1470 |

## Acknowledgments

We acknowledge TheRas, Inc. for use of compound A developed under a cCRADA with the Frederick National Laboratory. This project was funded in part with federal funds from the National Cancer Institute, National Institutes of Health under contract 75N91019D00024. The content of this publication does not necessarily reflect the views or policies of the Department of Health and Human Services, and the mention of trade names, commercial products, or organizations does not imply endorsement by the US government.

### References

1. Degorce F et al (2009) HTRF: a technology tailored for drug discovery–a review of theoretical aspects and recent applications. Curr Chem Genomics Transl Med 3:22–32
2. Downward J (2003) Targeting RAS signalling pathways in cancer therapy. Nat Rev Cancer 3: 11–22. https://doi.org/10.1038/nrc969
3. Vigil D, Cherfils J, Rossman KL, Der CJ (2010) Ras superfamily GEFs and GAPs: validated and tractable targets for cancer therapy? Nat Rev Cancer 10:842–857. https://doi.org/10.1038/nrc2960
4. Bos JL, Rehmann H, Wittinghofer A (2007) GEFs and GAPs: critical elements in the control of small G proteins. Cell 129:865–877. https://doi.org/10.1016/j.cell.2007.05.018
5. Hunter JC et al (2015) Biochemical and structural analysis of common cancer-associated KRAS mutations. Mol Cancer Res 13:1325–1335. https://doi.org/10.1158/1541-7786.Mcr-15-0203
6. Cox AD, Fesik SW, Kimmelman AC, Luo J, Der CJ (2014) Drugging the undruggable RAS: mission possible? Nat Rev Drug Discov 13:828–851. https://doi.org/10.1038/nrd4389
7. Moore AR, Rosenberg SC, McCormick F, Malek S (2020) RAS-targeted therapies: is the undruggable drugged? Nat Rev Drug Discov 19:533–552. https://doi.org/10.1038/s41573-020-0068-6

8. Scheffzek K et al (1997) The Ras-RasGAP complex: structural basis for GTPase activation and its loss in oncogenic Ras mutants. Science 277:333–338. https://doi.org/10.1126/science.277.5324.333
9. Ostrem JM, Peters U, Sos ML, Wells JA, Shokat KM (2013) K-Ras(G12C) inhibitors allosterically control GTP affinity and effector interactions. Nature 503:548–551. https://doi.org/10.1038/nature12796
10. Lanman BA et al (2020) Discovery of a covalent inhibitor of KRAS(G12C) (AMG 510) for the treatment of solid tumors. J Med Chem 63: 52–65. https://doi.org/10.1021/acs.jmedchem.9b01180
11. Hallin J et al (2020) The KRAS(G12C) inhibitor MRTX849 provides insight toward therapeutic susceptibility of KRAS-mutant cancers in mouse models and patients. Cancer Discov 10:54–71. https://doi.org/10.1158/2159-8290.Cd-19-1167
12. Hallin J et al (2022) Anti-tumor efficacy of a potent and selective non-covalent KRAS (G12D) inhibitor. Nat Med 28:2171–2182. https://doi.org/10.1038/s41591-022-02007-7
13. Lim SM et al (2014) Therapeutic targeting of oncogenic K-Ras by a covalent catalytic site inhibitor. Angew Chem Int Ed Engl 53:199–204. https://doi.org/10.1002/anie.201307387
14. Maurer T et al (2012) Small-molecule ligands bind to a distinct pocket in Ras and inhibit SOS-mediated nucleotide exchange activity. Proc Natl Acad Sci USA 109:5299–5304. https://doi.org/10.1073/pnas.1116510109
15. Sun Q et al (2012) Discovery of small molecules that bind to K-Ras and inhibit Sos-mediated activation. Angew Chem Int Ed Engl 51:6140–6143. https://doi.org/10.1002/anie.201201358
16. Zhang Z, Morstein J, Ecker AK, Guiley KZ, Shokat KM (2022) Chemoselective covalent modification of K-Ras (G12R) with a small molecule electrophile. J Am Chem Soc 144: 15916–15921
17. Zhang Z, Guiley KZ, Shokat KM (2022) Chemical acylation of an acquired serine suppresses oncogenic signaling of K-Ras (G12S). Nat Chem Biol 18:1177–1183

# Chapter 21

# A Method to Conditionally Measure Target Engagement at Intracellular RAS and RAF Complexes

**James D. Vasta, Ani Michaud, J. Aaron Crapster, and Matthew B. Robers**

## Abstract

Dysfunction of the RAS/mitogen-activated protein kinase (MAPK) pathway is a common driver of human cancers. As such, both the master regulator of the pathway, RAS, and its proximal kinase effectors, RAFs, have been of interest as drug targets for decades. Importantly, signaling within the RAS/MAPK pathway is highly coordinated due to the formation of a higher-order complex called the RAS/RAF signalosome, which may minimally contain dimers of both RAS and RAF protomers. In the disease state, RAS and RAF assemble in homo- and/or heterodimeric forms. Traditionally, drug development campaigns for both RAS and RAF have utilized biochemical assays of purified recombinant protein. As these assays do not query the RAS or RAF proteins in their full-length and complexed forms in cells, potency results collected using these assays have often failed to correlate with inhibition of the MAPK pathway. To more accurately quantify engagement at this signaling components, we present a bioluminescence resonance energy transfer (BRET)-based method to conditionally measure target engagement at individual protomers within the RAS/RAF signalosome in live cells.

**Key words** Target engagement, Intracellular affinity, BRET, NanoBRET™, RAS, RAF, MAPK

## 1 Introduction

The RAS/MAPK pathway is a master regulatory pathway controlling cell proliferation and is among the most frequently mutated pathways in cancer [1–3]. In normal cells, pathway agonism is tightly regulated by RAS-mediated GTP hydrolysis. In the GTP-bound state, RAS (KRAS, HRAS, NRAS) interacts with the primary pathway effectors, RAF kinases, which in turn facilitates RAF dimerization and formation of the active RAS/RAF heterotetrameric complex [3]. The RAF family of kinases comprise three paralogs: BRAF, CRAF, and ARAF, each with the capacity for homo- and heterodimerization. BRAF and CRAF are the most well-studied protomers in the system, and the potential role of ARAF in oncogenesis has gained attention in more recent years [2, 4]. Genetic lesions in this pathway are commonly observed in

Andrew G. Stephen and Dominic Esposito (eds.), *KRAS: Methods and Protocols*, Methods in Molecular Biology, vol. 2797, https://doi.org/10.1007/978-1-0716-3822-4_21,

cancers, most typically as somatic mutations in KRAS, NRAS, and BRAF [2, 3]. Mutations at codon 12 of KRAS (e.g., G12D, G12V, G12C) result in attenuated GTP hydrolysis and agonism of the active GTP-bound RAS/RAF signalosome. Historically, drug discovery efforts directly at KRAS were hampered due to the scarcity of binding pockets on this small GTPase [3]. The nucleotide binding pocket is considered undruggable owing to its strong affinity for GTP/GDP [3, 5]. Therefore, ATP-competitive inhibitors have been pursued at the proximal kinase effectors (RAF, MEK, ERK) for several decades. Among such ATP-competitive inhibitors, "RAF dimer inhibitors" are type II kinase inhibitors that stabilize RAF dimerization and occupy the nucleotide pocket of both RAF protomers within the RAS signalosome [4, 6–8]. As engagement of RAF dimer inhibitors is stabilized by RAS-GTP, RAF dimer inhibitors offer compelling utility for MAPK inhibition in RAS-driven tumors. However, RAF dimer inhibitors are relatively poor inhibitors of ARAF heterodimers [9, 10], indicating that ARAF avoidance may represent a clinical limitation in certain genetic contexts.

More recently, KRAS has proven druggable via the allosteric switch-2 pocket (S2P) [11]. The KRAS G12C allele was the first to be drugged at the S2P, exploiting the nucleophilicity of the cys-12 residue using a covalent strategy [11]. Building on the prototype KRAS (G12C) inhibitor, an expanded cohort of related drugs are now FDA approved or undergoing clinical investigation. Since the discovery of covalent KRAS (G12C) inhibitors, KRAS has proven more broadly vulnerable to reversible engagement at the S2P, enabling inhibition of more prevalent KRAS mutants (such as G12D) [5].

Recent successes in targeting RAS directly and the therapeutic potential of RAF dimer inhibitors together have galvanized drug discovery efforts at RAS-driven tumors. However, in the pursuit of novel medicines, methods to quantify drug-target engagement directly at protomers within this complex has represented a critical unmet. To date, the majority of assays utilizes pure protein fragments as proxies for these individual protomers in the RAS/RAF complex. However, as these cell-free methods do not query engagement of the full-length protein in the cellular milieu (e.g., PTMs conferring membrane localization, protein complexation, etc.), it is not surprising that drug potency with target fragments fails to correlate with intracellular target engagement or MAPK pathway inhibition. Herein we present an enzyme--complementation-based method to assemble a luminescent complex at RAS or RAF proteins in cells. This method utilizes protomers tagged with NanoBiT reporters (SmBiT, LgBiT), wherein luminescence originates conditionally upon the formation of specific protein dimers [10]. These complexes then enable the application of cell-permeable BRET probes to quantify engagement at select protomers within the complex via competitive displacement [5, 10, 12]. The conditional BRET reporter system generates potency data that is commensurate with pathway inhibition in disease-relevant cells [5], offering a more accurate platform over biochemical systems to query novel chemical matter at the RAS/RAF signalosome. Herein we describe a detailed method to validate this novel luminescence-based system for accurate assessments of target engagement at RAS/RAF complexes. This method should be adaptable to alternate protein complexes for future studies.

## 2 Materials

### *2.1 Determination of Specific Luminescence from RAS or RAF Complexes Using NanoBiT*

1. HEK293 cells.
2. Miscellaneous tissue culture reagents and plasticware.
3. Cell culture medium: 90% (v/v) DMEM, 10% (v/v) fetal bovine serum.
4. Assay medium: Opti-MEM without phenol red, 1% (v/v) fetal bovine serum.
5. Transfection reagent (FuGENE® HD).
6. BRAF, CRAF-SmBiT/LgBiT fusion vectors.
7. BRAF (A481F)-LgBiT fusion vector.
8. CRAF (A373F)-LgBiT fusion vector.
9. SmBiT-ARAF fusion vector.
10. SmBiT/LgBiT-KRAS (G12X) fusion plasmids.
11. EGFR-SmBiT/LgBiT and FRB-SmBiT/LgBiT fusion vector.
12. Untagged KRAS (G12C) expression vector.
13. Promoterless DNA (Transfection Carrier DNA).
14. White, tissue culture-treated 96-well plates.
15. White, nonbinding surface (NBS) 96-well plates, appropriately coated to avoid adsorptive properties of the fluorescent tracers.
16. Polypropylene plasticware (buffer troughs, serial dilution troughs, etc.).
17. Dimethylsulfoxide (DMSO).
18. BRET-compatible luminometer equipped with 450 nm (band-pass) and 600 nm (long pass) filters (e.g., GloMax Discover, PerkinElmer EnVision, or BMG CLARIOstar).
19. 3× complete live-mode detection solution: 1:166 dilution of the NanoBRET™ Nano-Glo® substrate plus a 1:500 dilution of the Extracellular NanoLuc® Inhibitor in Opti-MEM without serum or phenol red (*see* **Note 1**).
20. Digitonin stock solution: 50 mg/mL digitonin in DMSO.

21. 1000× HiBiT peptide (VSGWRLFKKIS) stock solution (100 μM in water).
22. 3× complete permeabilized-mode detection solution: 1:166 dilution of the NanoBRET™ Nano-Glo® substrate in assay medium (*see* **Note 1**).
23. 10× HiBiT peptide plus digitonin solution. Dilute 1 part (by volume) 50 mg/mL digitonin solution and 1 part (by volume) 1000× HiBiT peptide stock solution into 98 parts Opti-MEM.

### 2.2 Quantifying Target Engagement at Select Protomers Within RAF and RAS Complexes

1. Assay medium: Opti-MEM without phenol red, 1% (v/v) fetal bovine serum.
2. NanoBRET™ Tracer Dilution Buffer (Promega).
3. NanoBRET RAS tracer stock solution: 400 μM in DMSO (stored at −80 °C; stable to at least 5 freeze/thaw cycles).
4. 100× RAS NanoBRET tracer solution: 200 μM NanoBRET RAS tracer in DMSO prepared from the NanoBRET RAS tracer stock solution.
5. 20× RAS NanoBRET tracer working solution: 1 part (by volume) 100× RAS NanoBRET tracer solution, 4 parts (by volume) Tracer Dilution Buffer.
6. NanoBRET tracer K-4 stock solution: 400 μM in DMSO (stored at −80 °C; stable to at least 5 freeze/thaw cycles).
7. 100× BRAF NanoBRET tracer solution: 50 μM NanoBRET tracer K-4 in DMSO prepared from the NanoBRET tracer K-4 stock solution.
8. 20× BRAF NanoBRET tracer working solution: 1 part (by volume) 100× BRAF NanoBRET tracer solution, 4 parts (by volume) Tracer Dilution Buffer.
9. 100× CRAF NanoBRET tracer solution: 100 μM NanoBRET tracer K-4 in DMSO prepared from the NanoBRET tracer K-4 stock solution.
10. 20× CRAF NanoBRET tracer working solution: 1 part (by volume) 100× CRAF NanoBRET tracer solution, 4 parts (by volume) Tracer Dilution Buffer.
11. NanoBRET tracer K-10 stock solution: 400 μM in DMSO (stored at −80 °C; stable to at least 5 freeze/thaw cycles).
12. 100× ARAF NanoBRET tracer solution: 13 μM NanoBRET tracer K-10 in DMSO prepared from the NanoBRET tracer K-10 stock solution.
13. 20× ARAF NanoBRET tracer working solution: 1 part (by volume) 100× ARAF NanoBRET tracer solution, 4 parts (by volume) Tracer Dilution Buffer.

14. Test compound stock solution: 1000× desired final concentration in DMSO.
15. Test compound working solution (10×): 1 part (by volume) test compound stock solution diluted with 99 parts (by volume) Opti-MEM (*see* **Note 2**).

## 3 Methods

### *3.1 Determination of Specific Luminescence from RAS or RAF Complexes Using NanoBiT*

The first step in developing a conditional NanoBRET target engagement assay is verifying that the BRET donor is specific to the luminescent complex of the protein-protein interaction (PPI) of interest. In this section, protocols are provided for the detection of luminescence in HEK293 cells expressing NanoBiT fusions of KRAS or RAF. This protocol provides a starting point to express the targets of interest in cells such that the user can ensure the majority of the NanoBiT signals generated is due to specific interactions between the proteins of interest, and not due to spurious signal arising from spontaneous complementation of NanoBiT reporters. This is essential, as NanoBiT subunits possess intrinsic subunit affinities and will self-associate as a function of local intracellular concentration [13]. To verify that the majority of luminescence originates from specific PPIs, it is recommended that negative control proteins tagged with NanoBiT are included in the analysis. Ideally, these proteins have similar intracellular localization and expression levels to the proteins of interest. Therefore, to establish specific signal, iterative optimization may be required to ensure that expression levels are similar between the target and the control proteins. Figure 1 provides example specificity data for the KAS dimer and RAF dimer analysis. In each case, there is at least a tenfold relative increase in live cell luminescence for the intended KRAS or RAF dimer pairs compared to the live cell luminescence when using a control LgBiT fusion protein at a similar expression level, which confirms the specificity.

1. Cultivate HEK293 cells appropriately prior to assay, and resuspend cells into a single-cell suspension using cell culture medium.
2. Adjust the cell density to $2 \times 10^5$ cells/mL in pre-warmed (37 °C) assay medium in a sterile conical tube.
3. Prepare desired plasmid DNAs for transfection in sterile conical tubes:
   (a) For RAF dimer analysis using BRAF/CRAF dimers as an example:
      (i) For RAF dimer analysis, combine by mass 0.5 part the desired NanoBiT® fusion vectors (e.g., BRAF-SmBiT + CRAF-LgBiT, each 0.25 part) and 9.5 parts

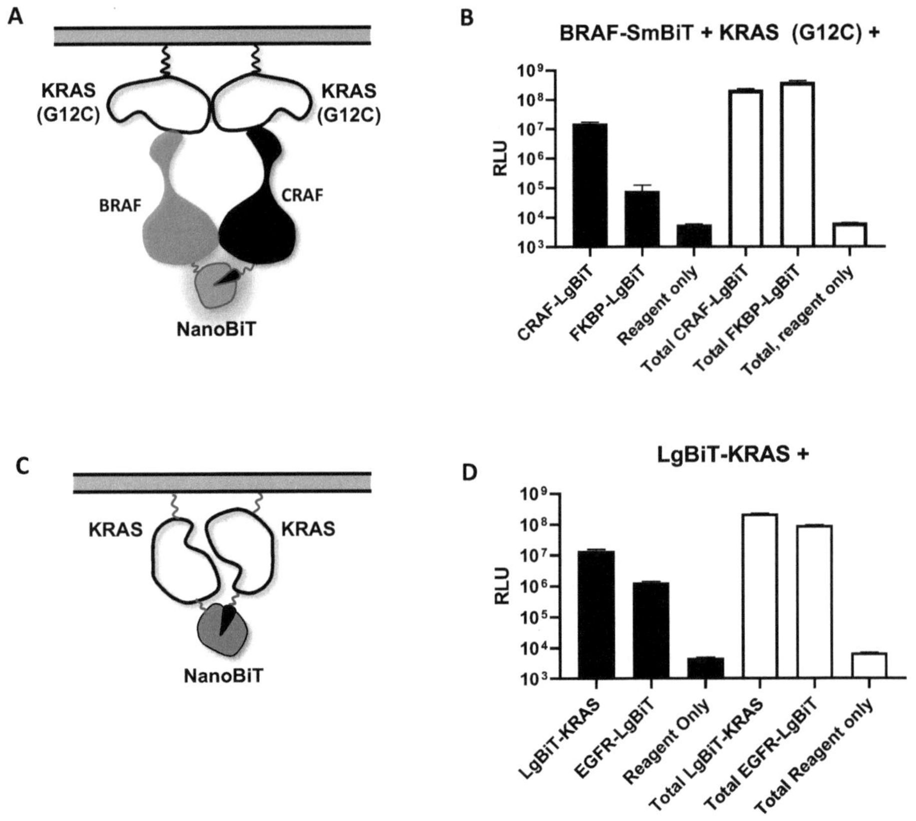


**Fig. 1** Conditional luminescence at complexes of RAFs and/or KRAS. (**a**) Illustration of the luminescent complex comprised of dimers of KRAS and dimers of RAF, tagged with NanoBiT reporters. (**b**) Luminescence observed at BRAF/CRAF heterodimers in complex with KRAS (G12C) using NanoBiT detection. (**c**) Illustration of the luminescent complex comprised of dimers of KRAS, tagged with NanoBiT reporters. (**d**) Luminescence observed at dimers of KRAS using NanoBiT. To demonstrate that the expression levels of the NanoBiTs were approximately similar, cells were lysed, and HiBiT peptide reagent was added at a saturating dose

untagged KRAS (G12C) expression vector in Opti-MEM at a total DNA concentration of 10 μg/mL (*see* **Note 3**).

(ii) For RAF negative controls, combine by mass 0.25 part BRAF-SmBiT fusion vector +2 parts FKBP-LgBiT fusion vector and 7.75 parts untagged KRAS (G12C) expression vector in Opti-MEM at a total DNA concentration of 10 μg/mL.

(b) For RAS dimer analysis, using KRAS wild-type dimers as an example:

(i) KRAS dimer analysis: combine by mass 2 parts KRAS WT/NanoBiT fusion vectors (1 part each) and 8 parts Transfection Carrier DNA in Opti-MEM at a total DNA concentration of 10 μg/mL (*see* **Note 4**).

(ii) For KRAS negative controls, combine by mass 1 part SmBiT-KRAS WT fusion vector and 1 part EGFR-LgBiT, with 8 parts Transfection Carrier DNA in Opti-MEM at a total DNA concentration of 10 μg/mL.

4. Initiate transfection complex formation by adding 3 parts (by volume) FuGENE® HD for each part (by mass) of DNA. Mix by inversion five times. Allow 20–30 min for complexation to occur (*see* **Note 5**).
5. Combine 1 part (by volume) of transfection complex with 20 parts (by volume) HEK293 cell suspension at $2 \times 10^5$ cells/mL in a sterile conical tube. Mix gently by inversion five times.
6. For RAF experiments: dispense 100 μL/well of the cell suspension plus transfection complex into each well of a white, tissue culture-treated 96-well plate. This will yield 20,000 cells per well. Incubate at least 20 h to allow expression to occur.
7. For RAS experiments: dispense cells/lipid-DNA complex into a sterile tissue culture flask, and incubate at least 20 h to allow expression to occur. After expression is allowed to occur, harvest the transfected cell population (by trypsinization and centrifugation), and resuspend the cell pellet in pre-warmed (37 °C) assay medium at a density of $2 \times 10^5$ cells/mL. Seed 85 μL/well into each well of a white, NBS plate.
8. Allow plates to re-equilibrate to ambient temperature for approximately 15 min.
9. To detect specific luminescence from NanoBiT reporters in live cells, add 50 μL of 3× complete live-mode detection solution, and incubate at ambient temperature for 3 min to allow production of luminescence.
10. To detect total LgBiT expression from permeabilized cells, add 50 μL/well of 3× complete permeabilized-mode detection solution and 15 μL/well of 10× HiBiT peptide plus digitonin solution. Mix on an orbital shaker for 3 min at 300–600 RPM.
11. Measure total luminescence on the BRET compatible luminometer.

### 3.2 Quantifying Target Engagement at Select Protomers Within RAF and RAS Complexes

Once specific signal can be determined from RAF or RAS dimers using NanoBiT, the system can be used to quantify target engagement using NanoBRET tracers [5, 10]. The SmBiT(q) variant should be utilized instead of the standard SmBiT tag on one of the protomers for target engagement studies (*see* **Notes 6** and 7). This section provides a protocol for the use of NanoBRET tracers for measuring target engagement at RAF dimers in complex with RAS, or at homodimers of KRAS, using the cell transfection and

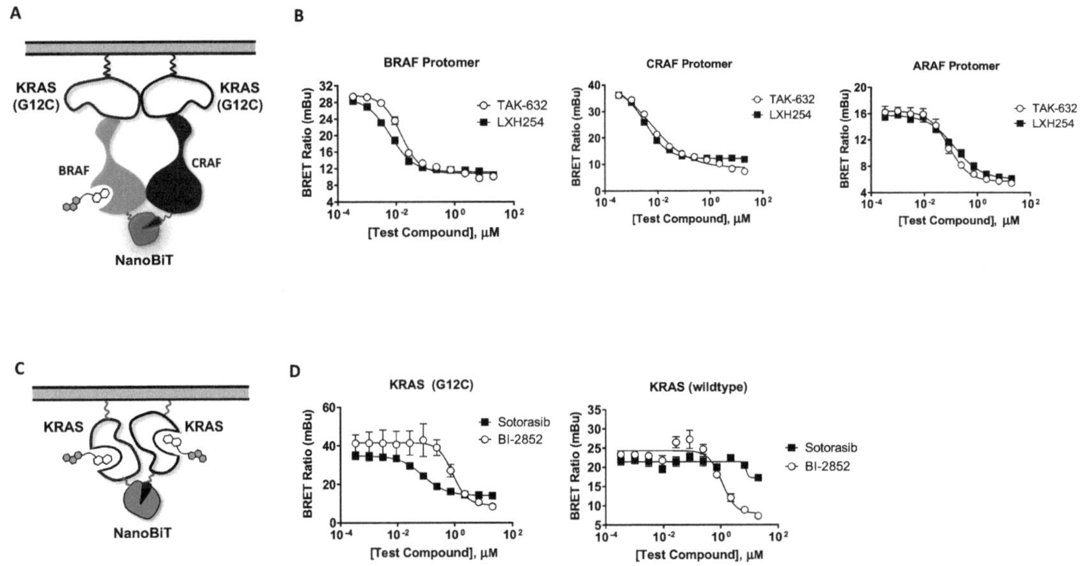


**Fig. 2** Conditional target engagement analysis at complexes of RAFs and/or KRAS. (**a**) Illustration of the BRET complex enabling target engagement analysis at BRAF protomers in complex with CRAF and KRAS (G12C). BRET is observed using a RAF kinase tracer. (**b**) BRET target engagement data observed at BRAF, CRAF, or ARAF dimers of BRAF in complex with KRAS (G12C). To query BRAF, the CRAF (A373F)-LgBiT and BRAF-SmBiT (q) were used. To query CRAF within BRAF/CRAF heterodimers, BRAF (A481F)-LgBiT and CRAF-SmBiT(q) are used. To query ARAF within BRAF/ARAF heterodimers, SmBiT(q)-ARAF and BRAF (A481)-LgBiT are used. (**c**) Illustration of the BRET complex comprised of NanoBiT-tagged dimers KRAS. BRET is observed using a KRAS tracer. (**d**) BRET target engagement data observed at dimers of KRAS wild type or KRAS (G12C)

seeding protocol provided in Subheading 3.1.6 or 3.1.7. Figure 2 provides example compound potency data collected using KRAS dimer or RAF dimer target engagement assays.

1. To each well of the 96-well plate, add 5 μL per well of the 20× NanoBRET tracer working solution for the protomer of interest (as described in Subheading 2.2 **items 5**, **8**, **10**, and **13**) (*see* **Notes 8**, **9**, and **10**).
2. Mix on an orbital shaker for 15 s at 900 RPM (*see* **Note 11**).
3. Add 10 μL per well of the 10× test compound working solution.
4. Mix on an orbital shaker for 15 s at 900 RPM.
5. Incubate the plate at 37 °C and 5% $CO_2$ for 2 h.
6. Allow plate to re-equilibrate to ambient temperature for approximately 15 min.
7. Add 50 μL of 3× complete live-mode detection solution, and incubate at ambient temperature for 3 min to allow production of luminescence.

8. Measure donor emission (450 nm) and acceptor emission (600 nm) using the BRET compatible luminometer.
9. Generate raw BRET ratio values by dividing the acceptor emission value by the donor emission value.
10. Convert raw BRET ratio values to milliBRET units (mBUs) by multiplying each raw BRET ratio value by 1000.
11. (Optional) If a background correction is desired, use the NanoBRET equation (Eq. 1):

$$\left(\frac{\text{Acceptor}_{\text{sample}}}{\text{Donor}_{\text{sample}}} - \frac{\text{Acceptor}_{\text{no tracer control}}}{\text{Donor}_{\text{no tracer control}}}\right) \times 1000 \qquad (1)$$

12. Determine the potency ($IC_{50}$) of your test compound by plotting the BRET ratio values (as mBUs) as a function of test compound concentration and fitting to the sigmoidal dose-response (variable slope) equation (Eq. 2):

$$\begin{aligned} Y &= \text{Bottom} \\ &+ (\text{Top} - \text{Bottom})/\left(1 + 10^{((\text{Log}(\text{IC}_{50}) - X)*\text{HillSlope})}\right) \end{aligned} \qquad (2)$$

where $X$ = test compound concentration and $Y$ = BRET ratio.

## 4 Notes

1. Prepare 3× complete detection solutions immediately prior to BRET measurement and use within 1 h. Omit the Extracellular NanoLuc® Inhibitor when preparing the permeabilized-mode reagent.
2. For preparation of test compound dilution series, we recommend first preparing a 1000× dilution series of the test compound in DMSO, after which 1 part (by volume) of this 1000× series can be diluted with 99 parts (by volume) assay medium to prepare a 10× test compound dilution series. This ensures that the same amount of DMSO is present in all assay wells.
3. For heterodimerization studies, the tag orientation should be evaluated empirically [13]. In a standardized workflow, this represents eight possible combinations (N- and C-terminal tagging, protein A, protein B, SmBiT and LgBiT in each variation). In some instances, only one tagging orientation is possible due to posttranslational modifications. For example, KRAS is known to tolerate only N-terminal reporter tags, owing to lipidation modifications that occur at the C-terminus. For the protein complexes in this study, these orientations have already been optimized [5, 10].

4. The NanoBRET RAS tracer is compatible with wild-type KRAS, all of the hotspot mutant alleles of KRAS (G12C, G12D, G12V, G13D, etc.), as well as wild-type HRAS. Similar target engagement protocols are used for these variants [5].
5. Add the FuGENE HD directly to the cell suspension.
6. To ensure that target engagement with NanoBRET tracers is only measured in intact cells, it is recommended that a version of NanoBiT is utilized that confers sensitivity to the Extracellular NanoLuc Inhibitor reagent. This reagent is designed to quench all target engagement signals in the extracellular debris, or in cells damaged during the cultivation step. To introduce the quenchable (q) variant of NanoBiT, the SmBiT peptide should be replaced with the SmBiT(q) peptide. The following amino acid sequence is used for SmBiT(q) in place of the standard SmBiT peptide: VTGYRLLEKIL.
7. To conditionally measure target engagement at BRAF within BRAF/CRAF heterodimers, the ATP pocket of CRAF must be blocked via mutation. To query BRAF, the CRAF (A373F)-LgBiT and BRAF-SmBiT(q) are used. To conditionally measure target engagement at CRAF within BRAF/CRAF heterodimers, BRAF (A481F)-LgBiT and CRAF-SmBiT(q) are used. To conditionally measure target engagement at ARAF within BRAF/ARAF heterodimers, SmBiT(q)-ARAF and BRAF (A481)-LgBiT are used.
8. The final concentration of NanoBRET™ tracers to use in the assay (and hence the concentration of the 100× stock solution) should be at or below the Kd-apparent between the tracer and the target protein. This analysis has been performed previously [5, 10].
9. Prepare 20× NanoBRET tracer reagents on the day of assay, and discard 20× NanoBRET tracer reagents after use, and do not store.
10. Dispense these solutions slowly owing to viscosity.
11. Mixing may vary between orbital shakers and may require optimization.

## References


1. Agianian B, Gavathiotis E (2018) Current insights of BRAF inhibitors in cancer. J Med Chem 61:5775–5793
2. Karoulia Z, Gavathiotis E, Poulikakos PI (2017) New perspectives for targeting RAF kinase in human cancer. Nat Rev Cancer 17: 676–691
3. Moore AR, Rosenberg SC, McCormick F et al (2020) RAS-targeted therapies: is the undruggable drugged? Nat Rev Drug Discov 19:533–552
4. Cook FA, Cook SJ (2021) Inhibition of RAF dimers: it takes two to tango. Biochem Soc Trans 49:237–251
5. Vasta JD, Peacock DM, Zheng Q et al (2022) KRAS is vulnerable to reversible switch-II pocket engagement in cells. Nat Chem Biol 18:596–604

6. Desideri E, Cavallo AL, Baccarini M (2015) Alike but different: RAF paralogs and their signaling outputs. Cell 161:967–970
7. Holderfield M, Deuker MM, McCormick F et al (2014) Targeting RAF kinases for cancer therapy: BRAF-mutated melanoma and beyond. Nat Rev Cancer 14:455–467
8. Jin T, Lavoie H, Sahmi M et al (2017) RAF inhibitors promote RAS-RAF interaction by allosterically disrupting RAF autoinhibition. Nat Commun 8:1211
9. Monaco KA, Delach S, Yuan J et al (2021) LXH254, a potent and selective ARAF-sparing inhibitor of BRAF and CRAF for the treatment of MAPK-driven tumors. Clin Cancer Res 27: 2061–2073
10. Vasta JD, Michaud A, Zimprich CA et al (2023) Protomer selectivity of type II RAF inhibitors within the RAS/RAF complex. Cell Chem Biol
11. Ostrem JM, Peters U, Sos ML et al (2013) K-Ras(G12C) inhibitors allosterically control GTP affinity and effector interactions. Nature 503:548–551
12. Vasta JD, Corona CR, Wilkinson J et al (2018) Quantitative, wide-spectrum kinase profiling in live cells for assessing the effect of cellular ATP on target engagement. Cell Chem Biol 25(206–214):e11
13. Dixon AS, Schwinn MK, Hall MP et al (2016) NanoLuc complementation reporter optimized for accurate measurement of protein interactions in cells. ACS Chem Biol 11:400–408

# Chapter 22

# FLAG-KRAS4B as a Model System for KRAS4B Proteoform and PTM Evaluation by Mass Spectrometry

**Robert A. D'Ippolito, Grace M. Scheidemantle, Brian P. Smith, Katie Powell, Scott Eury, Abigail Neish, Jennifer Mehalko, Lauren Beaumont, Nicole Fer, Vanessa Wall, William Burgan, Anna E. Maciag, Dominic Esposito, and Caroline J. DeHart**

## Abstract

Prior analysis of intact and modified protein forms (proteoforms) of KRAS4B isolated from cell lines and tumor samples by top-down mass spectrometry revealed the presence of novel posttranslational modifications (PTMs) and potential evidence of context-specific KRAS4B modifications. However, low endogenous proteoform signal resulted in ineffective characterization, making it difficult to visualize less abundant PTMs or perform follow-up PTM validation using standard proteomic workflows. The NCI RAS Initiative has developed a model system, whereby KRAS4B bearing an N-terminal FLAG tag can be stably expressed within a panel of cancer cell lines. Herein, we present a method for combining immunoprecipitation with complementary proteomic methods to directly analyze N-terminally FLAG-tagged KRAS4B proteoforms and PTMs. We provide detailed protocols for FLAG-KRAS4B purification, proteoform analysis by targeted top-down LC-MS/MS, and validation of abundant PTMs by bottom-up LC-MS/MS with example results.

**Key words** KRAS4B, Immunoprecipitation, Proteoform, Top-down mass spectrometry, Posttranslational modifications, Farnesylation, Hydroxyfarnesylation, Geranylgeranylation, FLAG tag

## 1 Introduction

Analysis of intact and modified protein forms (proteoforms) of KRAS4B by targeted top-down mass spectrometry (TDMS) enables the precise detection of KRAS4B posttranslational modifications (PTMs) within a variety of cell or cancer contexts [1]. Moreover, analysis of intact KRAS4B maintains connectivity between N-terminal oncogenic mutations (Gly 12, Gly 13, Gln 61) [2–5] and C-terminal PTMs within the hypervariable region (HVR) [5], particularly those resulting from prenylation of Cys 185 to promote membrane association [6–8]. This methodology provides a level of molecular detail that is simply unachievable by traditional

Andrew G. Stephen and Dominic Esposito (eds.), *KRAS: Methods and Protocols*, Methods in Molecular Biology, vol. 2797, https://doi.org/10.1007/978-1-0716-3822-4_22,

peptide-based proteomic methods. Previous studies have combined immunoprecipitation and TDMS (IP-TDMS) to analyze KRAS4B proteoforms within malignant cell lines and primary patient tumors, resulting in the discovery of 4 novel KRAS4B PTMs and the characterization of 28 KRAS4B proteoforms [1, 9]. However, attempts to validate these proteoforms by proteolytic digestion and subsequent peptide-based (bottom-up) LC-MS/MS workflows were not successful, likely due to the low endogenous abundance, high sequence identity with other RAS isoforms [5], difficulty maintaining labile PTMs throughout analysis, and challenging biochemical properties exhibited by endogenous KRAS4B. Efforts to characterize key endogenous KRAS4B PTMs within the C-terminal HVR proved especially challenging due to the high percentage of tryptic cleavage sites and the hydrophobicity of common prenylation PTMs [5–7].

To address this, the NCI RAS Initiative has developed N-terminal 3xFLAG-tagged KRAS4B expression constructs that can be stably and inducibly expressed within a range of cell line backgrounds. These FLAG-KRAS4B constructs were designed to incorporate a range of KRAS4B mutations, be expressed at high abundance, and exhibit endogenous C-terminal processing and PTMs [7]. These proteins are effectively purified by FLAG IP and have a final molecular weight highly amenable to top-down proteomic analysis (~25 kDa) [10]. Moreover, the selectivity of the IP enables the direct analysis of isolated FLAG-KRAS4B by proteolytic digestion and bottom-up LC-MS/MS, providing 100% sequence coverage, comprehensive PTM characterization, and the potential for direct comparison to complementary FLAG-KRAS4B proteoform data for evaluation of relative PTM abundance within a specific context or cell line background.

Herein, we present a detailed protocol for analysis of FLAG-KRAS4B by immunoprecipitation followed by TDMS or complementary bottom-up LC-MS/MS approaches (Fig. 1). We also present example data from FLAG-KRAS4B-G12D constructs isolated from four cell line backgrounds, showing the types of possible experiments and typical data achieved by each proteomic approach. Both methods present an opportunity for researchers to gain new insight into the identity and relative abundance of endogenous KRAS4B PTMs, particularly those associated with oncogenic mutations.

## 2 Materials

### *2.1 FLAG-KRAS4B Constructs, Stable Cell Lines, and Cell Pellets*

Lentivirus DNA and Virus Production

1. ViraPower™ Lentiviral Packaging Mix.
2. Lipofectamine™ 3000 Transfection Reagent.

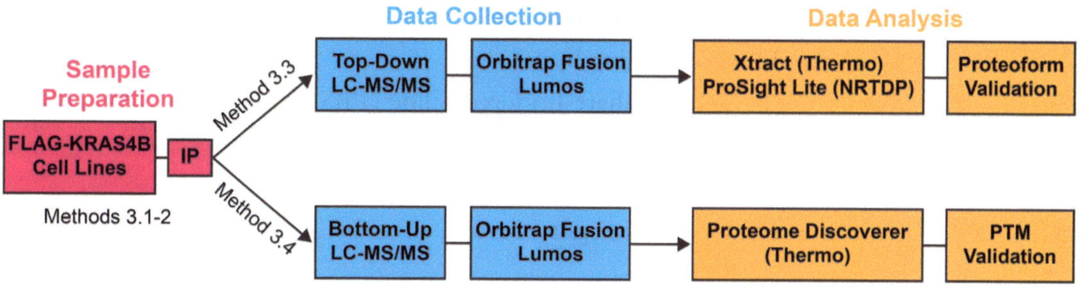


**Fig. 1** Method workflow for performing FLAG-KRAS4B IP-MS experiments. Following immunoprecipitation of FLAG-KRAS4B from a given cell line background, eluted proteins can be analyzed by top-down LC-MS/MS and bottom-up LC-MS/MS in parallel

3. DNA constructs containing the Tet-On 3G transactivator under a CMV promoter with blasticidin selection (e.g., Frederick National Laboratory R980-M38-658 [11]).
4. DNA constructs containing the Tet-inducible 3× FLAG-Hs. KRAS4B-G12D protein expressed under a TRE3G promoter with puromycin selection (e.g., Frederick National Laboratory R750-M17-663 and R728-M07-663 [11]).
5. Viral titering method (e.g., PCR assay in Support Protocol 3 from [12]).
6. HEK293, HeLa, Panc-1, and MiaPaca-2 cell lines.
7. Dulbecco's Modified Eagle Media (DMEM).
8. Fetal bovine serum (FBS).
9. Ham's F-12 Nutrient Mixture.
10. *Mycoplasma* contamination test (e.g., VenorGem Classic PCR test, Minerva Biolabs).
11. Cell line authentication assay (e.g., AmpFLSTR™ Identifiler™, Thermo Fisher Scientific).
12. Online reference database for cell line identities (e.g., www.cellosaurus.org).
13. 6-well plates.
14. Cell counter.
15. Polybrene transfection reagent.
16. Blasticidin.
17. Puromycin.
18. Cell freezing medium (e.g., Recovery™ Cell Culture Freezing Medium, Thermo Fisher Scientific).
19. Cell freezing apparatus.
20. Complete DMEM, containing 4.5 g/L D-glucose, 3.7 g/L sodium bicarbonate, 2 mM L-glutamine.

21. Tetracycline-free FBS.
22. Doxycycline hyclate.
23. 1× phosphate-buffered saline (PBS).
24. 0.25% trypsin-EDTA.
25. 50 mL conical tubes.
26. Liquid nitrogen.

### *2.2 Immunoprecipitation of FLAG-KRAS4B*

1. FLAG IP beads: ANTI-FLAG® M2 Affinity Gel (MilliporeSigma, cat# A2220-5ML).
2. Ice bucket or bath.
3. Water bath (30 °C).
4. Cold room or refrigerated case (4 °C).
5. 15 and 50 mL conical tubes.
6. 1.5 mL Protein LoBind tubes (Eppendorf, cat# 0030108442).
7. Shaking mixer with thermal control and a block to hold 1.5 mL microcentrifuge tubes.
8. Tube rotator.
9. Benzonase® nuclease (MilliporeSigma).
10. 1 M $MgCl_2$ in 18.2 Ω Milli-Q filtered water.
11. Lysis buffer: 50 mM Tris–HCl (pH 7.5), 150 mM NaCl, 2% NP-40 (IGEPAL CA-360), 2% Triton X-100 (electrophoresis grade), 1× HALT Protease and Phosphatase Inhibitor Cocktail (EDTA free, Thermo Fisher Scientific), 18.2 Ω Milli-Q filtered water.
12. Lysis buffer-NaCl: 50 mM Tris–HCl (pH 7.5), 2% NP-40 (IGEPAL CA-360), 2% Triton X-100, 1× HALT Protease and Phosphatase Inhibitor Cocktail (EDTA free), 18.2 Ω Milli-Q filtered water.
13. Wash buffer 1: 50 mM Tris–HCl (pH 7.5), 150 mM NaCl, 2% NP-40 (IGEPAL CA-360), 2% Triton X-100, 1× HALT Protease and Phosphatase Inhibitor Cocktail, 18.2 Ω Milli-Q filtered water.
14. Wash buffer 2: 50 mM Tris–HCl (pH 7.5), 150 mM NaCl, 18.2 Ω Milli-Q filtered water.
15. Elution buffer: 0.5% MS-grade trifluoracetic acid (MilliporeSigma) in LC-MS-grade water prepared immediately before use in a glass scintillation vial (*see* **Notes 1** and **2**).

### *2.3 Analysis of FLAG-KRAS4B by Targeted Top-Down LC-MS/MS*

1. 2 mg/mL bovine serum albumin (BSA).
2. Autosampler vials.
3. $C_4$ ZipTips (MilliporeSigma).

4. 1.5 mL Protein LoBind tubes.
5. LC-MS-grade acetonitrile, decanted into a glass scintillation vial.
6. 0.2% LC-MS-grade formic acid in LC-MS-grade water, prepared in a glass scintillation vial.
7. ZipTip elution buffer: 80% LC-MS-grade acetonitrile and 0.2% LC-MS-grade formic acid in LC-MS-grade water, prepared in a glass scintillation vial.
8. HPLC solvent A: 5% LC-MS-grade acetonitrile and 0.2% LC-MS-grade formic acid in LC-MS-grade water. Decant into a glass scintillation vial for use in sample preparation.
9. HPLC solvent B: 95% LC-MS-grade acetonitrile and 0.2% LC-MS-grade formic acid in LC-MS-grade water.
10. UltiMate 3000 ultrahigh-performance liquid chromatography system (Thermo Fisher Scientific).
11. Trap column: 150 μm ID × 6 cm L, packed in house with PLRP-S resin (5 μm particle size, 1000 Å pore size, Agilent Technologies).
12. Analytical column: 75 μm ID × 25 cm L, packed in house with PLRP-S resin (5 μm particle size, 1000 Å pore size, Agilent Technologies).
13. Commercially available components for column fabrication:
    (a) Molex 150 μm ID × 360 μm OD nanocapillary fused silica, DigiKey, cat# WM22006-ND.
    (b) Molex 75 μm ID × 360 μm OD nanocapillary fused silica, DigiKey, cat# WM19372-ND.
    (c) Materials and instructions to formulate and bake column frits, NextAdvance, cat# FRIT-KIT.
    (d) Pressure vessel for packing columns, NextAdvance, cat# PC77, with optional accessories for connecting pressure vessel to a compressed gas tank.
    (e) Ultra-High Purity 5.0 Grade Nitrogen Size 300 Cylinder, AirGas, cat# NI UHP300.
    (f) PLRP-S resin, Agilent Technologies, cat# PL1912–3802. Crack open column using a vice grip and adjustable wrench. Scrape PLRP-S resin into a scintillation vial. Wash and define resin in LC-MS-grade methanol in a ventilated fume hood, pouring supernatant to waste after resin has settled. Repeat two times. After the final wash, allow resin to air dry to completion in a ventilated fume hood. Store in a sealed scintillation vial in a desiccator.
14. Nanospray Flex™ ion source (Thermo Fisher Scientific).

15. Stainless steel emitters (Thermo Fisher Scientific, cat# ES542).
16. Extender line: 20 μm ID, 360 μm OD nanocapillary fused silica, cut to size (e.g., Molex P/N 1068150041).
17. Orbitrap Fusion Lumos mass spectrometer (Thermo Fisher Scientific).
18. Xcalibur Qual Browser with Xtract license (Thermo Fisher Scientific) or alternative software capable of monoisotopic intact mass determination from intact protein and fragment ion spectra.
19. ProSight Lite v1.4 (available for free download from http://prosightlite.northwestern.edu/) [13].

### 2.4 Analysis of FLAG-KRAS4B by Bottom-Up LC-MS/MS

1. 1.5 mL Protein LoBind tubes.
2. Shaking mixer with thermal control and a block to hold 1.5 mL microcentrifuge tubes.
3. ACS-grade acetone decanted into a clean glass bottle and stored at −20 °C until use.
4. 8 M urea, prepared fresh in LC-MS-grade water.
5. 1 M dithiothreitol (DTT), prepared in LC-MS-grade water, stored at −20 °C until use, and maintained on ice.
6. 100 mM ammonium bicarbonate (ABC), prepared fresh in LC-MS-grade water in a glass scintillation vial.
7. 100 mM DTT in 100 mM ABC, prepared in a glass scintillation vial.
8. 300 mM iodoacetamide (IAA) in 100 mM ABC, prepared in a 1.5 mL LoBind tube immediately prior to use.
9. Sequencing or LC-MS-grade proteases (e.g., trypsin or GluC).
10. $C_{18}$ columns for peptide concentration and desalting (e.g., Thermo Fisher Scientific, cat# 89873). Prepare all solutions with LC-MS-grade reagents in glass scintillation vials just prior to use for best results.
11. UltiMate 3000 ultrahigh-performance liquid chromatography system (Thermo Fisher Scientific).
12. Trap column: Acclaim™ PepMap™ 100 $C_{18}$ trap column 5 μm × 0.1 mm × 20 mm (Thermo Fisher Scientific).
13. Analytical column: Acclaim™ PepMap™ 100 $C_{18}$ analytical column 3 μm × 0.075 mm × 500 mm (Thermo Fisher Scientific), with included 20 μm ID extender line.
14. Nanospray Flex™ ion source (Thermo Fisher Scientific).
15. Stainless steel emitters (Thermo Fisher Scientific).
16. Orbitrap Fusion Lumos mass spectrometer (Thermo Fisher Scientific).

17. Proteome Discoverer v.2.4.1.15 (Thermo Fisher Scientific). Older or newer versions may also be used, as this workflow will be compatible with both.

## 3 Methods

### 3.1 FLAG-KRAS4B Constructs, Stable Cell Lines, and Cell Pellets

While alternative cell lines expressing N-terminally FLAG-tagged KRAS4B can be used for this experiment (e.g., as in [14]), the cell lines and cell pellets from which this protocol and example data were derived were generated as follows.

#### 3.1.1 Lentivirus DNA and Virus Production

1. Generate a VSV-G pseudotyped lentiviral DNA construct containing the Tet-On 3G transactivator expressed under a CMV promoter with blasticidin selection.
2. Generate a VSV-G pseudotyped lentiviral DNA construct containing the Tet-inducible 3× FLAG-Hs.KRAS4B-G12D protein expressed under a TRE3G promoter with puromycin selection.
3. Perform lentivirus production using the ViraPower™ Lentiviral Packaging Mix and Lipofectamine™ 3000 Transfection Reagent according to Invitrogen's application note "Improve lentiviral production using Lipofectamine 3000 reagent" (available here, https://assets.thermofisher.com/TFS-Assets/BID/Application-Notes/lentiviral-production-lipofectamine-reagent-app-note.pdf).
4. Determine viral titer for each lentiviral construct by quantitative PCR using Support Protocol 3 from [12].

#### 3.1.2 Cell Lines

1. Obtain HEK293, HeLa, Panc-1, and MiaPaca-2 cell lines from a reputable source.
2. Maintain HEK293, HeLa, and Panc-1 cell lines in DMEM supplemented with 10% FBS at 37 °C and 5% $CO_2$.
3. Maintain the MiaPaca-2 cell line in 10% DMEM supplemented with Ham's F-12 Nutrient Mixture and 10% FBS at 37 °C and 5% $CO_2$.
4. Verify that cell lines are not contaminated with *Mycoplasma* via PCR test.
5. Verify cell line identity and integrity by short tandem repeat (STR) profiling.

#### 3.1.3 Doxycycline-Inducible Cell Line Pool Generation

1. Seed cell lines into 6-well plates the day prior to transduction, at sufficient numbers to reach 30–50% confluency on the day of transduction.

2. Count one well to calculate the multiplicity of infection (MOI) for lentiviral transduction.
3. Reserve one well to evaluate cell death during the post-transduction selection phase.
4. Transduce cells in the remaining four wells with the Tet-On 3G transactivator lentiviral construct at an MOI of 0.05 (HEK293, HeLa) or 1 (Panc-1, MiaPaca-2). Remove media and replace with an equivalent amount containing 8 μg/mL polybrene and sufficient lentivirus to achieve the desired MOI. Incubate for 24 h at 37 °C and 5% $CO_2$.
5. Change media to remove residual lentivirus. Incubate for 24 h at 37 °C and 5% $CO_2$.
6. Add blasticidin to a final concentration of 8 μg/mL (HEK293), 4 μg/mL (HeLa), 15 μg/mL (Panc-1), or 8 μg/mL (MiaPaca-2) for antibiotic selection. Maintain antibiotic in cell media for all steps moving forward.
7. Repeat **steps 1–5** for second-stage transduction of the TRE3G-3xFLAG-KRAS4B-G12D lentiviral construct into the four cell lines.
8. Add puromycin to a final concentration of 2 μg/mL (HEK293), 1 μg/mL (HeLa), 6 μg/mL (Panc-1), or 1 μg/mL (MiaPaca-2) for antibiotic selection.
9. Perform antibiotic selection by maintaining both antibiotics in cell media for 2 weeks post-transduction.
10. Expand transduced cell lines and freeze stocks for future use.

#### *3.1.4 Cell Line Cultivation and Cell Pellet Generation*

1. Analyze at least $1 \times 10^8$ cells or greater from each cell population to ensure sufficient protein for proteomic analysis.
2. Expand all four stable cell lines (HEK293, HeLa, Panc-1, MiaPaca-2) in complete DMEM supplemented with tetracycline-free FBS at 37 °C and 5% $CO_2$.
3. 48 h prior to harvest, treat cells with 200 ng/mL doxycycline hyclate.
4. At 48 h posttreatment, remove media and wash cells with 1× PBS.
5. Dissociate cells with 0.25% trypsin-EDTA and transfer to a clean 50 mL conical tube.
6. Pellet cells by centrifugation (e.g., 5 min at 1500 rpm and 4 °C).
7. Wash cells with 1× PBS and pellet again by centrifugation.
8. Snap freeze cell pellets in liquid nitrogen and store at −80 °C before use.

### 3.2 Immunoprecipitation of FLAG-KRAS4B

Day 1:

1. Prepare autosampler vials for day 2. Add sufficient 2 mg/mL BSA to fill one vial per FLAG-KRAS4B sample. Incubate in a box with a lid at 4 °C until just prior to LC-MS/MS analysis (*see* **Note 3**).
2. Prepare lysis buffer and lysis buffer-NaCl. Add all components except HALT inhibitor cocktail to room temperature (RT) - $MQH_2O$ to facilitate incorporation of detergent. Chill all buffers on ice, and then add HALT inhibitor cocktail to all buffers just prior to use.
3. Prepare FLAG immunoprecipitation (IP) beads. Add 2.5 mL of chilled lysis buffer (with HALT inhibitor cocktail) to two 15 mL conical tubes (A, B; per IP). Swirl FLAG resin to resuspend, and then use a blunted pipet tip to transfer 300 μL 50% slurry (150 μL beads) to each 15 mL conical. Pipet gently to mix, and then allow resin to settle by gravity on ice during lysis.
4. Resuspend frozen cell pellets ($>1 \times 10^8$ cells recommended, prepared in a 50 mL conical tube) directly into 8 mL of chilled lysis buffer, pipetting gently until all cells are thawed.
5. Incubate lysate on ice for 20 min, pipetting to resuspend cells after 10 min. Do not vortex lysate.
6. Add 1 mL of 5 M NaCl to lysate to crack nuclei at a final concentration of 500 mM NaCl. Pipet to resuspend until genomic DNA becomes visible and lysate becomes viscous.
7. Add 9 mL of lysis buffer-NaCl to dilute lysate to a final concentration of 250 mM NaCl. Pipet to resuspend.
8. Add 18 μL 1 M $MgCl_2$ to lysate for a final concentration of 1 mM.
9. Add 2.5 μL benzonase nuclease. Pipet to resuspend.
10. Incubate lysate for 20 min at 30 °C in a water bath. Swirl to mix every 5 min.
11. Clarify lysate by centrifugation for 30 min at maximum speed and 4 °C.
12. Retain gel aliquots for subsequent Western blot (WB) analysis and store at −20 °C (*see* **Note 4**).
13. Remove supernatant from FLAG IP resin, leaving just enough liquid to cover the resin surface.
14. Add 9 mL of lysate to each 15 mL conical with FLAG IP resin. Add 5 mL of chilled lysis buffer to each conical to bring the final volume to 14 mL. Seal the conical tubes, and then wrap a strip of Parafilm around the caps to prevent leaks.
15. Incubate lysates with FLAG IP resin for 16–20 h at 4 °C with constant rotation.

Day 2:

1. Prepare wash buffer 1 and wash buffer 2. Add all components except HALT inhibitor cocktail to wash buffer 1 at room temperature (RT) to facilitate incorporation of detergent. Chill all buffers on ice, and then add HALT inhibitor cocktail to wash buffer 1 just prior to use.
2. Collect IP flowthrough by centrifugation for 10 min at 800 rpm and 4 °C.
3. Retain gel aliquots for subsequent WB analysis and store at −20 °C.
4. Remove supernatant from FLAG resin, leaving enough liquid to visibly cover the resin (*see* **Note 5**).
5. Add 5 mL of wash buffer 1, starting near the resin bed and drawing the pipet up the side of the conical tube to gently resuspend resin. Do not pipet to resuspend.
6. Collect wash flowthrough by centrifugation for 5 min at 800 rpm and 4 °C.
7. Retain gel aliquots for subsequent WB analysis and store at −20 °C.
8. Remove supernatant from FLAG resin, leaving enough liquid to visibly cover the resin.
9. Repeat **steps 5–8** two additional times.
10. Add 7.5 mL of wash buffer 2, starting near the resin bed and drawing the pipet up the side of the tube to gently resuspend resin. Do not pipet to resuspend.
11. Collect wash flowthrough by centrifugation for 5 min at 800 rpm and 4 °C.
12. Retain gel aliquots for subsequent WB analysis and store at −20 °C.
13. Remove supernatant from FLAG resin, leaving enough liquid to visibly cover the resin.
14. Repeat **steps 10–13** four additional times.
15. For the final wash, remove supernatant until ~1.5 mL of liquid remains. Use a blunted pipet tip to gently resuspend FLAG resin and transfer to a pre-chilled 1.5 mL Eppendorf LoBind tube. Perform this step for each IP, using a separate LoBind tube for each. Collect final wash flowthrough by centrifugation for 5 min at 800 rpm and RT. Allow tubes to sit on ice for ~5 min for FLAG resin to settle by gravity.
16. Remove as much supernatant as possible from FLAG resin, leaving the surface exposed. Add 500 μL of elution buffer, starting near the resin bed and drawing the pipet up the side of the tube to gently resuspend resin. Do not pipet to resuspend.

17. Incubate the resin in a shaking mixer for 20 min at 900 rpm and RT.
18. Collect eluted FLAG-KRAS4B (E1) by centrifugation for 5 min at 800 × *g* and RT. Allow tubes to sit on ice for ~5 min for FLAG resin to settle by gravity.
19. Retain gel aliquots for subsequent WB analysis and store at −20 °C.
20. Transfer supernatant to a pre-chilled 1.5 mL LoBind tube, taking care to not transfer any FLAG resin. Keep eluted protein on ice.
21. Add an additional 500 μL of elution buffer, starting near the resin bed and drawing the pipet up the side of the tube to gently resuspend resin. Add 100 μL of LC-MS-grade acetonitrile. Do not pipet to resuspend resin.
22. Incubate the resin in a shaking mixer for 20 min at 900 rpm and RT.
23. Collect eluted FLAG-KRAS4B (E2) by centrifugation for 5 min at 800 rpm and RT. Allow tubes to sit on ice for ~5 min for FLAG resin to settle by gravity.
24. Retain gel aliquots for subsequent WB analysis and store at −20 °C.
25. Transfer supernatant to a pre-chilled 1.5 mL LoBind tube, taking care to not transfer any FLAG resin. Keep eluted protein on ice.
26. Elutions (E1, E2) from both halves of the IP (A, B) can be combined for a single experiment (e.g., proteoform analysis by targeted top-down LC-MS/MS) or kept divided for parallel experiments (e.g., complementary PTM analyses by top-down and bottom-up LC-MS/MS).
27. Add 200 μL gel sample buffer to the remaining FLAG resin from each IP. Mix by centrifugation at 800 rpm for 5 min at RT. Store at −20 °C for optional analysis by WB or in-gel digestion and subsequent bottom-up LC-MS/MS (not covered by this protocol).

### 3.3 Analysis of FLAG-KRAS4B by Targeted Top-Down LC-MS/MS

#### 3.3.1 Concentration and Desalting

1. Prepare one 1.5 mL LoBind tube per sample by adding 10 μL of $C_4$ ZipTip elution buffer and pre-chilling on ice.
2. Prepare for the final sample dilution and vial preparation by pre-chilling HPLC solvent A on ice.
3. Perform concentration and desalting of eluted FLAG-KRAS4B protein by $C_4$ ZipTip. Several protocols are available, but the one observed to work best for this purpose entails the following:

(a) Apply a clean $C_4$ ZipTip to a p10 pipet.

(b) Activate resin by drawing 10 μL of LC-MS-grade acetonitrile through the ZipTip and dispensing to waste. Repeat five additional times.

(c) Equilibrate resin by drawing 10 μL of 0.2% formic acid in LC-MS-grade water and dispensing to waste. Repeat five additional times.

(d) Bind FLAG-KRAS4B to the resin by transferring the ZipTip to the end of a p200 tip, inserting the ZipTip into the IP elution, and pipetting 200 μL of elution slowly up and down for 30 times. Take extreme care to not allow any air or bubbles to pass through the resin (*see* **Note 6**).

(e) Return the ZipTip to the p10 pipet. Wash bound protein by drawing 10 μL of 0.2% formic acid in LC-MS-grade water through the ZipTip and dispensing to waste. Do not allow air or bubbles to pass through the resin during this step. Repeat nine additional times.

(f) Elute FLAG-KRAS4B into the prepared LoBind tube by pipetting the 10 μL of chilled ZipTip elution buffer up and down at least ten times, again taking extreme care to prevent air from passing through the resin.

4. Multiple IP elutions can be combined, or additional protein can be recovered from a given fraction by performing the ZipTip elution step within the same LoBind tube. As an example, two ZipTips from E1 and two ZipTips from E2 can be combined into a single sample by this method to boost final FLAG-KRAS4B proteoform signal.
5. Add 40 μL of ice-cold HPLC solvent A to the final ZipTip elution (1:5 final dilution), and pipet to resuspend. Keep the sample on ice.
6. Remove BSA solution from a prepared vial, and wash well with 200 μL of ice-cold HPLC solvent A, dispensing to waste. Repeat five times.
7. Transfer sample from **step 5** to washed autosampler vial and immediately proceed with LC-MS/MS analysis (*see* **Note 7**).

#### *3.3.2 LC-MS/MS Parameters*

1. Resolve FLAG-KRAS4B proteoforms by reverse-phase liquid chromatography (LC) prior to mass spectrometry (MS) analysis. The parameters provided herein are for an UltiMate 3000 UHPLC system coupled to an Orbitrap Fusion Lumos mass spectrometer (Thermo Fisher Scientific).
2. The nanocapillary trap and analytical columns (described in Methods) should be coupled to a vented tee setup, a 20 μm ID fused silica extender line, and a NanoFlex ion source equipped with a stainless steel emitter (Thermo Fisher

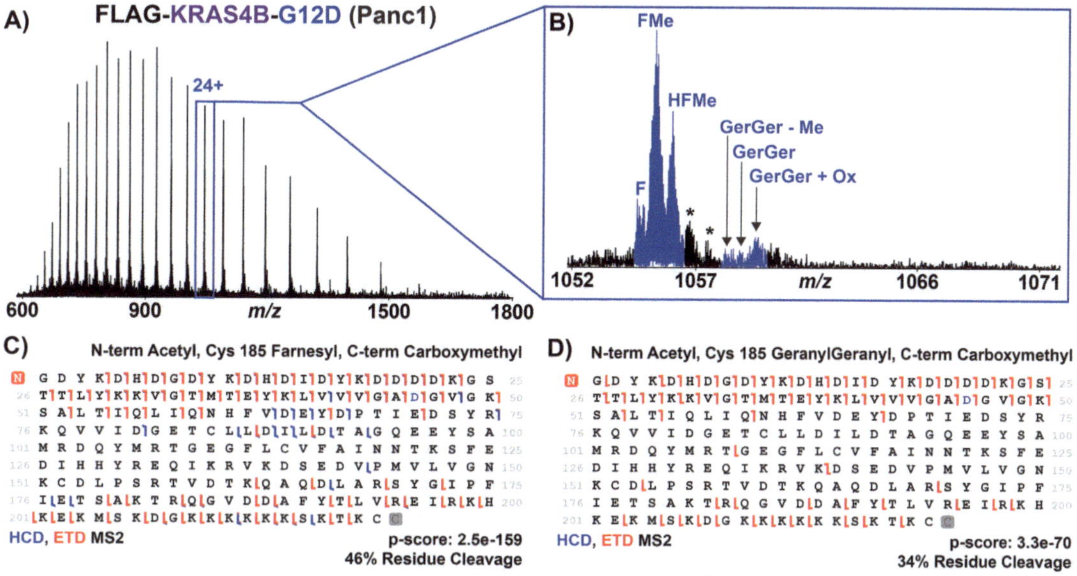


**Fig. 2** Representative data of FLAG-KRAS4B-G12D analyzed by IP-TDMS from Panc1 cells. Full MS1 scan (**a**) and zoom in on the 24+ charge state (**b**) are used to visualize abundant proteoforms, obtain intact masses, and select charge states for subsequent tMS2 experiments. Example fragment ion maps by tMS2 confirming the proteoforms of N-term acetyl, Cys185 farnesyl, and C-term carboxymethyl (**c**) and N-term acetyl, Cys185 geranylgeranyl, and C-term carboxymethyl (**d**)

Scientific). Alternative configurations (as in [1, 9, 15]) will also work well for LC separation.

3. The temperature of the LC should be maintained at 45 °C for the duration of the experiment.
4. Inject 2.5 μL of sample onto the trap column, and wash in HPLC solvent A for 10 min at a flow rate of 3 μL/min.
5. Elute FLAG-KRAS4B proteoforms onto the analytical column and into the mass spectrometer with the following gradient at a flow rate of 0.3 μL/min: 5% HPLC solvent B at 0 min, 25% HPLC solvent B at 1 min, 45% HPLC solvent B at 40 min, 90% HPLC solvent B from 41 to 43 min, and 5% HPLC solvent B from 45 to 60 min.
6. Obtain a full intact mass (MS1) spectrum of FLAG-KRAS4B to visualize the location of proteoforms within the chromatogram, determine their intact mass, and identify the most abundant charge states for subsequent fragmentation (MS2) and targeted proteoform characterization. An example full MS1 spectrum for FLAG-KRAS4B-G12D is shown in Fig. 2a.
7. Full MS1 scan method parameters:
   (a) Intact protein mode.
   (b) FT resolving power (r.p.): 120,000 (at 200 *m/z*).

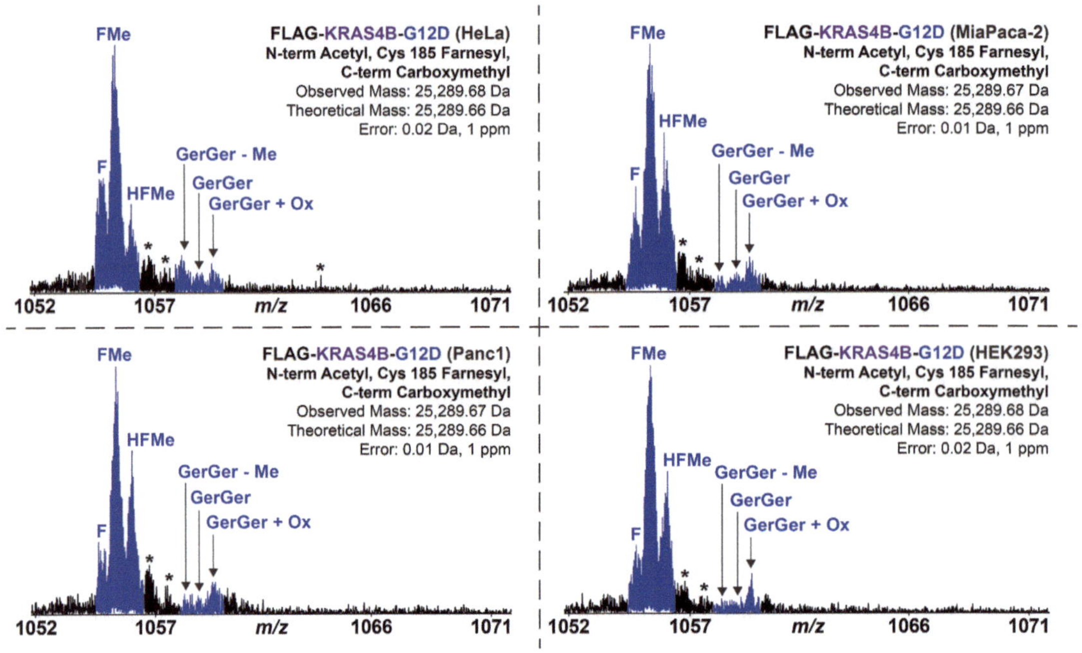


**Fig. 3** Comparison of the intact mass 24+ charge states from FLAG-KRAS4B-G12D by IP-TDMS. Data are from HeLa (top left), MiaPaca-2 (top right), Panc1 (bottom left), and HEK293 (bottom right). Abundant proteoforms are labeled

   (c) Scan window: 500–2000 *m/z*.
   (d) Average of 20 microscans.
   (e) Automatic gain control (AGC) target: 5E+05.
   (f) Maximum ion injection time: 50 ms.

8. Mass shifts due to FLAG-KRAS4B PTMs can be better visualized by obtaining a targeted MS1 (tMS1) spectrum focusing on individual protein charge states within a 10–50 *m/z* window. By focusing the ion current within this narrow isolation window, tMS1 can markedly increase sensitivity and reveal lower abundance proteoforms not detectable within the full MS1 scan. This can then enable the visualization and direct comparison of abundant proteoform populations between samples. An example tMS1 spectrum for FLAG-KRAS4B-G12D is shown in Fig. 2b, and an example comparison of abundant FLAG-KRAS4B-G12D proteoforms between four cell line backgrounds facilitated by tMS1 is shown in Fig. 3.
9. Targeted MS1 (tMS1) scan parameters:
   (a) Intact protein mode.
   (b) FT r.p.: 120,000 (at 200 *m/z*).
   (c) Isolation window: 10–50 *m/z* (e.g., *m/z* 1053–1063 for the 24+ charge state of FLAG-KRAS4B).

(d) 1 microscan

(e) AGC target: 1E+05.

(f) Maximum ion injection time: 50 ms.

10. FLAG-KRAS4B proteoform sequences can be confirmed and PTMs localized by applying targeted MS2 (tMS2) fragmentation to selected abundant charge states. Similarly to tMS1, by focusing an MS2 fragmentation method on one or more individual charge states rather than a data-dependent method, the signal from the resulting fragment ions can be dramatically improved, thus resulting in improved proteoform characterization. Moreover, different charge states can be targeted over successive retention time windows, allowing for the characterization of multiple proteoform populations within a single experiment. Example applications of two tMS2 methods to confirm the presence of farnesylation and geranylgeranylation on KRAS4B Cys 185 within the FLAG-KRAS4B-G12D construct are shown in Fig. 2c and Fig. 2d, respectively.

11. Targeted MS2 (tMS2) method parameters:

(a) Intact protein mode.

(b) Default charge state: targeted charge state (e.g., 24+) or average of targeted charge states (e.g., 36+ for targeting 37+, 36+, and 35+ charge states).

(c) FT r.p.: 60,000 (at 200 *m/z*).

(d) Isolation window: 3 *m/z*.

(e) Average of 4 microscans.

(f) AGC target: 1E+06.

(g) Maximum ion injection time: 800 ms.

(h) High-energy collisional dissociation (HCD) MS2 parameters: assisted, normalized collision energies of 21%, 22%, 23%.

(i) Electron transfer dissociation (ETD) MS2 parameters: 10 ms reaction time, 7E+05 reagent ion target, 200 ms max reagent injection time.

12. Additional MS parameters:

(a) Heated transfer capillary temperature: 275 °C.

(b) S-lens RF amplitude: 50%.

(c) 15 V in-source dissociation.

#### 3.3.3 Data Analysis

1. Prepare by installing ProSight Lite software [13], which is available for free download from http://prosightlite.northwestern.edu/.

2. To analyze FLAG-KRAS4B top-down LC-MS/MS data in ProSight Lite, the monoisotopic mass of the precursor (MS1 spectrum), the monoisotopic masses of the fragment ions (MS2 spectrum), and the protein sequence will all be needed.
3. Monoisotopic masses can be obtained by performing deconvolution via the Xtract algorithm in Xcalibur Qual Browser (as in [13, 15]) or using an alternative software of choice. The MS1 monoisotopic precursor mass can be typed directly into the Precursor dialog box in the Modify Experimental Data pane of ProSight Lite. MS2 fragment ion masses can be exported directly from Xcalibur (Xtract Result Pane #2, Monoisotopic Mass) into a spreadsheet program (e.g., Microsoft Excel). The values in the Mass column can then be copied and pasted directly into the Fragments dialog box in the Modify Experimental Data pane in ProSight Lite. The type of precursor (monoisotopic), mass mode (M-neutral), and fragmentation method (e.g., HCD) can be selected, and the acceptable mass error can be defined (recommended values are 10 ppm or below) within this pane. To combine MS2 data from two fragmentation types (e.g., HCD and ETD), paste both lists of monoisotopic fragment ion masses into the dialog box, and select a fragmentation mode incorporating all possible ion types for both methods (e.g., EThcD). Finally, select Save to retain data for graphical fragment ion map generation.
4. The FLAG-KRAS4B protein sequence can be obtained by downloading the endogenous KRAS4B protein sequence from https://www.uniprot.org/uniprot/P01116-2.fasta. The sequence can be directly pasted into the Sequence dialog box in the Modify Candidate Sequence pane in ProSight Lite. To reflect endogenous processing, the last three amino acid residues (-VIM) of the KRAS4B sequence will need to be deleted. To account for the FLAG tag, the sequence of the tag and linker will need to be added N-terminal to the initial methionine of KRAS4B. Any mutations (e.g., G12D) will need to be manually entered within the candidate sequence to account for their respective mass shifts. Fixed modifications (cysteine, methionine) should be left as no modification unless the protein has been treated with the appropriate reagents. Finally, select Save to retain this information for graphical fragment ion map generation.
5. The graphical fragment ion map (examples shown in Fig. 2c, d) will contain a series of flags, each representing a fragment ion mass that has been matched to the candidate proteoform sequence within the stated error tolerance. The color and shape of the flags are determined by convention. The quality of the overall match between the monoisotopic fragment ion masses and the candidate proteoform sequence is designated by

a p-score, with lower values representing better matches. The quality of the overall match between the observed precursor monoisotopic mass and the theoretical mass of the candidate proteoform is designated by a mass discrepancy and resulting ppm error.

6. PTMs can be manually assigned to given residues by selecting the residue, then selecting one of the available PTMs from the Modification pane on ProSight Lite. Typical FLAG-KRAS4B PTMs include cleavage of the initial methionine and acetylation of the N-terminus, along with C-terminal farnesylation, hydroxyfarnesylation, and geranylgeranylation (all including the mass of the accompanying carboxymethylation when selected in ProSight Lite). Custom mass shifts (e.g., farnesylation, 204.187801 Da, UniMod) can be added to residues using the Custom dialog box within the Modification pane of ProSight Lite.
7. For additional guidance on data analysis and interpretation, please refer to [13] for a detailed walkthrough of ProSight Lite and [15] for example applications to KRAS4B proteoform data.

### 3.4 Analysis of FLAG-KRAS4B by Bottom-Up LC-MS/MS

#### 3.4.1 Concentration and Desalting

1. Prepare for protein concentration by dividing each FLAG-KRAS4B IP elution (e.g., E1, E2) equally between two 1.5 mL LoBind tubes. Maintain tubes on ice. It is also recommended to store a sufficient volume of ACS-grade acetone at −20 °C for this purpose to facilitate protein precipitation (*see* **Note 8**).
2. Add 1.25 mL ice-cold ACS-grade acetone to each divided IP elution. Incubate at −20 °C for at least 4 h (or −80 °C for at least 1 h) to precipitate eluted proteins. This step can also be performed overnight to maximize recovery of lower-abundance FLAG-KRAS4B proteoforms.
3. Pellet precipitated proteins by centrifugation at 21,000 ×*g* for 10 min at 4 °C. This step can also be performed at RT with minimal protein loss.
4. Remove supernatant, taking care to not disturb protein pellet.
5. Add 1 mL of ice-cold acetone to the side of the tube. Do not resuspend the protein pellet.
6. Wash pellet by centrifugation at 21,000 ×*g* for 10 min at 4 °C.
7. Remove supernatant, taking care to not disturb protein pellet.
8. Air dry pellets for ~1 min at RT to allow residual acetone to evaporate. Proceed immediately to in-solution digestion.

*3.4.2 In-Solution Digestion*

1. Resuspend protein pellet in 20 μL of fresh 8 M urea.
2. Reduce proteins by adding 2 μL of 100 mM DTT (in fresh 100 mM ABC). Pipet to evenly mix and incubate for 20 min at RT.
3. Alkylate proteins by adding 3 μL of 300 mM fresh iodoacetamide (in freshly prepared 100 mM ABC). Pipet to evenly mix and incubate for 30 min in the dark at RT.
4. Quench alkylation reaction by adding 1 μL of 1 M DTT. Pipet to evenly mix.
5. Add sufficient 100 mM ammonium bicarbonate to dilute urea below 1 M final concentration (135 μL or greater).
6. Add 1 μg of trypsin (Promega) or GluC (Promega), each prepared according to the manufacturer's protocol. Pipet to evenly mix, and then incubate overnight (>16 h) in a shaking mixer at 900 rpm and 37 °C.
7. Peptides should be concentrated and desalted prior to LC-MS/MS analysis. While many options are available, Pierce $C_{18}$ spin columns (Thermo Fisher Scientific) are recommended for this specific experiment. The manufacturer's protocol may be followed, with the following modifications: elute peptides in 80% acetonitrile and 0.2% formic acid in LC-MS-grade water ($C_4$ ZipTip elution buffer), using twice the stated volume (i.e., 40 μL) for both elution steps. Dry eluted peptides completely via SpeedVac at RT and store at −20 °C until analysis by LC-MS/MS.
8. Immediately prior to LC-MS/MS analysis, resuspend peptides in 36 μL of pre-chilled solvent A. Centrifuge for 2 min at 21,000 ×*g* and RT to pellet any debris. Transfer supernatant to a clean autosampler vial. BSA conditioned and washed vials may also be used if the FLAG-KRAS4B C-terminal PTMs are of particular interest (*see* Subheading 3.3).

*3.4.3 LC-MS/MS Parameters*

1. Resolve FLAG-KRAS4B peptides by reverse-phase liquid chromatography (LC) prior to mass spectrometry (MS) analysis. The parameters provided herein are for an UltiMate 3000 UHPLC system coupled to an Orbitrap Fusion Lumos mass spectrometer (Thermo Fisher Scientific).
2. The nanocapillary trap and analytical columns (described in Methods) should be coupled to a trap-in-valve setup, a 20 μm ID fused silica extender line, and a NanoFlex ion source equipped with a stainless steel emitter (Thermo Fisher Scientific).
3. The temperature of the LC should be maintained at 60 °C for the duration of the experiment.

4. Inject 2 μL of sample onto the trap column and wash in HPLC solvent A for 10 min at a flow rate of 3 μL/min.
5. Elute FLAG-KRAS4B peptides onto the analytical column and into the mass spectrometer with the following gradient at a flow rate of 0.3 μL/min: 2% HPLC solvent B at 0 min, 25% HPLC solvent B at 90 min, 90% HPLC solvent B from 91 to 93 min, and 5% HPLC solvent B from 95 to 110 min. This gradient can be altered to elute more hydrophobic peptides or obtain better separation of peptide species of interest as needed.
6. Obtain precursor (MS1) spectra of eluted FLAG-KRAS4B peptides to determine their intact masses and identify the most abundant species within each scan for subsequent data-dependent MS2 fragmentation.
7. MS1 scan method parameters:
   (a) Peptide mode.
   (b) FT resolving power (r.p.): 120,000 (at 200 *m/z*).
   (c) Scan window: 375–1500 *m/z*.
   (d) 1 microscan
   (e) Automatic gain control (AGC) target: 4E+05.
   (f) Maximum ion injection time: 50 ms.
8. FLAG-KRAS4B peptide sequences can be confirmed and PTMs localized by applying multiple MS2 fragmentation methods to selected abundant peptides within the MS1 scan. Moreover, employing dynamic exclusion can ensure that more peptides can be fragmented within a sample, generating deeper protein and PTM coverage. Example MS2 fragmentation spectra of GluC peptides confirming the presence of and localizing abundant FLAG-KRAS4B-G12D C-terminal PTMs are shown in Fig. 4.
9. Data-dependent (top20) MS2 method parameters:
   (a) Peptide mode.
   (b) Default charge state: 2+.
   (c) Include charge states: 2–7+.
   (d) Dynamic exclusion: exclude after $n = 1$ times, for 60 s, with a high/low mass tolerance of 10 ppm, and all isotopes also excluded.
   (e) FT r.p.: 30,000 (at 200 *m/z*).
   (f) Isolation window: 1.6 *m/z*.
   (g) 1 microscan
   (h) AGC target: 5E+04.
   (i) Maximum ion injection time: 54 ms.

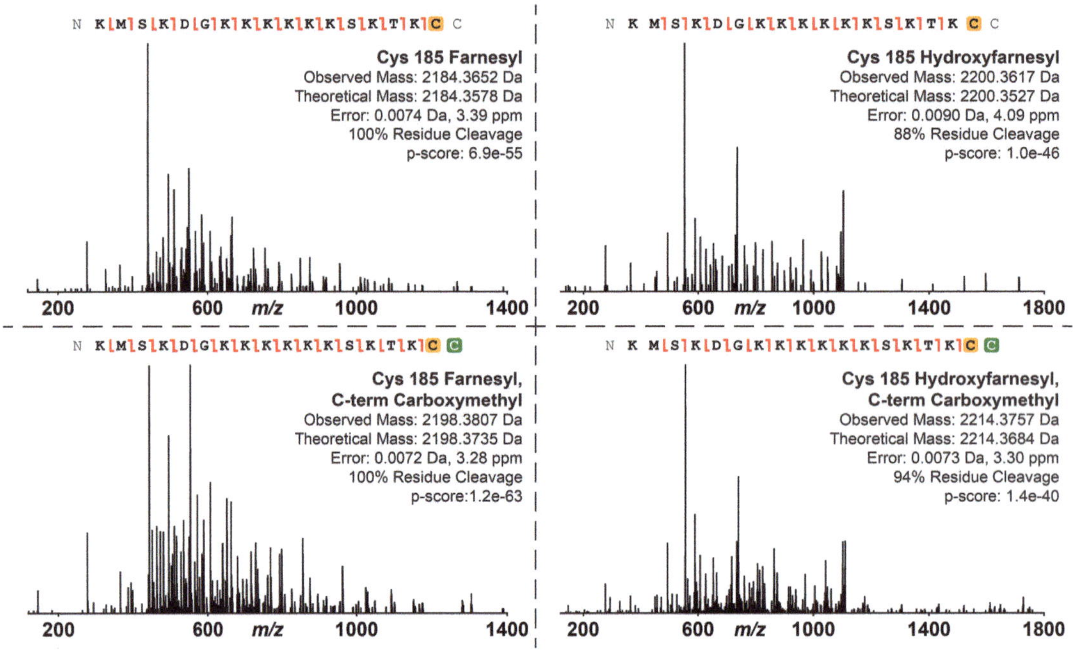


**Fig. 4** Bottom-up confirmation of abundant C-terminal modifications of FLAG-KRAS4B-G12D from Panc1 cells. Cys 185 farnesylation (top left), Cys 185 hydroxyfarnesylation (top right), Cys 185 farnesylation and C-term carboxymethylation (bottom left), and Cys 185 hydroxyfarnesylation and C-term carboxymethylation (bottom right) were confidently localized using ETD fragmentation

(j) High-energy collisional dissociation (HCD) MS2 parameters: 30% normalized collision energy.

(k) Electron transfer dissociation (ETD) MS2 parameters: calibrated charge-dependent.

(l) Electron transfer/high-energy collisional dissociation (EThcD) MS2 parameters: calibrated charge-dependent; ETD supplemental activation type of EThcD with 15% SA collision energy.

10. Additional MS parameters:

(a) Heated transfer capillary temperature: 275 °C.

(b) S-lens RF amplitude: 30%.

#### 3.4.4 Data Analysis

1. Raw data files will need to be processed and analyzed for protein identification and PTM localization. While many options are available, we have employed Proteome Discoverer v.2.4.1.15 (Thermo Fisher Scientific), for which we will provide the steps and parameters below.
2. Prepare by generating a .fasta database containing the FLAG-KRAS4B protein sequence. Include any relevant mutations and manual removal of the final 3 amino acid residues (-VIM) to

reflect endogenous processing. Add to the Proteome Discoverer database list via the Maintain Fasta Files pane within the Administration tab.

3. Prepare by generating a processing workflow incorporating the following parameters:
   (a) Spectrum files RC node: select the custom FLAG-KRAS4B database and additional relevant databases (e.g., UniProt filtered and reviewed *Homo sapiens* (human) database, known-contaminants database), along with the relevant enzyme (e.g., GluC).
   (b) Spectrum selector node: select Use MS1 Precursor, with a mass range of 150 Da to 15,000 Da, and 1 minimum peak. Select Is FTMS for the mass analyzer, with the relevant MS2 fragmentation type (e.g., ETD) and a signal-to-noise ratio threshold of 1.5. Select FTMS, MS2, the relevant fragmentation method, positive polarity, and the respective MS1 or MS2 FT resolving power for unrecognized replacements.
   (c) SEQUEST HT node: select the custom FLAG-KRAS4B database and additional relevant databases (e.g., UniProt filtered and reviewed *Homo sapiens* (human) database, known-contaminants database), along with the relevant enzyme and fragment ion types (e.g., c/z for ETD MS2). Allow for 2 missed cleavages, a maximum peptide length of 144 residues, a precursor mass tolerance of 10 ppm, a fragment mass tolerance of 0.02 Da, and 3 modifications per peptide. Select a range of expected dynamic modifications, for example, oxidation of Met, acetylation of Lys/N-terminus, phosphorylation of Ser/Thr/Tyr, or expected modifications of KRAS4B Cys 185 (i.e., farnesylation, geranylgeranylation, and carboxymethylation).
   (d) MS Amanda 2.0 node: select the custom FLAG-KRAS4B database and additional relevant databases (e.g., UniProt filtered and reviewed *Homo sapiens* (human) database, known-contaminants database), along with the relevant enzyme and fragment ion types. Allow for 2 missed cleavages, a precursor mass tolerance of 5 ppm, a fragment mass tolerance of 0.02 Da, and 3 modifications per peptide. Select a range of expected dynamic modifications.
   (e) Percolator node: select a concatenated target/decoy strategy, validation based on q-value, a maximum delta Cn of 0.05, a target strict FDR of 0.01, and a target relaxed FDR of 0.05.
   (f) Save the workflow in the Common folder for ease of future access.

4. Apply the processing workflow to search input data files, selecting the preferred option for consensus workflow for data processing, result file generation, and protein annotation.
5. Open the Analysis Results pane to view search results, including peptide coverage of FLAG-KRAS4B and successfully localized PTMs.
6. Scan numbers for individual peptide spectral matches can be located in the PSMs tab and used to locate the MS2 fragmentation spectrum of interest within the raw data file. The monoisotopic masses of these fragment ions can be used to generate a high-quality peptide fragmentation map in ProSight Lite, as described above. This step is particularly useful for the confirmation and localization of KRAS4B Cys 185 PTMs (Fig. 4), as these modifications tend to violate the one-PTM-per-residue rule incorporated within the SEQUEST and MS Amanda search algorithms. Confident assignment of MS2 fragment ions to this C-terminal region also enables close analysis of which PTMs may or may not be present within the hypervariable region (HVR) and direct correlation with complementary proteoform data obtained by targeted top-down LC-MS/MS analysis.

## 4 Notes

1. Use only LC-MS- or Optima-grade reagents for the IP elution buffer and all subsequent steps. Prepare all solutions in glass scintillation vials and do not store reagents in plastic. This will minimize the leaching of polymers into the sample and subsequent interference with FLAG-KRAS4B proteoform detection.
2. Use only fresh TFA when preparing the IP elution buffer, and perform this step immediately before use. Single-use glass ampules of concentrated TFA are recommended. Do not store concentrated TFA in plastic tubes or bottles.
3. Coat all autosampler vials in 2 mg/mL BSA prior to use. Overnight is preferable, but a minimum of 4 h is recommended. This will prevent loss of hydrophobic FLAG-KRAS4B proteoforms due to adhering to the glass surface of the vial.
4. It is possible that the FLAG IP may fail. Should this occur, utilize reserved gel aliquots from the FLAG IP and perform a Western blot as described in [15], which provides a detailed explanation of KRAS4B Western blot strategies and interpretation. An αFLAG and/or α-RAS antibody may be used, with the expected molecular weight at ~25 kDa.

5. Keep the FLAG resin covered by liquid at all times during the IP to minimize oxidization of FLAG-KRAS4B proteoforms from exposure to air. Oxidization will be prominently visible by top-down LC-MS/MS analysis, particularly when isolating individual charge states, and can impair proteoform characterization by targeted MS2 fragmentation.
6. Take great care to not allow air through the $C_4$ ZipTip resin at any point after protein has been bound. This will help to further minimize oxidation of FLAG-KRAS4B proteoforms.
7. Endogenous KRAS4B proteoforms (including FLAG-KRAS4B) degrade rapidly in HPLC solvent A, resulting in the loss of key proteoform species of interest and increased difficulty in proteoform characterization by targeted MS1/2. It is therefore highly recommended that top-down LC-MS/MS analysis be performed immediately after the IP elutions have been collected to preserve FLAG-KRAS4B proteoform quality and maximize available proteoform signal. If that is not feasible, samples should be stored in pre-chilled, preconditioned, and prewashed vials on ice at 4 °C until they can be run. Do not freeze samples, as they will not recover after thawing.
8. ACS-grade acetone should be decanted into a clean glass bottle and stored at −20 °C prior to use. This will facilitate protein precipitation and reduce wait times due to cooling requirements. Acetone waste is hazardous and should be retained in a dedicated container. Store and dispose of the acetone waste container in accordance with local hazardous waste handling policies.

## Acknowledgments

This project has been funded in full or in part with federal funds from the National Cancer Institute, National Institutes of Health, under contract no. 75N91019D00024. The content of this publication does not necessarily reflect the views or policies of the Department of Health and Human Services nor does mention of trade names, commercial products, or organizations imply endorsement by the US government.

### References

1. Ntai I, Fornelli L, DeHart CJ, Hutton JE, Doubleday PF, LeDuc RD, van Nispen AJ, Fellers RT, Whiteley G, Boja ES, Rodriguez H, Kelleher NL (2018) Precise characterization of KRAS4b proteoforms in human colorectal cells and tumors reveals mutation/modification cross-talk. Proc Natl Acad Sci USA 115(16):4140–4145. https://doi.org/10.1073/pnas.1716122115
2. Prior IA, Hood FE, Hartley JL (2020) The frequency of Ras mutations in cancer. Cancer

Res 80(14):2969–2974. https://doi.org/10.1158/0008-5472.CAN-19-3682

3. Moore AR, Rosenberg SC, McCormick F, Malek S (2020) RAS-targeted therapies: is the undruggable drugged? Nat Rev Drug Discov 19:533–552. https://doi.org/10.1038/s41573-020-0068-6
4. Hobbs GA, Der CJ, Rossman KL (2016) RAS isoforms and mutations in cancer at a glance. J Cell Sci 129(7):1287–1292. https://doi.org/10.1242/jcs.182873
5. Schubbert S, Shannon K, Bollag G (2007) Hyperactive RAS in developmental disorders and cancer. Nat Rev Cancer 7(4):295–308. https://doi.org/10.1038/nrc2109
6. Cox AD, Der CJ, Philips MR (2015) Targeting RAS membrane association: back to the future for anti-RAS drug discovery? Clin Cancer Res 21(8):1819–1827. https://doi.org/10.1158/1078-0432.Ccr-14-3214
7. Wright LP, Philips MR (2006) Thematic review series: lipid posttranslational modifications. CAAX modification and membrane targeting of Ras. J Lipid Res 47(5):883–891. https://doi.org/10.1194/jlr.R600004-JLR200
8. Willumsen BM, Christensen A, Hubbert NL, Papageorge AG, Lowy DR (1984) The p21 ras C-terminus is required for transformation and membrane association. Nature 310(5978):583–586. https://doi.org/10.1038/310583a0
9. Adams LM, DeHart CJ, Drown BS, Anderson LC, Bocik W, Boja ES, Hiltke TM, Hendrickson CL, Rodriguez H, Caldwell M, Vafabakhsh R, Kelleher NL (2023) Mapping the KRAS proteoform landscape in colorectal cancer identifies truncated KRAS4B that decreases MAPK signaling. J Biol Chem 299(1):102768. https://doi.org/10.1016/j.jbc.2022.102768
10. Compton PD, Zamdborg, Thomas PM, Kelleher NL (2011) On the scalability and requirements of whole protein mass spectrometry. Anal Chem 83(17):6868–6874. https://doi.org/10.1021/ac2010795
11. Hartley JL, Temple GF, Brasch MA (2000) DNA cloning using in vitro site-specific recombination. Genome Res 10(11):1788–1795. https://doi.org/10.1101/gr.143000
12. Salmon P, Trono D (2007) Production and titration of lentiviral vectors. Curr Protoc Hum Genet 54:12.10.1–12.10.24. https://doi.org/10.1002/0471142905.hg1210s54
13. DeHart CJ, Fellers RT, Fornelli L, Kelleher NL, Thomas PM (2017) Bioinformatics analysis of top-down mass spectrometry data with ProSight Lite. Methods Mol Biol 1558:381–394. https://doi.org/10.1007/978-1-4939-6783-4_18
14. Miller SP, Maio G, Zhang X, Badillo Soto FS, Zhu J, Ramirez SZ, Lin H (2022) A proteomic approach identifies isoform-specific and nucleotide-dependent RAS interactions. Mol Cell Proteomics 21(8):100268. https://doi.org/10.1016/j.mcpro.2022.100268
15. Adams LM, DeHart CJ, Kelleher NL (2021) Precise characterization of KRAS4B proteoforms by combining immunoprecipitation with top-down mass spectrometry. Methods Mol Biol 2262:47–64. https://doi.org/10.1007/978-1-0716-1190-6_3

# Chapter 23

# Creation of an Isogenic H/N/KRAS-Less Mouse Embryonic Fibroblast Cell Line Panel Derived from a Size-Sorted Diploid Clone

**William Burgan and Nicole Fer**

## Abstract

Cell line panels have proven to be an invaluable tool for investigators researching a range of topics from drug mechanism or drug sensitivity studies to disease-specific etiology. The cell lines used in these panels may range from heterogeneous tumor populations grown from primary tumor isolations to genetically engineered clonal cell lines which express specific gene isoforms. Mouse embryonic fibroblast (MEF) cells are a commonly used cell line for biological research due to their accessibility and ease of genetic manipulation. This chapter will describe the process of creating a size-sorted diploid (SSDC) clonal cell panel expressing specific RAS isoforms from a previously engineered RAS-less MEF cell line pool.

**Key words** Cell line panel, Clone, Size-sorted diploid clone (SSDC), Lentiviral vector, Single cell cloning, Mouse embryonic fibroblast, RAS-less MEF

## 1 Introduction

Cell line panels used in research are broadly available from several sources including cell line repositories and commercial biopharmaceutical companies [1–3]. The National Cancer Institute and its collaborators have a well-established record of creating, using, and distributing cell line panels ranging from the NCI-60 cell line panel produced over two decades ago to the recent transition to a patient-derived xenografts (PDX) panel which were first released in 2016 [4, 5]. In 2013 the RAS Initiative was created to stimulate the research community to apply new techniques and technology in a renewed effort to pursue new treatment options for cancers caused by mutations in the Ras gene. Shortly after the formation of the

The original version of the chapter has been revised. A correction to this chapter can be found at https://doi.org/10.1007/978-1-0716-3822-4_26

Andrew G. Stephen and Dominic Esposito (eds.), *KRAS: Methods and Protocols*, Methods in Molecular Biology, vol. 2797, https://doi.org/10.1007/978-1-0716-3822-4_23,

NCI RAS Initiative, a H/N/KRAS-less mouse embryonic fibroblast (MEF) cell line panel was generated, validated, and shared internationally to aid in this research effort. While this panel has been beneficial to the RAS research community, an inherent variability, due in part to the ploidy of the parental pool from which the panel was derived, complicated comparisons between isolates and alleles. This impact has been minimized by utilizing a diploid clone derived from the DU1473 MEF; $Hras^{-/-}$; $Nras^{-/-}$; $Kras^{lox/lox}$; $RERT^{ert/ert}$ -parental cell line [6, 7]. A diploid cell population was sorted and cloned, and clones were cultured continuously for up to 20 passages to identify those that retained a diploid signature. These clones were further screened for p53 mutations and mouse exome sequenced to determine which final clone was selected to develop the improved cell line panel. The selected clone was treated with 600 nM 4-hydroxytamoxifen (4-OHT) to activate the Cre-recombinase which excised a portion of the endogenous KRAS located between two loxP sites effectively arresting cells in a senescent G1 state. A proliferation assay of the tamoxifen-treated cells showed no proliferation which is a telltale sign of KRAS removal, and the cells were then transduced with a third-generation VSV-G pseudotyped lentiviral vector which delivered different versions of the human Ras transgene as well as an antibiotic selection marker. This transduction restored the proliferative ability of the cells allowing for expansion and selection of the transduced pool. The pools were single cell cloned, and subsequent clones underwent a systematic, rigorous quality control evaluation culminating with another round of exome sequencing before isoform/allele clones were selected to be part of the MEF 2.0 panel. Clones contained in this new panel were confirmed to be diploid, had fewer integrations than the initial panel, and performed similarly on proliferation assays as well as epidermal growth factor stimulation signaling assays. Quality control testing will be discussed in further detail in a later chapter (Xu and Powell).

## 2 Materials

### 2.1 General Cell Culture

1. Cell line DU1473 mouse embryonic fibroblast (MEF); $Hras^{-/-}$; $Nras^{-/-}$; $Kras^{lox/lox}$; $RERT^{ert/ert}$.
2. Tissue culture-treated plates and flasks.
3. Complete MEF media: Dulbecco's Minimum Essential Media (DMEM) high-glucose media +50 mL fetal bovine serum (FBS).
4. 1× phosphate-buffered saline (1× PBS) pH 7.4.
5. 0.25% trypsin-EDTA.
6. Trypan blue 0.4%.
7. Bio-Rad TC20 and counting slides.
8. Recovery™ cell culture freeze media (Thermo Fisher Scientific).

9. Mr. Frosty controlled rate freeze container.
10. Vybrant DyeCycle Green (Thermo Fisher Scientific).
11. Calcein-AM viable cell stain (*see* **Note 1**).
12. Greiner 384-well tissue culture-treated plates—black wall, clear bottom.
13. FxCycle™ PI/RNase Solution (Thermo Fisher Scientific).

### 2.2 Endogenous KRAS Removal and Generation of RAS Expression Cell Lines Via Lentiviral Transduction

1. 600 nM tamoxifen (4-hydroxytamoxifen) (*see* **Note 2**).
2. Incucyte imaging system with compatible 96-well plate (Sartorius).
3. Third-generation, VSV-pseudotyped lentiviral DNA expressing Ras transgene.
4. Tamoxifen-treated size-sorted diploid clonal MEF cell line.
5. Transduction media: complete media, 8 μg/mL hexadimethrine bromide (Polybrene).
6. Blasticidin: 4 μg/mL final concentration.

## 3 Methods

All manipulations performed with cell lines and lentivirus should be performed in a class 2 biosafety cabinet (BSC) using appropriate personal protective equipment (gloves, lab coat/gown, and safety glasses). All materials used for tissue culture should be sterile, and flasks and plates should be tissue culture treated. Institutional-specific requirements regarding safety and proper waste handling should be strictly adhered to. General cell culture procedures outlined in Subheading 3.1 will be followed while handling cells unless otherwise noted.

### 3.1 General Cell Culture Techniques

#### 3.1.1 Thawing a Viable Frozen Cell Line

1. Turn on water bath, set at 37 °C, and add an appropriate volume of pre-warmed complete MEF media to an appropriately sized flask for the cell number to be seeded. Generally, one million cells are added to a T75. Keep flask in 37 °C/5% $CO_2$ incubator until vial is thawed.
2. Retrieve viable frozen vial of cells from the liquid nitrogen freezer and wrap the top of the vial with parafilm to seal the area where the cap and vial meet, thus preventing this seam from encountering water bath water.
3. Without submerging the vial completely, gently swirl the vial in the water bath for approximately 1 to 2 min. Check the vial every 15 s, and remove the vial when a small pellet (pea-sized) of ice is present in the vial.

4. Dry the vial with a Kim-wipe, and decontaminate the outside using a disinfectant such as a quaternary ammonium chloride detergent disinfectant (or similar) prior to removing the parafilm.
5. Place the vial in the BSC and unscrew the vial lid. Gently pipet the vial contents up and down 2–3 times with a 2 mL pipet before transferring all the contents to the pre-warmed flask described in **step 1** of this procedure. Gently rock the flask to disperse the media and cells (*see* **Note 3**).
6. Return flask to the incubator for roughly 6–18 h to allow the cells to attach.
7. After the incubation in the previous step, pre-warm an appropriate volume of media for at least 30 min in the incubator before performing a media exchange on the newly attached cells.

*3.1.2 Culture and Expansion of Cell Lines*

1. Monitor cells daily for general condition and level of confluency. Media should be changed every 2–3 days depending on cell density (*see* **Note 4**).
2. When flask has reached 80–90% confluency, cells should be passaged to a new flask or plate.
3. Remove and dispose of the media in the 80–90% confluent flask.
4. Gently add an appropriate volume of 1× PBS to the flask, and gently rock the flask to wash any remaining medium off the cell monolayer.
5. Remove and dispose of the PBS wash.
6. Add an appropriate volume of 0.25% trypsin-EDTA to the flask, and gently rock to fully cover the cell monolayer.
7. Return the flask to the incubator, and incubate for 5–15 min until cells have lifted off from the surface of the flask on their own or after a gentle tap on the side of the flask.
8. Once cells are free floating, neutralize the trypsin by adding complete media at a minimum of two times the amount of the original amount of trypsin added (e.g., 1 mL trypsin then 2 mL complete media).
9. Remove 20 μL of single cell suspension, and add to a microcentrifuge tube containing 20 μL 0.4% trypan blue. Incubate at room temp for 3–5 min.
10. Remove 10 μL of cell/trypan blue mixture and load into counting slide for Bio-Rad TC20 or equivalent cell counter to determine cell viability and number of viable cells.

11. If passaging the cells, then transfer required number of viable cells to new cell culture-treated vessel pre-warmed with fresh complete media.
12. Repeat the above steps to carry the cells in culture.

#### *3.1.3 Generation of a Viable Cell Line Freeze*

1. Transfer cells to freeze to new tube and pellet cells by centrifuging at 800 rpm for 5 min at 4 °C.
2. Discard the supernatant, gently resuspend the cell pellet in Recovery™ cell culture freezing media, and aliquot 0.5 to 1 mL to 1.8 mL Nunc Cryovials (*see* **Note 5**).
3. Place aliquots into Mr. Frosty controlled rate freeze container, and place in −80 °C freezer for a minimum of 4 h before transferring to vapor-phase liquid nitrogen freezer for long-term storage (*see* **Note 6**).

### *3.2 Cell Cycle Determination and Diploid Cell Sorting Using Vybrant DyeCycle Green*

1. Start parental DU1473 cells in culture and harvest as described in Subheadings 3.1.1 and 3.1.2.
2. Transfer 1 mL of cell suspension at a concentration of 1e6 cells/mL to a new tube.
3. Add 2 μL of Vybrant DyeCycle Green stain to the tube and gently mix.
4. Incubate at 37 °C for 30 min, protected from light.
5. Analyze cells on a flow cytometer or fluorescent activated cell sorter using 488 nm excitation and green emission to determine cell ploidy.
6. Setup parameters to perform the cell sorting will vary depending on the instrument being used; however, sorting should be based on a known diploid control under sorting conditions that have been previously optimized to protect the viability of the mouse embryonic fibroblast cells.
7. Once a sufficient number of diploid cells have been accumulated, remove from the cell sorter, and immediately place into a treated cell culture vessel (suitable for the cell number obtained) containing pre-warmed complete media, and transfer to a 37 °C/5% $CO_2$ incubator to grow.
8. Culture the size-sorted diploid pool until sufficient cell number is achieved to viably freeze multiple vials (*see* Subheading 3.1.3), and proceed to single cell cloning which requires at least two million cells to perform (*see* **Note 7**).

### *3.3 Single Cell Cloning and Ploidy Stability Verification*

1. Pre-warm 20 mL each of serum free and complete media in 50 mL conical tubes to 37 °C in the incubator.
2. Resuspend the Calcein-AM stain in 25.1 μL DMSO to make 2 mM stock solution.

3. Make a working stock solution of the Calcein-AM stain by diluting 1:10 in PBS (200 μM).
4. Harvest cells when they are between 80 and 90% confluent.
5. Count cells with hemocytometer or automated cell counter.
6. Transfer 2e6 viable cells to new tube, centrifuge at 100 ×*g* for 5 min at 4 °C, and gently resuspend pellet in 2 mL serum-free medium (concentration should be 1e6 cells/mL) (*see* **Note 8**).
7. Add Calcein-AM stain from working stock to cells for final concentration of 0.25 μM (2.5 μL in 2 mL).
8. Resuspend cells that have settled by swirling the tube or gently tapping the bottom with your finger.
9. Protect cells from light and incubate at 37 °C for 30 min flicking the tube every 5 min to mix.
10. Wash cells with 10 mL of warm serum-free medium and pellet at 100 ×*g* for 5 min at 4 °C.
11. Resuspend pellet in complete media *without* selection antibiotic.
12. Count cells with hemocytometer or automated cell counter, and determine amount needed for 400 cells **total**.
13. Add 400 cells to 15 mL of complete media in a 50 mL conical and mix gently, but well.
14. Fill the outer ring of wells of the 384-well plate with 100 μL of media or PBS (*see* **Note 9**).
15. Pipet diluted cells into a sterile v-bottom reservoir.
16. Carefully add 25 μL of cells to each well of the 384-well plate (*see* **Note 10**).
17. Overseed wells B2, H12, and O23 to use as a focusing reference.
18. Centrifuge the plate at 100 ×*g* for 2 min to bring cells to the bottom of the wells for imaging.
19. Image the plate using a fluorescent microscope or automated plate reader with the ability to detect GFP expressing cells.
20. Identify the wells containing a single green cell and then return plate to a 37 °C/5% $CO_2$ incubator.
21. Monitor the identified wells for cell growth over the course of the next 2 weeks. Wells should be fed with 25 μL complete media at approximately 1 week post seeding.
22. When cells have reached 50% or more confluency, transfer (using procedure in Subheading 3.1.2) the colony to tissue culture plates of increasing surface area commensurate with the number of cells present (*see* **Note 11**).

23. When sufficient cell numbers are present to test for ploidy, transfer 1e6 cells to a 15 mL conical tube for ploidy testing (*see* **Note 12**).

### 3.4 Ploidy Verification and Stability Testing

1. Centrifuge cells at 800 rpm for 5 min at 4 °C to pellet.
2. Discard supernatant and resuspend pellet in ice-cold 70% ethanol. Cells can be further processed for ploidy determination or stored at −20 °C for testing later.
3. Incubate resuspended cells in freezer or ice bucket for 15 min.
4. Centrifuge cells at 800 rpm for 5 min at 4 °C to pellet.
5. Discard supernatant.
6. Wash pellet once by adding 5 mL 1× PBS and repeating **steps 4** and **5**.
7. Add 0.5 mL of FxCycle™ PI/RNase solution to the cell pellet.
8. Vortex gently to resuspend the pellet.
9. Incubate for 15–30 min at room temperate in the dark.
10. Analyze cells using a flow cytometer capable of 488 nm excitation, and collect data using 585/42 bandpass filter or equivalent.
11. Clones that are diploid should be viably frozen (*see* Subheading 3.1.3) then continuously passaged for at least 20 passages (*see* Subheading 3.1.2). Cells should be sampled every five passages to verify that they have remained diploid. If the ploidy has changed, then discard the clone.
12. Clones that remain diploid after 20 passages in culture are determined to be stable and should be submitted for quality control testing to verify at a minimum that p53 has remained unmodified (sequencing verification) and no deleterious mutations have developed (exome sequencing) (*see* **Note 13**).
13. The final clone selected for panel creation will be one that has remained diploid during the extended passaging tests, is p53 wild type, and grows well (doubling time between 18–36 h) and exome sequencing has not identified any deleterious mutations that affect the RAS/MAPK signaling members, tumor suppressors, or known oncogenes.

### 3.5 Endogenous KRAS Removal Via Tamoxifen Treatment

1. Obtain a vial of the chosen stable diploid clone and thaw according to the instructions in Subheading 3.1.1.
2. Following the second passage, seed a T150 flask with either $5 \times 10^5$ cells if tamoxifen will be added within 24 h or $1 \times 10^5$ cells if tamoxifen will be added within 72 h in 20 mL MEF complete media (*see* **Note 14**).

3. Twenty-four or 72 h after seeding, remove the media and replace with 20 mL of fresh pre-warmed MEF complete media containing 600 nM 4-hydroxytamoxifen (4-OHT).
4. Change the media with fresh pre-warmed MEF complete media containing 600 nM 4-OHT every 2–3 days for 11 days.
5. Around day 6, or when cells have reached 70–80% confluency, split the cells into additional T150 flasks using a 1:6 split.
6. On the 11th day, harvest and count the cells according to the procedure in Subheading 3.1.2. before proceeding to the verification of senescence using a proliferation assay (*see* **Note 15**).

### 3.6 Verification of Senescence Using an Incucyte® Imaging System

1. Determine concentration of tamoxifen-treated cells (cells/mL) following the instructions for harvest and counting in Subheading 3.1.2.
2. Generate cell dilutions at two or more of the following concentrations:
   (a) 6250 cells/mL which is equivalent to 1250 cells/200 μL/well (this is equivalent to approximately 300,000 cells if seeding in a T75)
   (b) 12,500 cells/mL which is equivalent to 2500 cells/200 μL/well (this is equivalent to approximately 590,000 cells if seeding in a T75)
   (c) 25,000 cells/mL which is equivalent of 5000 cells/200 μL/well (this is equivalent to approximately 1.2 million cells if seeding in a T75).
3. Obtain an Incucyte-compatible 96-well plate.
4. Seed a full column or row for each dilution chosen. Include a PBS ring around the wells that have been seeded to prevent evaporation.
5. Setup the Incucyte to scan the wells every 4–6 h for a period of 7 days.
6. Cells successfully treated with tamoxifen appear to be larger and more spread out relative to non-treated MEF cells. The cell images obtained over the course of 7 days should appear similar to the tamoxifen-treated cells in Fig. 1, and a review of the confluency chart should show a nearly horizontal line at the end of 7 days indicating no cell proliferation (*see* **Note 16**).

### 3.7 Generation of Specific RAS-Expressing Cell Lines Via Lentiviral Transduction

1. Turn on water bath and set temperature at 37 °C.
2. Pre-warm 14 mL of MEF complete media to 37 °C in the incubator.
3. Add 200,000 tamoxifen-treated MEF cells in 2 mL warm complete media to each well of a 6-well plate.

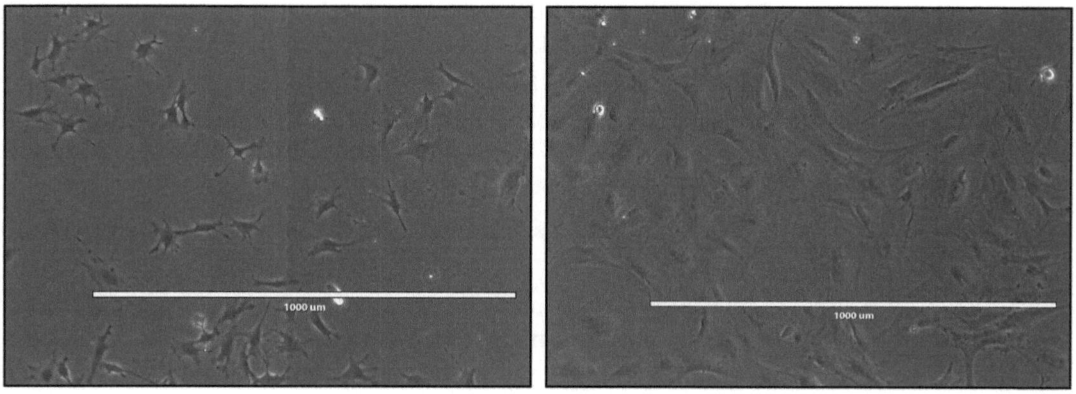


**Fig. 1** The image on the left shows mouse embryonic fibroblasts after attachment on day zero prior to addition of tamoxifen. The image on the right shows the mouse embryonic fibroblasts following 11 days of tamoxifen treatment displaying the characteristic flattened "fried egg" appearance

4. Place the 6-well plate in a 37 °C incubator supplemented with 5% $CO_2$ overnight. Transduction can be performed 16–18 h post seeding.
5. Count one well of cells:
   (a) Remove all media from a single well with a 5 mL pipet.
   (b) Add 1 mL of room temperature sterile 1× PBS and gently swirl around plate.
   (c) Remove the PBS wash from the well.
   (d) Add 250 μL of 0.25% trypsin-EDTA to the well, and return the plate to the 37 °C incubator for 5 min.
   (e) Gently resuspend the cells in 750 μL of pre-warmed MEF complete media (insuring that a single cell suspension is present), and perform a cell count.
6. Using the count from above and the lentivirus titer (**infectious particles per mL**), determine the volume of virus needed to transduce at the specified MOI per cell.
   (a) Calculate the # of viable cells per well: (total volume of cell stock (mL) × # viable cells counted per mL) = total # viable cells per well.
   (b) Calculate the volume of lentivirus for MOI: (total # viable cells × MOI)/(lentiviral titer/1000) = volume (μL) of lentivirus to add. For values <2 μL a dilution may be necessary.
7. Remove the growth media from the remaining wells of the 6-well plate, and using a P-1000, gently add 1 mL transduction media per well.

8. Add the previously calculated, volume of lentivirus for each MOI to be used. Dilute virus stock if necessary, so that total volume added is greater than 2 μL.
9. Gently rock the plate to evenly distribute virus/media, and incubate in the “infected”-labeled incubator at 37 °C/5% $CO_2$ for 24 h.
10. Remove the virus-containing media 24 h later, and replace with 3 mL of pre-warmed MEF complete media.
11. Twenty-four hours after the media change, perform another media change on all wells with fresh pre-warmed MEF complete media containing the appropriate antibiotic for selection:
    (a) Blasticidin = 4 μg/mL.
12. Monitor the control well (antibiotic selection verification) for cell death, and watch for resumption of proliferation in the transduced wells (*see* **Note 17**).
13. Transduced cell lines are expanded, using the procedure described in Subheading 3.1.2, from 6-well plate through T75 as determined by cell growth.
14. Cell line pools should be frozen as described in Subheading 3.1.3 before proceeding to single cell cloning (*see* Subheading 3.1.3).
15. RAS isoform-specific clones will be generated as described in Subheading 3.3 and further evaluated using a robust suite of quality control assays which will be described in Chapter 24 to make the final selections for the final isogenic panel.

## 4 Notes

1. Calcein-AM should be reconstituted by adding 25.1 μL DMSO to a single vial to make a 2 mM stock. This stock solution can be frozen and thawed *once* for use within a single month. The working concentration is made by diluting the stock solution 1:10 with 1× PBS to create a working stock of 200 μM. *Always protect from light* while using this dye.
2. Tamoxifen is solubilized in ethanol and heated in a 37 °C water bath for a minimum of 30 min to ensure complete dissolution of crystals. Cells should be approximately 80% confluent and at low passage number (<5) to start the tamoxifen treatment. Tamoxifen waste should be collected separately from other cell culture waste and safely disposed of according to your institution’s regulations.
3. When thawing viably frozen cells, we prefer to transfer the thawed viable cells directly to a flask containing pre-warmed complete media. Allow the cells to attach, and perform a media

exchange anywhere from 6 to 16 h post-attachment. The media change removes the DMSO (freeze media) from the flask. We have experienced no issues with this approach. There are other techniques for thawing and removing the freeze media from viably frozen cells such as pelleting the freshly thawed cells via a gentle spin in the centrifuge, discarding the supernatant, and resuspending the cells in fresh media before transferring to a flask. Although this technique is common, we have found that MEFs are very sensitive to this technique and can lead to substantially lower attachment and cell death.

4. Mouse embryonic fibroblasts (MEFs), and particularly those that express a single RAS isoform, are very sensitive to cell density when grown in culture. Generally, we passage 3–500,000 cells to start cells in a T75 flask in culture. Cells should not be allowed to grow to over 90% confluency. MEFs that are initially seeded at too low a concentration could also see no growth. We do not use MEF cells for experiments once they have reached 20 passages.
5. Cells should be kept in the freeze media for less than 30 min prior to freezing. Make sure that cells are gently resuspended/mixed completely before aliquoting to labeled freeze vials. Failure to properly mix the cells can lead to low viability when thawed. Likewise, aggressive pipetting can damage the cells leading to low viability when thawed.
6. Viable cells can be stored in the −80 °C for generally less than a month with no adverse effects to viability. Longer storage times will gradually see a reduction in viability. Vapor-phase liquid nitrogen is the preferred long-term storage solution for viably frozen cells with a storage time measured in years with no loss in viability.
7. Be sure to freeze at least two vials of the sorted population. We freeze viable cells at each step of the production process. This allows us to return to each step of the production process in case there's an issue in the future.
8. Cells will be lost in the washes, so it's recommended to use more than 1e6 cells total. We typically use 2e6 cells at a minimum. The higher cell number will ensure that cells will be remaining following the washes, and it'll make it easier to see the pellet after each centrifugation although doing this may add an extra dilution when diluting to the 400 cell level for seeding the single cell cloning plate.
9. Since the 384-well plate will be in the incubator for an extended period (up to 2 weeks), we place 100 μL 1× PBS or media into the outer ring of the 384-well plate and only add cells to the inner portion of the plate. This prevents evaporation

of the media on the wells containing cells allowing for uniform growth conditions across the seeded areas of the plate.

10. The low 25 μL volume allows the single MEF cell to condition the media faster and leads to a higher success rate at forming clones. We then gradually add more media to the wells as the cell number increases—an additional 25 μL initially and then possibly a full media exchange if cell growth and number warrant.
11. Generally, MEFs are moved progressively into larger plates based on the cell number which can be misleading due their variable size and surface area coverage. We generally move using a split ratio range of 1:3 or 1:6 once sufficient cell numbers are present. This typically means moving a clone from 384- to 48-well to 12-well to 6-well or T25 and then onward from there depending on the quality control and culture needs. When moving from the 384-well plate to the 96- or 48-well plate, we recommend allowing the cells to attach and then performing a media change to remove the media containing trypsin. Performing this procedure tends to increase the number of clones that survive the transfer out of the 384-well plate. With MEFs, a common mistake is to overestimate the number of cells present; therefore a fine balance needs to be struck between confluency and cell number.
12. We recommend freezing at least two vials of the clone when ploidy is checked. This removes the cells from active culture lessening the burden on the cell culture staff and, more importantly, keeping the passage number low on the clone.
13. It may be advantageous to viably freeze a few vials at the higher passage for future comparison studies. Again, we generally try to freeze at major milestones or transition points.
14. Initiation of tamoxifen treatment and subsequent scheduling need to be carefully coordinated both in terms of cell density and staff in the laboratory. The effectiveness of tamoxifen treatment and its downstream implications can be easily affected by inconsistent handling of the cells during the treatment period. The different cell concentrations provide starting points that take scheduling and cell line density into consideration.
15. The original parental pool DU1473 can be viably frozen following tamoxifen treatment and used later without issue. However, the size-sorted diploid clone does not perform equally as well in the transductions following the viable freeze; therefore, we typically move directly to transduction following the tamoxifen treatment. Tamoxifen treatment can yield unpredictable genetic anomalies, so we try to batch

transductions to create cell line pools together as much as possible to minimize lot to lot variation following tamoxifen treatment.

16. Tamoxifen-treated cells tend to spread out following tamoxifen treatment, almost giving the appearance of a fried egg, and therefore cover more surface area on the plate (*see* Fig. 1). If imaging is only relying on confluency, then lower seeding densities yield a more realistic growth curve with the caveat that too low of a seeding density can adversely affect cell growth. The Incucyte platform also offers a passive cell dye which stains the nucleus red and can allow the number of cells to be quantitated as opposed to relying on confluency alone. Use of the passive cell dye is particularly useful when calculating a doubling time.
17. Be cautious about passaging too quickly from the 6-well plate. Unlike typical tumor cells, MEF morphology can make it appear that there are far more cells present than there actually are. At the end of 7 days of selection treatment the cells can be trypsinized and reseeded back into the 6-well plate to remove cells that were attached but were not viable. Incremental expansion to 2 wells on a 6-well plate or to a T25 is a possible way to move forward depending on the growth of the cells following selection.

## Acknowledgments

The author would like to thank members of the NCI RAS Initiative who supported this effort, specifically the current and past members of the Cell Development Group within the Protein Expression Laboratory who work diligently to generate research reagents in support of the RAS Initiative. This project has been funded in full or in part with federal funds from the National Cancer Institute, National Institutes of Health, under contract no. 75N91019D00024. The content of this publication does not necessarily reflect the views or policies of the Department of Health and Human Services nor does mention of trade names, commercial products, or organizations imply endorsement by the US government.

## References

1. Kim H, Sung Y, Paik S (2015) Cancer cell line panels empower genomics-based discovery of precision cancer medicine. Yonsei Med J 56(5): 1186–1198
2. American Type Culture Collection (2023) Cancer panels. https://www.atcc.org/cell-products/human-cells/cancer-panels#t=productTab&numberOfResults=24. Accessed 13 June 2023
3. Crown Biosciences (2023) Cell line panels. https://www.crownbio.com/model-systems/in-vitro/cell-line-panels. Accessed 13 June 2023

4. Alley M, Scudiero D, Monks A et al (1988) Feasibility of drug screening with panels of human tumor cell lines using a microculture tetrazolium assay. Cancer Res 48(3):589–601
5. Ledford H (2016) US cancer institute overhauls cell lines. Nature 25(530):91
6. Drosten M, Dhawahir E, Sum YM et al (2010) Genetic analysis of Ras signaling pathways in cell proliferation, migration, and survival. EMBO 29(6):1091–1103
7. Nissley D, McCormick F (2022) RAS at 40: update from the RAS initiative. Cancer Discov 12(4):895–898

# Chapter 24

# Quality Control of an Isogenic H/N/KRAS-Less Mouse Embryonic Fibroblast Cell Line Panel


**Bingfang Xu and Katie Powell**



## Abstract

Isogenic H/N/KRAS-less mouse embryonic fibroblast (MEF) cell lines have been developed to assist in cell-based assays of RAS inhibitors. The quality control assessment of a panel of these isogenic MEFs is described here, with a focus on ensuring the proper insertion of the desired mutant RAS transgene, a determination of gene copy number, and an investigation of potential off-target mutations which could lead to phenotypes which are undesired in downstream experiments. Using this suite of quality control tools, a MEF cell line can be readily validated, and researchers can be assured of the rationale for an observed phenotype.



**Key words** Transgene, Sanger sequencing, Capillary electrophoresis, Droplet digital PCR (ddPCR), Copy number analysis, Next-generation sequencing (NGS) analysis, Vector integration site assay (VISA), Whole exome sequencing (WES)


## 1 Introduction

The panel of cell lines in the isogenic H/N/KRAS-less mouse embryonic fibroblast (MEF) 2.0 (*see* Chap. 23 of this volume) contains 12 different individual transgenes. To validate these lines, five sequential quality control steps were developed and implemented. First, the cell lines were sequenced using the Sanger method to ensure the intended transgene and desired mutations were present. Second, the H/N/KRAS-less MEF cell lines exhibit large fragment deletions in the Nras and Hras genes and contain a pair of loxp sites on Kras exon2 [1]. By employing tamoxifen-induced CRE activation, the endogenous mouse Kras is removed, resulting in a "ras-less" background for the cell lines. To verify the deletion of endogenous Kras exon2 in the individual cell lines of the MEF panel 2.0, we conducted copy number analysis of Kras to ensure the transgene is the exclusive source of RAS signaling.


Andrew G. Stephen and Dominic Esposito (eds.), *KRAS: Methods and Protocols*, Methods in Molecular Biology, vol. 2797, https://doi.org/10.1007/978-1-0716-3822-4_24,

Single nucleotide mutation, coupled with the loss of heterozygosity due to a large fragment deletion in Trp53 gene, can result in the complete loss of tumor suppression function of Trp53 [2], granting a growth advantage to immortal cells. We originally detected approximate 400 and 600 kb large fragment deletions surrounding Trp53 in subpopulations of our original H/N/KRAS-less MEF version 1 cell line. The parental diploid clone for the current panel has an approximate 200 kb deletion surrounding Trp53, leaving one wild-type (WT) allele to provide Trp53 regulation of cell growth. It is not uncommon for the descendants of this line to acquire additional mutations that lead to complete loss of Trp53 function. Therefore, we additionally conducted mutation screening for Trp53 coding regions and splicing junction sites for each individual cell line of the MEF panel 2.0.

As the lentivirus used to insert the transgene randomly integrates into the host genome, the integration site can serve as a unique molecular barcode to track the genomic location and abundance of integration events. Lentivector integration site analysis was performed using linker-mediated PCR, as previously reported [3–5]. For the MEF 2.0 panel, we selected cell lines that had one or two copies of integrations with intact 5′ and 3′ end LTRs. In addition, these integrations must not be located within any exons or critical gene regulatory regions.

Finally, whole exome sequencing (WES) provides comprehensive information about the protein-coding regions of the genome. The cell lines in MEF panel 2.0 have undergone complicated engineering involving growth bottlenecks caused by the lentiviral transduction and selection as well as the Cre recombination. Therefore, we conducted WES to validate the presence of intended modifications but also to confirm the absence of off-target mutational effects. Mutations identified in any important genes or genes related to the cancer pathways being studied immediately disqualify those cell lines from entry into the panel.

## 2 Materials

### 2.1 Transgene Sequencing Analysis

1. Genomic DNA from cell lines (*see* **Note 1**).
2. Phusion™ Plus PCR Master Mix.
3. Oligonucleotides for amplification and sequencing (*see* **Note 2**).
4. PCR-grade water.
5. Thermal cycler for PCR.
6. Agilent Bioanalyzer and DNA 7500 kit.
7. ExoSAP-IT™ PCR Product Cleanup Reagent, Thermo Fisher Scientific.

8. BigDye™ Terminator v3.1 Cycle Sequencing Kit, Thermo Fisher Scientific.
9. BigDye XTerminator™ Purification Kit, Thermo Fisher Scientific.
10. Digital Vortex-Genie.
11. Capillary electrophoresis instrument.

### 2.2 Endogenous KRAS Copy Number Verification

1. ddPCR Supermix for probes (No dUTP), Bio-Rad.
2. Primers (*see* **Note 3**).
3. TaqMan MGB probe with FAM, Thermo Fisher Scientific.
4. TaqMan Copy Number Reference Assay, mouse, Tfrc, Thermo Fisher Scientific.
5. QX200 Droplet Digital PCR System, Bio-Rad.
6. QuantaSoft™ software, Bio-Rad.
7. Thermal cycler, Bio-Rad.

### 2.3 Trp53 Verification

1. Platinum Hot Start PCR Master Mix (2×), Thermo Fisher Scientific.
2. M13-tagged PCR primers (*see* Table 1).
3. Thermal cycler.
4. AMPure XP Reagent, Beckman Coulter Life Sciences.
5. Exonuclease I.
6. Qubit assay, Thermo Fisher Scientific.
7. Nanodrop spectrophotometer.
8. Agilent TapeStation System.
9. Illumina MiSeq sequencer.
10. Illumina MiSeq V2 Reagent Kits 500 cycles.
11. PhiX Sequencing Control V3 from Illumina.
12. Illumina BaseSpace software.
13. SnapGene software.
14. Other software: Python 3, Trimmomatic, Seqtk.

### 2.4 Vector Integration Site Assay (VISA)

1. NEBNext® Ultra™ II FS DNA Library Prep Kit for Illumina.
2. Platinum Hot Start PCR Master Mix (2×), Thermo Fisher Scientific.
3. PCR primers (*see* Table 2).
4. Thermal cycler.
5. AMPure XP Reagent, Beckman Coulter Life Sciences.
6. Exonuclease I.
7. Qubit assay, Thermo Fisher Scientific.

**Table 1**
**List of primer sequences for Trp53 mutation screen**

| Name | Primer sequence | Specificity |
|---|---|---|
| mTRP53-E2-1F | GTAAAACGACGGCCAGTATTCTACCCTTTCCTATAAGCC | exon2 |
| mTRP53-E2-2R | GGAAACAGCTATGACCATGCCATACCATGTTTGAACAC TAC | exon2 |
| mTRP53-E3-1F | GTAAAACGACGGCCAGTCATTGACTACATAGCAAGTTGG | exon3 |
| mTP53_Ex3_Rnew | GGAAACAGCTATGACCATGCAAGTCCCTTTCTGCTCTGG | exon3 |
| mTP53_Ex4_Fnew | GTAAAACGACGGCCAGTACACAGTCCTGAGGGTTCTTC | exon4 |
| mTP53_Ex4_Rnew | GGAAACAGCTATGACCATGAAAGGTCACACGAAAGACAA CTC | exon4 |
| mTRP53-E5-1F | GTAAAACGACGGCCAGTTAGTTCCCCACCTTGACAC | exon5 |
| mTP53_Ex5_Rnew | GGAAACAGCTATGACCATGCCACAGGCGGTGTTGAG | exon5 |
| mTP53_Ex6_Fnew | GTAAAACGACGGCCAGTGGCTTCTGACTTATTCTTGCTC | exon6 |
| mTP53_Ex6_Rnew | GGAAACAGCTATGACCATGCAGCTTGCACCTCTAAGC | exon6 |
| mTRP53-E7-1F | GTAAAACGACGGCCAGTGGTGGAATATCCCTACTCTACA | exon7 |
| mTRP53-E7-3R | GGAAACAGCTATGACCATGGAAACAGGCTAACCTAACC TAC | exon7 |
| mTRP53-E8-1F | GTAAAACGACGGCCAGTCCTAGTTTACACACAGTCAGGA | exon8 |
| mTP53_Ex8_Rnew | GGAAACAGCTATGACCATGCTCCGCCTCCTTGGT | exon8 |
| mTP53_Ex9_Fnew | GTAAAACGACGGCCAGTTCACCTCTTGCTCTCTCCT | exon9 |
| mTP53_Ex9_Rnew | GGAAACAGCTATGACCATGCAAGCTAATGTCACGGCT AGA | exon9 |
| mTRP53-E10-1F | GTAAAACGACGGCCAGTCAAAAACCTGTAAGTGGAGC | exon10 |
| mTRP53-E10-2R | GGAAACAGCTATGACCATGAGGTCTGGGTAGAGCACC | exon10 |
| mTRP53-E11-1F | GTAAAACGACGGCCAGTAGCCCAAACTGCTAGCTC | exon11 |
| mTRP53-E11-1R | GGAAACAGCTATGACCATGAGTCATAAGACAGCAAGGA GAG | exon11 |
| Nested primer_F | AATGATACGGCGACCACCGAGATCTACACNNNNNNNNA CACTCTTTCCCTACACGACGCTCTTCCGATCTNNNNN NNNGTAAAACGACGGCCAGT | |
| Nested primer_R | CAAGCAGAAGACGGCATACGAGATNNNNNNNNGTGAC TGGAGTTCAGACGTGTGCTCTTCCGATCTNNNNNN NNGGAAACAGCTATGACCATG | |

Primer sequences are listed from 5′ to 3′, and specificity of the given primer for various Trp53 exons is noted. NNNNNNNN represents i5, i7, and inline barcode sequences

8. Nanodrop spectrophotometer.
9. Agilent TapeStation System.

**Table 2**
**List of primer sequences for integration site analysis**

| Name | Sequence | Specificity |
|---|---|---|
| HIV_P2 | TGTGACTCTGGTAACTAGAGATCCCTC | 3LTR |
| RAS_5LTR_P1 | CAGATCTGGTCTAACCAGAGAGA | 5LTR |
| Linker_P1 | GTAATACGACTCACTATAGGGC | Linker |
| PE1_3LTRnest | AATGATACGGCGACCACCGAGATCTACACTCTTTCCC TACACGACGCTCTTCCGATCTNNNNNNNCCCTTTTAG TCAGTGTGGAAAATC | 3LTR |
| PE1_5LTRnest | AATGATACGGCGACCACCGAGATCTACACTCTTTCCC TACACGACGCTCTTCCGATCTNNNNNNNCACTGTTGTC TTTTCTGGGAGTGAACTAGCC | 5LTR |
| PE2_Linkernest | CAAGCAGAAGACGGCATACGAGATCGGTCTCGGCATTCC TGCTGAACCGCTCTTCCGATCTNNNNNNNAGGGCTCCGC TTAAGGGAC | Linker |

Primer sequences are listed from 5′ to 3′, and specificity of the given primer for various regions of the integrated lentivirus is noted. NNNNNNNN represents barcode sequences

10. Illumina MiSeq sequencer.
11. Illumina MiSeq V2 Reagent Kits 300 cycles.
12. PhiX Sequencing Control V3 from Illumina.
13. Illumina BaseSpace software.
14. Software: Python 3, Perl, Bash.

### 2.5 Exome Sequencing Analysis

1. Agilent SureSelect XT Mouse All Exon Kit.
2. Thermal cycler.
3. AMPure XP Reagent, Beckman Coulter Life Sciences.
4. Qubit assay, Thermo Fisher Scientific.
5. Agilent TapeStation System.
6. Covaris E220 Focused-ultrasonicator.
7. Vacuum concentrator.
8. Automated Liquid Handling Platform.
9. Dynabeads MyOne Streptavidin T1 magnetic beads, Thermo Fisher Scientific.
10. Illumina NextSeq 2000 sequencer.
11. NextSeq 1000/2000 P2 Reagents (300 cycles) v3.
12. Illumina NovaSeq 6000 sequencer.
13. NovaSeq 6000 S1 Reagent Kit v1.5 (300 cycles).

14. PhiX Sequencing Control V3 from Illumina.
15. Dragen v3.10 software.
16. Exome seek/XAVIER software (*see* **Note 4**).
17. High-performance computing system.
18. Integrated Genome Viewer (IGV) from the Broad Institute.
19. Ensembl Variant Effect Predictor (VEP).
20. GATK4 from the Broad Institute.
21. Other software: Python 3, R.

## 3 Methods

### 3.1 Transgene Sequencing Analysis

#### 3.1.1 PCR Amplification

1. Add 1 μL of each genomic DNA to a well.
2. Add 1 μL of PCR-grade water to one well to serve as a no template control (NTC).
3. Make master mix in a 1.5 mL microcentrifuge tube adjusting for sample number and adding a reasonable extra amount (*see* **Note 5**).
4. Add 24 μL of master mix to each well for a total volume of 25 μL.
5. Centrifuge tubes or plate briefly to collect the volume at the bottom.
6. PCR amplify at the following cycling conditions: 30 s at 98 °C, 35 cycles of 98 °C for 10 s, 60 °C for 10 s, and 72 °C for 2 min and then 5 min at 72 °C, and hold at 4 °C.
7. Run 1 μL of each amplified, unpurified genomic DNA product and master mix on an Agilent Bioanalyzer DNA 7500 chip to confirm the presence of an amplified product band.
8. Remove ExoSAP-IT™ reagent from −20 °C freezer and keep on ice throughout this procedure.
9. Mix 5 μL of a post-PCR reaction product with 2 μL of Exo-SAP-IT™ reagent for a combined 7 μL reaction volume.
10. Incubate at 37 °C for 15 min to degrade remaining primers and nucleotides.
11. Incubate at 80 °C for 15 min to inactivate ExoSAP-IT™ reagent.

#### 3.1.2 Perform Cycle Sequencing

1. Completely thaw the contents of the BigDye™ Terminator v3.1 Cycle Sequencing Kit and your primers, and store on ice.
2. Vortex the tubes for 2 to 3 s, and then centrifuge briefly (2 to 3 s) with a benchtop microcentrifuge to collect contents at the bottom of the tubes.

3. Add reaction components as follows to each well: 8 μL BigDye™ Terminator 3.1 Ready Reaction Mix, 3.2 pmol forward or reverse primer, and template from PCR amplification, and bring to 20 μL with PCR-grade water (*see* **Note 6**).
4. Seal the plate with MicroAmp™ Clear Adhesive Film.
5. Vortex the plate for 2 to 3 s, and then centrifuge in a swinging-bucket centrifuge to collect contents to the bottom of the wells (5 to 10 s) at 1000 ×*g*.
6. Place the tubes or plate(s) in a thermal cycler and set the correct volume.
7. Perform cycle sequencing under the following conditions: 1 min at 96 °C, 25 cycles of 96 °C for 10 s, 50 °C for 5 s, and 60 °C for 4 min, and then hold at 4 °C.
8. Briefly centrifuge the reactions (1000 ×*g* for 1 min).
9. Prepare fresh premix BigDye XTerminator™ solution for each processing run.
10. Dispense 110 μL of bead mix (BigDye XTerminator™ bead solution and SAM solution) to each sample.
11. Seal the plate using MicroAmp™ Clear Adhesive Film.
12. Vortex the 96-well plate for 20 min at 1800 rpm (for the Digital Vortex-Genie™ 2).
13. Centrifuge the plate at 1000 ×*g* in a swinging-bucket centrifuge for 2 min.

#### *3.1.3 Perform Capillary Electrophoresis*

1. Remove the adhesive film, and then place a septum on the plate.
2. Load the reaction plate in the instrument.
3. Set up an instrument run using the appropriate conditions for your instrument. Select a BDx run module to use the direct injection method.
4. Start the electrophoresis run.
5. When the run is complete, check data quality and export the data.
6. Use analysis software to perform base calling, trim low-quality bases at the 5′ and 3′ ends, and assign a quality value for each base.
7. Compare sequence results to expected sequence to confirm presence of transgene in cell line.

### 3.2 Copy Number Analysis of Endogenous Mouse KRAS

#### 3.2.1 ddPCR Assay

1. Extract genomic DNA using the Qiagen AllPrep protocol.
2. Measure the DNA concentration using a nanodrop spectrophotometer.
3. Prepare 20× mouse Kras/FAM copy number assay by combining a pair of primers and a TaqMan probe with a dye label (FAM) on the 5′ end and a minor groove binder (MGB) and nonfluorescent quencher (NFQ) on the 3′ end and using low-TE to adjust volume. In the 20× assay, the primer concentration is 20 μM, and the probe concentration is 5 mM (*see* **Note 3** for primer sequences).
4. Combine 25 ng genomic DNA, the mouse Kras/FAM copy number assay, the Tfrc/VIC copy number reference assay, and 2× ddPCR Supermix for probes (No dUTP) for the PCR reaction, and adjust the final volume using $H_2O$.
5. Generate droplets using Bio-Rad QX200 Droplet Generator and Droplet Generation Oil for probes (*see* **Note 7**).
6. Transfer droplets to a 96-well plate, and conduct PCR amplification using the following protocol: preheating at 95 °C for 10 min, 40 cycles at 94 °C for 30 s, and 60 °C for 1 min, and a final cycle at 98 °C for 10 min.

#### 3.2.2 ddPCR Data Analysis

1. Analyze the samples using Bio-Rad QX200 droplet reader following PCR amplification of the targeted Kras and Tfrc fragments. The reader utilizes the FAM and VIC detection system, enabling multiplexed analysis of both targets within the same sample (*see* **Note 7**).
2. Utilize the QuantaSoft™ software to count the number of PCR-positive and PCR-negative droplets. Adjust the threshold for positive counts manually based on the positive and negative controls. Fit the fraction of positive droplets to a Poisson distribution to determine the absolute initial copy number of the target DNA molecule in the input reaction mixture.
3. Calculate the ratio of Kras vs. Tfrc and determine the copy number of endogenous mouse Kras gene.

### 3.3 Trp53 Mutation Screen

#### 3.3.1 PCR Amplification of Trp53 Exons

1. Amplify the coding regions and splicing junctions of 10 exons (exon 2 to exon 11) of Trp53. Set up the PCR reaction for each exon separately for each sample. For each PCR reaction, combine a primer pair (*see* Table 1) with a final concentration of 1 μM each, 25 ng genomic DNA, and 2× Platinum Hot Start PCR Master Mix. Adjust the final volume to 25 μl using $H_2O$.
2. Conduct PCR amplification using the following protocol: an initial preheating step at 95 °C for 2 min, 36 cycles at 94 °C for 15 s, 58 °C for 15 s, and 68 °C for 1 min, and finally, a single cycle at 68 °C for 1 min.

3. Quantify the resulting amplicons using the Agilent TapeStation.
4. Pool ten PCR products from each sample by equal molarity to serve as the template for the barcode PCR.

*3.3.2 Barcode PCR for Each Sample*

1. For each barcode PCR reaction, combine a pair of barcode primers (*see* Table 1) with a final concentration of 1 μM, 2 μL DNA template obtained from the PCR described above, and 2× Platinum Hot Start PCR Master Mix, and adjust the final volume to 25 μL using $H_2O$.
2. Conduct the barcode PCR using the following protocol: an initial preheating at 95 °C for 2 min, 15 cycles at 94 °C for 15 s, 56 °C for 15 s, and 68 °C for 1 min, and finally, a single cycle at 68 °C for 1 min.

*3.3.3 Preparation for MiSeq*

1. Treat the products from the barcode PCR with Exonuclease I at 37 °C for 15 min.
2. Purify the treated products using AMPure XP Reagent.
3. Quantify the purified amplicons using a nanodrop spectrophotometer, and pool the amplicons together based on equal molarity.
4. Measure the length of the final pooled sample by using the Agilent TapeStation, and determine the final concentration of the pooled sample using a Qubit assay.

*3.3.4 MiSeq Sequencing and Data Analysis*

1. Perform next-generation sequencing on the Illumina MiSeq platform (*see* **Note 8**) using MiSeq V2 Reagent Kits 500 cycles. We spike in 10% PhiX Sequencing Control (*see* **Note 9**) and set read1 length as 272 bp and read2 length as 244 bp to have the best coverage on coding regions.
2. Monitor the sequencing quality on the BaseSpace website.
3. Download the raw FASTQ files from BaseSpace.
4. Trim the sequence reads using Trimmomatic (*see* **Note 10**).
5. Identify the top 100–300 most common reads for each sample, and detect sequence reads containing mutations with a frequency greater than 4% using custom-developed Python scripts (*see* **Note 10**).
6. Visualize the sequence reads with mutations by aligning them with the wild-type mouse Trp53 gene using SnapGene.
7. Calculate the mutation frequency using a custom-developed Python script (*see* **Note 10**).

### 3.4 Vector Integration Site Analysis (VISA)

#### 3.4.1 Library Preparation

1. Utilize NEBNext Ultra II FS DNA Library Prep Kit for Illumina for library preparation. For each sample, randomly shear 500 ng DNA to 200–450 bp fragments. The sheared DNA fragments undergo end repair, and a single dA is added to the 3′ ends of genome fragments (*see* **Note 11**). 3 ng DNA from a cell line carrying a single known HIV provirus was added to each sample to monitor the PCR and sequencing reactions.
2. Ligate the dA-tailing genomic fragments to a partial double-stranded linker (5′GTAATACGACTCACTATAGGGCTCCG CTTAAGGGACT 3′, 5′–PO4-GTCCCTTAAGCGGAG-3′-C6) with 3′ T overhang following the protocol of NEBNext Ultra II FS DNA Library Prep Kit (*see* **Note 11**).
3. Purify the ligated DNA fragments using AMPure XP Reagent. Quantify the DNA fragments using a nanodrop spectrophotometer.
4. Set up the first PCR reaction to selectively amplify the integration junctions. Combine linker-ligated DNA fragments, three primers with a final concentration of 1 μM each, and 2× Platinum Hot Start PCR Master Mix. Adjust the final volume to 50 μL using $H_2O$. The three primers are specific to the linker, the lentivector 5LTR, and 3LTR, respectively (*see* Table 2).
5. Conduct PCR amplification using the following protocol: an initial preheating step at 95 °C for 2 min, 31 cycles at 94 °C for 15 s, 56 °C for 15 s, and 68 °C for 1 min, and finally, a single cycle at 68 °C for 1 min.
6. Purify the first PCR products using AMPure XP Reagent. Quantify the amplicons using a nanodrop spectrophotometer.
7. Set up the nested PCR reaction to amplify the 5LTR and 3LTR integration junctions separately. Combine the first PCR products with an inner primer pair, each with a final concentration of 1 μM, and 2× Platinum Hot Start PCR Master Mix. Adjust the final volume to 25 μL using $H_2O$. The primer pairs of PE1_3LTRnest and PE2_Linkernest are used to amplify the 3LTR integration junctions, while the primer pairs of PE1_5LTRnest and PE2_Linkernest are used to amplify the 5LTR integration junctions (*see* Table 2). This nest PCR amplification increased the specificity of amplification and was used to introduce the Illumina sequencing adapters and barcodes.
8. Conduct the nested PCR amplification using the following protocol: an initial preheating step at 95 °C for 2 min, 30 cycles at 94 °C for 15 s, 56 °C for 15 s, and 68 °C for 1 min, and finally, a single cycle at 68 °C for 1 min.
9. Treat the products from the barcode PCR with Exonuclease I at 37 °C for 15 min.
10. Purify the treated products using AMPure XP Reagent.

11. Quantify the purified amplicons using a nanodrop spectrophotometer. Pool the amplicons together based on equal molarity.
12. Measure the length of the final pooled sample by using the Agilent TapeStation, and determine the final concentration using a Qubit assay.

#### 3.4.2 Integration Site Identification and Quantification

1. Perform next-generation sequencing on the Illumina MiSeq platform (*see* **Note 8**) using MiSeq V2 Regent Kits 300 cycles. Five percent PhiX was spiked in as the sequencing control (*see* **Note 9**). Paired-end sequencing was performed, and the read length for read1 and 2 was 158 bp. Read1 sequences include the LTR-host DNA junction, and read2 sequences include the broken end of the host DNA.
2. Monitor the sequencing quality on BaseSpace website, and download the raw FASTQ files from BaseSpace.
3. Trim the sequences and map them to mouse genome mm10 with BLAT. Subsequently, identify genomic location and copy number of integration sites using custom-developed pipeline published previously [3–5]. The modified pipeline has been uploaded to GitHub (*see* **Note 12**).

### 3.5 Exome Sequencing Analysis

#### 3.5.1 Library Preparation

1. Extract genomic DNA using the Qiagen AllPrep protocol.
2. Measure DNA concentration using a Qubit assay.
3. Shear genomic DNA (500 ng per sample) to ~300 bp fragments using Covaris E220 Focused-ultrasonicator (*see* **Note 13**).
4. Utilize the Agilent SureSelect XT Mouse All Exon protocol to prepare gDNA libraries (*see* **Note 13**).
5. Perform hybridization and exon capture steps on Bravo Automated Liquid Handling Platform (*see* **Note 14**).
6. Pool samples together based on equal molarity for multiplexed sequencing.
7. Measure the length of the final pooled libraries using Agilent TapeStation, and quantify the final concentration of the pooled libraries using a Qubit assay.

#### 3.5.2 Next-Generation Sequencing

1. Perform paired-end sequencing on the Illumina NovaSeq 6000 platform with NovaSeq 6000 S1 Reagent Kit v1.5 (300 cycles). Target 50 M paired reads for each sample.
2. Alternatively, perform paired-end sequencing on the Illumina NextSeq 2000 platform with NextSeq 1000/2000 P2 Reagent kit (300 Cycles) v3. Target 50 M paired reads for each sample.
3. Monitor sequencing quality and download the FASTQ files from sequencers.

*3.5.3 Exome Analysis*

1. Run Exome seek v1.0.3 or XAVIER v2.0 pipeline on a high-performance computing system. Exome seek/XAVIER pipeline uses Trimmomatic to trim FASTQ files and uses bwa-mem to conduct alignment and generate BAM files (*see* **Note 4**).
2. Utilize Exome seek v1.0.3 or XAVIER v2.0 pipeline to conduct germline and somatic calling and generate VCF files (*see* **Note 4**).
3. Alternatively, use Dragen v3.10 pipeline to conduct the secondary analysis on a Dragen server.
4. Annotate variants using Ensembl Variant Effect Predictor (VEP).
5. Filter single nucleotide variants and insertions or deletions based on sequencing quality, depth of coverage, allele frequency, and variant effect prediction. These filtered variants are compiled into a variant list (*see* **Note 15**).
6. For further analysis, a subset of the compiled variants should undergo manual review using the Integrated Genome Viewer (IGV) software from the Broad Institute.

## 4 Notes

1. Genomic DNA concentration should be between 50 and 150 ng/μL. Two reactions can be done for lower concentration DNAs.
2. Forward CMV51 oligonucleotide (5′-CGGGACTTTCCTACTTGGCA) and reverse WRE oligonucleotide (5′-AAGCAGCGTATCCACATAGC).
3. Primers and probe sequences for copy number ddPCR assay: mKRAS_F (20 μM, 5′-GGCCTGCTGAAAATGACTGAG), mKras_R (20 μM, 5′-GAGCAGCGTTACCTCTATCGTAGG), mKras_probe_ex2 (5 μM, 5′-TACGCCACCAGCTC).
4. The document for the Exome seek/XAVIER pipeline is on the GitHub website: https://ccbr.github.io/XAVIER/usage/run/.
5. Forward oligo (0.5 μM final), reverse oligo (0.5 μM final), to 25 μL PCR-grade water, 12.5 μL Phusion™ Plus PCR Master Mix.
6. Amount of template from PCR amplification is dependent on size. Use 30 ng for 1–2 kb and 40 ng if over 2 kb.
7. A digital droplet PCR guide is on Bio-Rad website: https://www.bio-rad.com/webroot/web/pdf/lsr/literature/Bulletin_6407.pdf.

8. MiSeq system guide is on https://support.illumina.com/content/dam/illumina-support/documents/documentation/system_documentation/miseq/miseq-system-guide-for-local-run-manager-15027617-05.pdf.
9. The technical note about PHIX control is on the Illumina website https://knowledge.illumina.com/library-preparation/general/library-preparation-general-reference_material-list/000001545.
10. The data analysis workflow of mouse Trp53 mutation screen is uploaded to GitHub: https://github.com/bingfang/P53-mutation-screen.git.
11. The product manual of NEBNext® Ultra™ II FS DNA Library Prep Kit for Illumina includes further details for how to use the products (https://www.neb.com/media/nebus/files/manuals/manuale6177-e7805.pdf).
12. The data analysis workflow of the integration site analysis is uploaded to GitHub: https://github.com/bingfang/Integration.git.
13. The protocol of SureSelect XT HS/SureSelect XT Low Input Target Enrichment with Pre-Capture Pooling and Preparation of Multiplexed NGS Libraries for the Illumina Sequencing Platform can be found on Agilent website: https://www.agilent.com/cs/library/usermanuals/public/G9702-90005.pdf.
14. The manual of Bravo Automated Liquid Handling Platform can be found on Agilent website: https://www.agilent.com/cs/library/usermanuals/Public/G7550-90000.pdf.
15. The downstream analysis workflow of exome sequencing is uploaded to GitHub: https://github.com/bingfang/exomeSeq_downstream_analysis.git.

## Acknowledgments

This project was funded in part with federal funds from the National Cancer Institute, National Institutes of Health under contract 75N91019D00024. The content of this publication does not necessarily reflect the views or policies of the Department of Health and Human Services, and the mention of trade names, commercial products, or organizations does not imply endorsement by the US government.

## References

1. Drosten M, Dhawahir A, Sum EY et al (2010) Genetic analysis of Ras signalling pathways in cell proliferation, migration and survival. EMBO J 29(6):1091–1104. https://doi.org/10.1038/emboj.2010.7
2. Alexandrova EM, Mirza SA, Xu S et al (2017) p53 loss-of-heterozygosity is a necessary prerequisite for mutant p53 stabilization and gain-of-function in vivo. Cell Death Dis 8(3):e2661. https://doi.org/10.1038/cddis.2017.80
3. De Ravin SS, Wu X, Moir S et al (2016) Lentiviral hematopoietic stem cell gene therapy for X-linked severe combined immunodeficiency. Sci Transl Med 8(335):335ra357. https://doi.org/10.1126/scitranslmed.aad8856
4. Maldarelli F, Wu X, Su L et al (2014) HIV latency. Specific HIV integration sites are linked to clonal expansion and persistence of infected cells. Science 345(6193):179–183. https://doi.org/10.1126/science.1254194
5. Wells DW, Guo S, Shao W et al (2020) An analytical pipeline for identifying and mapping the integration sites of HIV and other retroviruses. BMC Genomics 21(1):216. https://doi.org/10.1186/s12864-020-6647-4. (Wells et al., 2020)

# Chapter 25

# A Facile Method to Append a Bio-ID Tag to Endogenous Mutant Kras Alleles

**Konstantin Budagyan, Alexa C. Cannon, and Jonathan Chernoff**

## Abstract

KRAS mutations occur in approximately ~50% of colorectal cancers (CRCs) and are associated with poor prognosis and resistance to therapy. While these most common mutations found at amino acids G12, G13, Q61, and A146 have long been considered oncogenic drivers of CRC, emerging clinical data suggest that each mutation may possess different biological functions, resulting in varying consequences in oncogenesis. Currently, the mechanistic underpinnings associated with each allelic variation remain unclear. Elucidating the unique effectors of each KRAS mutant could both increase the understanding of KRAS biology and provide a basis for allele-specific therapeutic opportunities. Biotinylation identification (BioID) is a method to label and identify proteins located in proximity of a protein of interest. These proteins are captured through the strong interaction between the biotin label and streptavidin bead and subsequently identified by mass spectrometry. Here, we developed a protocol using CRISPR-mediated gene editing to generate endogenous $BioID_2$-tagged *Kras*$^{G12D}$ and *Kras*$^{G12V}$ isogenic murine colon epithelial cell lines to identify unique protein proximity partners by BioID.

**Key words** BioID, BirA, Proximity labeling, KRAS, Small G proteins, Signal transduction, Gene editing, Colorectal cancer

## 1 Introduction

RAS is one of the most commonly mutated proto-oncogenes in cancer [1]. Of the three RAS isoforms (HRAS, KRAS, and NRAS), KRAS mutations possess greater allelic diversity, with common mutation sites at codons G12, G13, Q61, and A146 [2]. Several studies support differences in downstream biological signaling; however, whether the interacting partners differ between these various KRAS mutants KRAS has yet to be investigated [3]. Further examination into allele-specific downstream effectors may offer valuable mechanisms into target-specific drug response and

Konstantin Budagyan and Alexa C. Cannon contributed equally with all other contributors.

Andrew G. Stephen and Dominic Esposito (eds.), *KRAS: Methods and Protocols*, Methods in Molecular Biology, vol. 2797, https://doi.org/10.1007/978-1-0716-3822-4_25,

resistance. Enzymatic biotinylation (BioID) is one approach to identify potential binding partners. In this method, a BirA (*Escherichia coli* biotin ligase) tag is linked to the protein of interest (KRAS) and introduced in cells, resulting in an overexpressed BirA-KRAS fusion protein that will add a biotin label on proteins within an approximate 10 nm radius. These biotin-labeled proteins can then be identified through streptavidin pull down and mass spectrometry [4, 5]. However, this method is subject to high background and artifactual results due to overexpression of a recombinant protein. Furthermore, overexpression of RAS can lead to mislocalization and artificial signaling in cellular compartments where endogenous KRAS is not expressed, consequentially labeling proteins KRAS would otherwise not interact with [3]. To avoid nonspecific labeling, new gene editing techniques can be employed to generate knock-ins of a BirA-tagged KRAS mutant allele that will be expressed at endogenous levels. Here we report a CRISPR-mediated gene editing method to create a series of endogenous KRAS-mutant alleles with a BirA tag.

This assay is based off our previously reported method of fluorescent co-selection for CRISPR-driven KRAS editing [6]. Similarly, this assay utilizes a fluorescent marker to allow for screening of cells that have undergone successful homologous recombination events. This assay involves simultaneous delivery of two single-guide RNAs (sgRNAs) to two different endogenous loci, along with two templates for homologous recombination (HR). One sgRNA and HR template targets and edits a fluorescent protein, while the other one targets mutant *Kras G12D* allele. This co-selection scheme is based on the idea that a cell that is competent to undergo homology-directed repair (HDR) at one site is more likely to also undergo HDR at additional sites [6, 7]. In this protocol, we first introduce Cas9-blue fluorescent protein (BFP) fusion into a mouse colon epithelial cell line with deleted *Apc* and mutant $Kras^{G12D/WT}$ ($Apc^{-/-}$; $Kras^{G12D/WT}$). Once Cas9-BFP is stably expressed, we next introduce two sgRNAs targeting either *BFP* or the $Kras^{G12D}$ allele in addition to two HR templates to edit (1) *BFP* to *green fluorescent protein (GFP)* and (2) $Kras^{G12D/WT}$ to $BirA\text{-}Kras^{G12V}$ with an N-terminus BirA tag (i.e., $Kras^{\mathrm{BirA\text{-}G12V}}$), while sparing the *WT Kras allele*. Selection for cells that are GFP-positive enriched for those with successful insertion of BirA tag into the N-terminus of mutant *Kras* allele. The result is endogenous expression of a BirA-tagged mutant *Kras* allele, while the wild-type allele remains unedited. Subsequent biotin labeling followed by streptavidin affinity capture and mass spectrometry can identify proteins located in proximity to mutant Kras, revealing potential unique mutant-specific binding partners (Fig. 1).

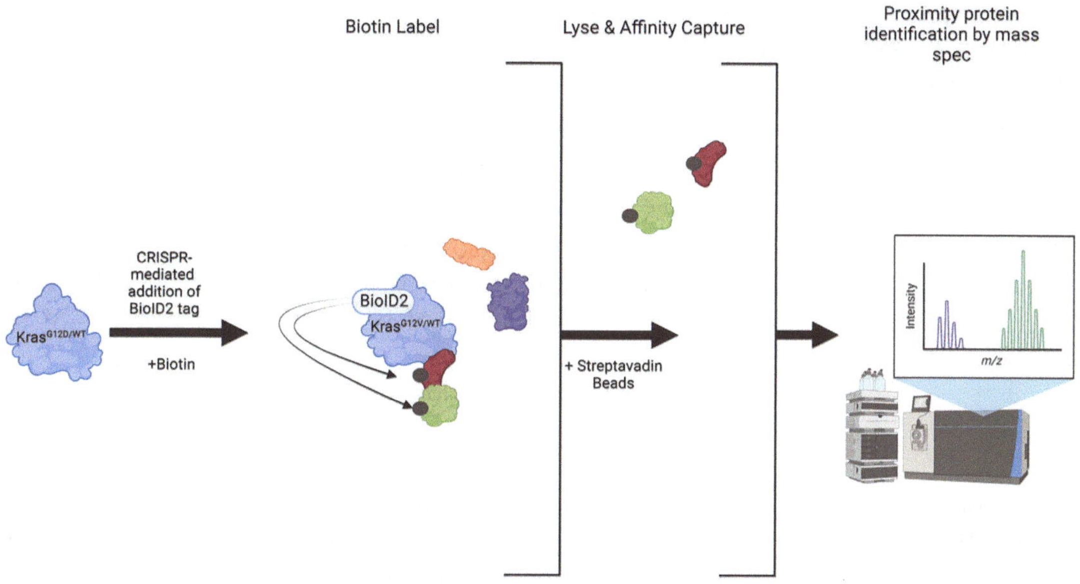


**Fig. 1** Overview of BioID. First, CRISPR is utilized to endogenously generate a BioID$_2$-tagged *Kras*$^{G12V}$ allele. After addition of biotin, BirA will biotinylate proteins within a 10 nm radius. These proteins can be pulled down with streptavidin beads and identified via mass spectrometry

## 2 Materials

### *2.1 Lentiviral Infection Components*

1. Lenti-X 293 T cells.
2. TransDUx MAX virus transduction reagent.
3. Opti-MEM.
4. DMEM.
5. Lenti-X 293 T cell medium: DMEM with high glucose (4.5 g/ L). 4 mM L-glutamine, and sodium bicarbonate, 10% fetal bovine serum (FBS), 100 units/mL penicillin G sodium, 100 mg/mL streptomycin sulfate, 1 mM sodium pyruvate.
6. Mouse colon epithelium medium: RPMI 1640, 10% FBS, 1% antibiotic-antimycotic, 4 mM L-glutamine.
7. Lipofectamine 3000 transfection kit.
8. 45 μm syringe filter.
9. Second-generation packaging plasmids: psPAX2 (Addgene, 12260), pMD2.G (Addgene 12259).
10. Lentiviral plasmid containing Cas9-P2A-BFP flanked by long terminal repeats: pLentiCas9-P2A-BFP (Addgene 149447).
11. 10 cm and 24-well culture plates.

### 2.2 CRISPR/Cas9-Mediated BirA Knock-In in Mouse Colon Epithelial Cells

#### 2.2.1 Electroporation Components

1. Neon Transfection System.
2. Single-strand DNA oligonucleotides for GFP, BirA-Kras$^{G12D}$, and BirA-Kras$^{G12V}$.
3. Trypsin-EDTA (0.25%).
4. sgRNA plasmid U6::sgBFP-U6::sgKrasG12D-pX333-red plasmid (Addgene 149448).
5. GFP ssODN: 5′-CACCACCGGCAAGCTGCCCGTGCCCT GGCCCACCCTCGTGACCACCCTGACCTACGGCGTGC AGTGCTTCAGCCGCTACCCCGACCACATGAAGCAGC AC-3′.
6. BirA-Kras G12V DNA donor plasmid: (Addgene xxx).
7. Fluorescent microscope.

#### 2.2.2 Clonal Selection Components

1. NarI restriction enzyme and rCutSmart™ Buffer.
2. NucleoSpin Tissue genomic DNA extraction kit or any other compatible genome DNA extraction kit.
3. DNA polymerase and accompanying buffers.
4. TAE buffer: 40 mM Tris–HCl (pH 7.6), 20 mM acetic acid, 1 mM EDTA.
5. 1.5% agarose gel in TAE buffer.
6. Ethidium bromide.
7. Ras G12V mutant-specific antibody (Cell Signaling Technologies 14412).
8. Ras G12D mutant-specific antibody (Cell Signaling Technologies 14429).
9. Anti-HA.11 Epitope Tag Clone: 16B12 antibody (BioLegend 901513).
10. BioID2 antibody (Abcam ab232733).
11. Ras antibody (Cell Signaling Technologies 91054).
12. BirA LHA forward primer: 5′- GGCTTCAGGCACAGGG TAAGGA-3′.
13. BirA LHA reverse primer: 5′-GATGCCCACGATCAGCTTG TCC-3′.
14. BirA RHA forward primer: 5′-GGAGGTGCTGCTGAAGGT-3′.
15. BirA RHA reverse primer: 5′-TTAATCGCAGCATCTGGG CAGC-3′.
16. BirA LHA sequencing primer: 5′-GCGGGTGTGTCCA-CAGGGTATA-3′.
17. Kras sgRNA top primer: 5′-CACCGTGGTTGGAGCTG ATGGCGT-3′.

18. Kras sgRNA bottom primer: 5′-AAACACGCCATCAG CTCCAACCAC-3′.
19. BFP sgRNA top primer: 5′-CACCGCACTGCACCCCGTGG CTCA-3′.
20. BFP sgRNA bottom primer: 5′-AAACTGAGCCACGGGG TGCAGTGC-3′.

#### 2.2.3 Biotin Labeling and Affinity Capture Components

1. Biotin solution: 1 mM biotin dissolved in serum-free DMEM.
2. 1× PBS.
3. Cell lysis buffer: 50 mM HEPES pH 7.5, 150 mM NaCl, 0.5% Triton X-100, 1 mM EDTA, 1 mM EGTA, 10 mM NaF, 2.5 mM $NaVO_4$, Protease Inhibitor Cocktail, Phosphatase Inhibitor Cocktail 2, Phosphatase Inhibitor Cocktail 3.
4. BSA blocking buffer: 5% BSA in 1× TBSTX (1× Tris-buffered saline; 0.1% Triton X-100).
5. Cell scrapers.
6. Table-top centrifuge.
7. 1.5 mL and 2 mL tubes.
8. Pierce BCA Protein Assay.
9. 90 °C heat block.
10. DynaMag-2 magnetic stand.
11. Rotator mixer.
12. Pierce High Sensitivity Streptavidin-HRP.
13. Wash buffer A: 2% SDS.
14. Wash buffer B: 0.5% Triton X-100, 1 mM EDTA, 500 mM NaCl, 50 mM Tris–HCl pH 7.5.
15. Wash buffer C: 0.5% IPEGAL-CA630, 1 mM EDTA, 50 mM Tris–HCl pH 7.5.
16. Dynabeads MyOne Streptavidin C1.
17. Elution buffer: 1% SDS, 50 mM Tris–HCl 7.5, 2 mM biotin.
18. 2× Laemeli sample buffer (65.8 mM Tris–HCl, pH 6.8, 2.1% SDS, 26.3% (w/v) glycerol, 0.01% bromophenol blue).

## 3 Methods

### 3.1 Design and Generation of CRISPR/Cas9 Vectors

This assay involves simultaneous introduction of two sgRNAs and two ssODNs to endogenously express *BirA-Kras*$^{G12V/WT}$ and allow in vitro labeling and identification of unique effector proteins in proximity. In this protocol, biotin ligase from *Aquifex aeolicus*, called BioID2, is utilized. This is a smaller ligase than BioID, yet functionally comparable, and consists of 233 amino acids [8] (*see*

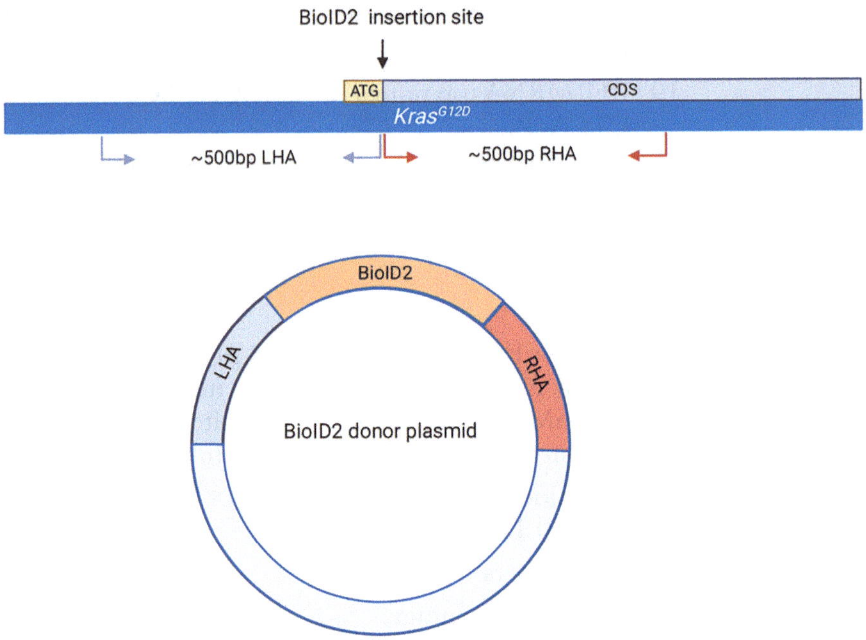


**Fig. 2** Introduction of BioID-tagged *Kras*$^{G12V}$ to the endogenous locus. A plasmid is used to incorporate the BioID$_2$ tag into the *Kras* coding region. The RHA (right homology arm) of the plasmid encodes the *Kras*$^{G12V}$ edit as well as silent mutations to aid in screening for successful edits and disrupt subsequent Cas9 DNA cleavage and further editing events

**Note 1**). The BioID2 insert also includes an HA-tag, a FLAG-tag, and a polyhistidine tag for various potential downstream applications.

As part of this protocol, a cell line was first established to stably express Cas9 and BFP. The BFP is used as a fluorescent marker to enrich cells that underwent BFP to GFP conversion and successful HDR. Once the target cell stably expresses both Cas9 and BFP, a plasmid containing two sgRNAs, one targeting BFP and one targeting the mutant *Kras* allele, is transfected into the cell. ssODN and donor plasmid encoding the desired edits (*BFP* → *GFP* and *Kras*$^{G12D}$ → *BirA-Kras*$^{G12V}$) are co-transfected into the cell with the dual sgRNA plasmid. The donor plasmid introduces BioID2 on the N-terminal domain of *Kras* (Fig. 2). The donor plasmid contains three silent mutations located near the protospacer adjacent motif (PAM) site in the RHA (right homology arm) to prevent guide RNA from binding to the edited sequence. A short flexible linker, GGS, was inserted between BioID and Kras, and polyhistidine, FLAG, and HA-tags were added for downstream applications (*see* **Note 2**). As a result, the edited cells will express polyhistidine, FLAG, and HA-tagged BioID2 linked to *Kras*$^{G12V}$ (Fig. 3a–c).

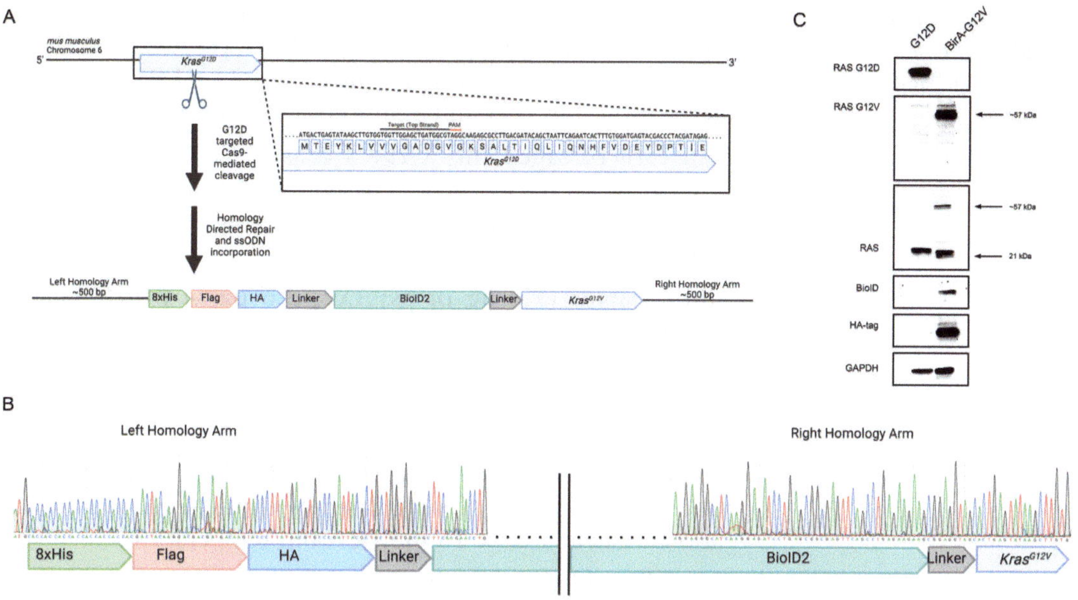


**Fig. 3** Resulting editing sequences and protein expression. (**a**) Schematic showing sgRNA target sequence and editing steps resulting in knock-in of the $BioID_2$ tag and edited *Kras*. (**b**) DNA sequence of cells after incorporation of edits. (**c**) Immunoblot displaying editing from $Kras^{G12D}$ to $BioID_2$-$Kras^{G12V}$

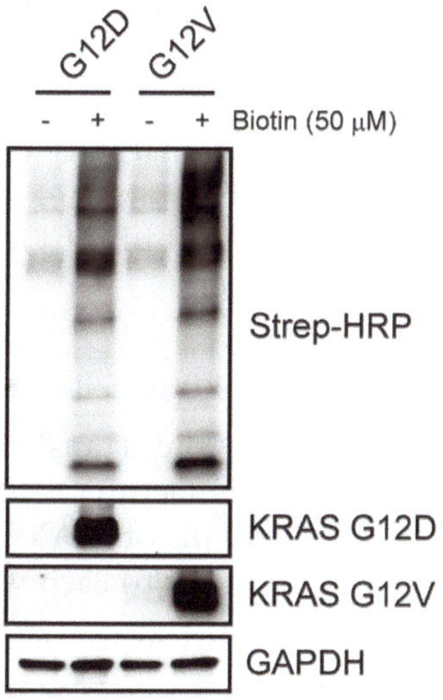


**Fig. 4** Biotinylation of cellular lysates. Confirmed biotinylation of cellular lysates after 16-h incubation with 50 μM biotin

After successful identification of desired edited clones, these cell lines can be used to perform biotin labeling of proteins located within ~10 nm proximity. This assay involves incubation of cells with biotin followed by affinity capture of biotin-labeled proteins by streptavidin beads (Fig. 4). These proteins can then be diluted and subsequently identified by mass spectrometry.

### 3.2 Establishment of Stable Cell Lines Expressing Cas9 and BFP

1. Generate lentivirus using a second-generation lentiviral assembly system. Packaging plasmids (pMD2.G, psPAX2) along with a lentivirus shuttle vector (pLentiCas9-P2A-BFP) were transfected using Lipofectamine 3000 transfection reagent.
2. Plate early passage Lenti-X 293 T cells in a 10 cm culture plate in DMEM with 10% FBS. These cells should be about 70% confluent. Prior to transfection, change the media to antibiotic-free media.
3. Add plasmids (7.5 μg of psPAX2, 2.5 μg of pMD2.G, and 10 μg of pLentiCas9-P2A-BFP) with 500 μL of Opti-MEM in an Eppendorf tube, and then add 40 μL of Lipofectamine 3000 reagent. In a separate tube, mix 500 μL of Opti-MEM with 45 μL of Lipofectamine 3000 reagent.
4. Combine the tube with the DNA/Opti-MEM mix with the tube containing the Lipofectamine/Opti-MEM mix at 1:1 ratio and incubate for 15 min at room temperature.
5. Carefully add the combined transfection mix to the plate of Lenti-X-293 T cells dropwise.
6. Sixteen hours later, remove the medium, and replace with fresh complete medium (DMEM +10% FBS), and incubate for 24 h.
7. The next day (~48 h post-transfection), collect the supernatant containing the virus. Centrifuge at 200 ×*g* for 5 min at 4 °C to remove any cells. Filter the supernatant through a 0.45 μm syringe filter.
8. (Optional) Concentrate the virus using an ultracentrifuge at 10,500 ×*g* at 4 °C overnight. Resuspend the viral pellet in 50–200 μL of Opti-MEM. Discard the supernatant into 10% bleach.
9. Plate $Apc^{-/-}$; $Kras^{G12D}$ colon epithelial cells in a 24-well plate (~50,000 cells per well) the day before infection.
10. Use the TransDux MAX transduction reagent according to manufacturer's protocol. Combine 2.5 μL of TransDux and 100 μL of MAX Enhancer to 400 μL of culture medium, and transfer to each well.
11. Add virus to each well and gently mix.
12. Incubate for 72 h. Check for BFP fluorescence using a fluorescent microscope.
13. Sort for BFP-positive cells (*see* **Note 3**).
14. Expand the BFP-positive cells and confirm expression of Cas9 via Western blot.

### 3.3 Electroporation of Mouse Colon Epithelial Cells with sgRNAs, ssODN, and Donor Plasmids

We recommend referring to the Neon Transfection System protocol on the Thermo Fisher webpage: http://tools.thermofisher.com/content/sfs/manuals/neo_device_man.pdf.

1. On the day of transfection, trypsinize BFP+ mouse colon epithelial $Apc^{-/-}$; $Kras^{G12/WT}$ cells with 0.25% trypsin, and rinse cells in phosphate-buffered saline (PBS) without $Ca^{2+}$ and $Mg^{2+}$.
2. Resuspend the cells in Resuspension Buffer R at a final density of 0.5–1 × $10^5$ cells per well in a 24-well plate.
3. Prepare a 24-well plate by adding 500 μL of pre-warmed RPMI 1640 with 10% FBS (without antibiotics) to each well that will be used for transfection.
4. Prepare DNA for the transfection (1 μg U6::sgBFP-U6::sgKrasG12D-pX333-HcRed plasmid, 1 μL of 10 uM ssODN (GFP), and 500 ng of donor plasmid (BirA-KrasG12D or BirA-KrasG12V)).
5. With a 10 μL Neon Transfection tip, electroporate according to Neon Transfection System protocol at the following settings:
   (a) Pulse voltage (v): 1300.
   (b) Pulse width (ms): 20.
   (c) Pulse number: 2.
6. Pipette the electroporated cells into a well with medium in the prepared 24-well plate.
7. Sixteen hours post electroporation, aspirate the medium, and refeed with complete medium (with antibiotics).
8. Three days post transfection, examine cells under a fluorescent microscope for GFP (a marker of HR events).
9. Sort and expand GFP-positive cells.
10. Once cells have been expanded, isolate genomic DNA for sequencing analysis.
11. Use primers to amplify regions around the left homology arm (LHA) and the right homology arm (RHA) to confirm successful incorporation of donor DNA (*see* **Note 4**).
12. Perform single-cell cloning of GFP-positive cells. Verify successful clones by isolating, amplifying, and sequencing individual clones.
13. Confirm the expression of BirA using an immunoblotting technique, by probing for BirA, or HA-tag, or KRAS G12V. The total size of the insert is ~26 kDa; therefore immunoblots should have bands at ~57 kDa indicating successful inserts.

### *3.4 Biotin Incubation and Cell Lysis*

1. Culture newly generated cells in DMEM +10% dialyzed FBS for 2–3 passages (*see* **Note 5**).
2. Plate *Apc*$^{-/-}$; *BirA-Kras*$^{mut/WT}$ mouse colon epithelial cells on a 10 cm plate, and allow cells to adhere to plate overnight.
3. The following day, add 50 μM of 1 mM biotin solution to cell culture medium, and incubate 16–24 h (*see* **Note 6**).
4. After biotin incubation, rinse the plate with 1× PBS three times to remove excess biotin.
5. Add 300 μL of cell lysis buffer to each 15 cm plate, and swirl plate for uniform covering of cells. Let incubate for 5 min.
6. Scrape lysed cells and transfer into a 1.5 mL Eppendorf tube. Put lysed cells on ice.
7. Incubate the mixture by rotating for 30 min at 4 °C.
8. Once incubation is complete, add equal volume of cold 50 mM Tris–HCl pH 7.5, and pipette to mix the solution.
9. Centrifuge the lysate at 12,500 ×*g* for 20 min at 4 °C.
10. Transfer supernatant to new 2 mL centrifuge tube on ice.
11. Quantify protein using BCA Assay.

### *3.5 Streptavidin Bead Preparation and Pull Down*

1. Pipette 50 μL of streptavidin magnetic beads into a 1.5 mL Eppendorf tube per sample. Incubate tubes on a magnetic stand for 3 min at room temperature.
2. Remove and discard the clear solution. Add 1 mL of lysis buffer to each tube and rotate for 5 min at room temperature.
3. Incubate tubes in a magnetic stand for 3 min and discard the clear solution. Add the lysate with the equilibrated beads to a new 2 mL tube and rotate overnight at 4 °C.
4. The next day, incubate the bead-lysate mixture on a magnetic stand for 5 min at room temperature.
5. Carefully discard the supernatant without disturbing the beads.
6. Add 1 mL of wash buffer A to wash the beads and rotate for 5 min.
7. Place tubes on a magnetic stand for 3 min. Carefully discard wash buffer.
8. Add 1 mL of wash buffer B to wash the beads and rotate for 5 min.
9. Place tubes on a magnetic stand for 3 min. Carefully discard wash buffer.
10. Add 1 mL of wash buffer C to wash the beads and rotate for 5 min.
11. Place tubes on a magnetic stand for 3 min. Carefully discard wash buffer.

12. Add 1 mL of 50 mM Tris–HCl pH 7.5 to wash the beads and rotate for 5 min.
13. Place tubes on a magnetic stand for 3 min. Carefully discard wash buffer.
14. Pipette 50 μL elution buffer to the beads and mix.
15. Place mixture on a 90 °C heat block, and incubate for 5 min to release the bound proteins from the beads.
16. Place the tubes on a magnetic stand for 5 min, transfer eluate to a new microcentrifuge tube, and immediately freeze on dry ice.
17. Analyze the eluate for biotin labeling prior to performing MS analysis.

### 3.6 Biotin Labeling Analysis Using Immunoblot

1. Load ~5 μl of the eluate into 4–20% SDS-polyacrylamide gels, and separate proteins by electrophoresis. Two gels are needed: one for detecting labeled proteins and another for expression of HA, BioID2, $Kras^{G12V}$.
2. Transfer proteins to a PVDF membrane for detecting biotin-labeled proteins.
3. Block the PVDF membrane for 1 h at room temperature using BSA blocking buffer.
4. Incubate the membrane in Strep-HRP antibody diluted in BSA blocking buffer (1:20,000) for 1 h at room temperature.
5. Wash the membrane three times for 10 min with 1× TBST.
6. Observe the biotinylated proteins using enhanced chemiluminescence (ECL).
7. Process the second membrane for expression of fusion proteins according to standard protocols.

## 4 Notes

1. Smaller ligase is preferred in this protocol to reduce the size of DNA insert as small as possible for efficient homologous recombination. The downside of BioID2 is that takes at least ~16 h to label. Alternative option would be to use miniTurbo, which is an engineered ligase that can label in ~10 min [9]. It is only slightly larger than BioID2 (28 kD vs 26 kD) and therefore should not dramatically affect the rate of CRISPR-mediated knock-in.
2. A short linker was used in this protocol, which has a practical labeling radius of BioID2 of ~10 nm. Alternatively, the biotinylation range of BIoID2 can be increased using a longer linker, thus enhancing the identification of proteins beyond the 10 nm range.

3. Sorting cells in bulk for a BFP$^+$ signal will enrich the population for BFP$^+$ cells. The enriched BFP$^+$ population will allow more robust use of the florescent co-selection method and increased visualization of successful BFP → GFP editing events once the sgRNAs are introduced.
4. Primers are designed to amplify regions around RHA and LHA. BirA LHA forward primer is located on the endogenous *Kras* gene, ~600 bp away from LHA, while the BirA LHA reverse primer is located on the donor plasmid. Similarly, BirA RHA forward is located on the donor plasmid, and BirA RHA reverse primer is located on the endogenous *Kras* gene, ~100 bp away from RHA.
5. This step is important to reduce the accumulation of basal biotinylation in normal medium due to endogenous biotin levels in FBS. These are reduced by culturing cells in biotin-depleted serum (dialyzed FBS) for >72 h.
6. Biotinylation reached maximum levels ~16 h after addition of 50 μM biotin. Labeling time can be optimized to avoid non-specific binding. Additionally, BioID2 has enhanced sensitivity for biotin and maintained maximum biotinylation at lower biotin concentrations. Therefore, a 0.1–50 μM biotin dose titration can be performed to determine optimal concentrations depending on cell type.

## References


1. Zehir A, Benayed R, Shah RH et al (2017) Mutational landscape of metastatic cancer revealed from prospective clinical sequencing of 10,000 patients. Nat Med 23:703–713
2. Serebriiskii IG, Connelly C, Frampton G et al (2019) Comprehensive characterization of RAS mutations in colon and rectal cancers in old and young patients. Nat Commun 10:3722
3. Haigis KM (2017) KRAS alleles: the devil is in the detail. Trends Cancer 3:686–697
4. Sears RM, May DG, Roux KJ (2019) BioID as a tool for protein-proximity labeling in living cells. Methods Mol Biol 2012:299–313
5. Adhikari H, Counter CM (2021) Using BioID to characterize the RAS interactome. Methods Mol Biol 2262:271–280
6. Budagyan K, Chernoff J (2021) A facile method to engineer mutant Kras alleles in an isogenic cell background. Methods Mol Biol 2262:323–334
7. Agudelo D, Duringer A, Bozoyan L et al (2017) Marker-free coselection for CRISPR-driven genome editing in human cells. Nat Methods 14:615–620
8. Kim DI, Jensen SC, Noble KA et al (2016) An improved smaller biotin ligase for BioID proximity labeling. Mol Biol Cell 27:1188–1196
9. Branon TC, Bosch JA, Sanchez AD et al (2018) Efficient proximity labeling in living cells and organisms with TurboID. Nat Biotechnol 36: 880–887

# Correction to: Creation of an Isogenic H/N/KRAS-Less Mouse Embryonic Fibroblast Cell Line Panel Derived from a Size-Sorted Diploid Clone

**William Burgan and Nicole Fer**

Correction to:
Chapter 23 in: Andrew G. Stephen and Dominic Esposito (eds.), *KRAS: Methods and Protocols*, Methods in Molecular Biology, vol. 2797,
https://doi.org/10.1007/978-1-0716-3822-4_23

The author's name (Nicole Fer) that was missed in chapter 23 has now been included in the new version. The newly added author's details are as follows.

**Nicole Fer** Nicole.fer@nih.gov
NCI RAS Initiative,
Cancer Research Technology Program,
Frederick National Laboratory for Cancer Research,
Frederick, MD, USA

The updated version of this chapter can be found at
https://doi.org/10.1007/978-1-0716-3822-4_23

Andrew G. Stephen and Dominic Esposito (eds.), *KRAS: Methods and Protocols*, Methods in Molecular Biology, vol. 2797,
https://doi.org/10.1007/978-1-0716-3822-4_26,

# Index




Andrew G. Stephen and Dominic Esposito (eds.), *KRAS: Methods and Protocols*, Methods in Molecular Biology, vol. 2797, https://doi.org/10.1007/978-1-0716-3822-4,

MIX
Papier aus verantwortungsvollen Quellen
Paper from responsible sources
FSC® C105338

If you have any concerns about our products,
you can contact us on
**ProductSafety@springernature.com**

In case Publisher is established outside the EU,
the EU authorized representative is:
**Springer Nature Customer Service Center GmbH**
**Europaplatz 3, 69115 Heidelberg, Germany**

Printed by Libri Plureos GmbH
in Hamburg, Germany